In recent years there has been a great deal of interaction among game theorists, philosophers, and logicians with respect to certain foundational problems concerning rationality, the formalization of knowledge and practical reasoning, and models of learning and deliberation.

This unique volume brings together the work of some of the preeminent figures in their respective disciplines, all of whom are engaged in research at the forefront of their fields. Together they offer a conspectus of the interaction of game theory, logic, and epistemology in the formal models of knowledge, belief, deliberation, and learning and in the relationship between Bayesian decision theory and game theory, as well as between bounded rationality and computational complexity.

Knowledge, belief, and strategic interaction

Cambridge Studies in Probability, Induction, and Decision Theory

General editor: Brian Skyrms

Advisory editors: Ernest W. Adams, Ken Binmore, Jeremy Butterfield, Persi Diaconis, William L. Harper, John Harsanyi, Richard C. Jeffrey, Wolfgang Spohn, Patrick Suppes, Amos Tversky, Sandy Zabell

This new series is intended to be a forum for the most innovative and challenging work in the theory of rational decision. It focuses on contemporary developments at the interface between philosophy, psychology, economics, and statistics. The series addresses foundational theoretical issues, often quite technical ones, and therefore assumes a distinctly philosophical character.

Other titles in the series
Ellery Eells, *Probabilistic Causality*
Richard Jeffrey, *Probability and the Art of Judgment*
Robert Koons, *Paradoxes of Strategic Rationality*

Forthcoming
J. Howard Sobel, *Taking Chances*
Patrick Maher, *Betting on Theories*
Patrick Suppes and Mario Zanotti, *Foundations of Probability with Applications*
Cristina Bicchieri, *Rationality and Coordination*
Clark Glymour and Kevin Kelly (eds.), *Logic, Confirmation, and Discovery*

Knowledge, belief, and strategic interaction

Edited by

Cristina Bicchieri
Carnegie Mellon University

Maria Luisa Dalla Chiara
University of Florence

CAMBRIDGE
UNIVERSITY PRESS

Published by the Press Syndicate of the University of Cambridge
The Pitt Building, Trumpington Street, Cambridge CB2 1RP
40 West 20th Street, New York, NY 10011-4211, USA
10 Stamford Road, Oakleigh, Victoria 3166, Australia

First published 1992

Printed in the United States of America

Library of Congress Cataloging-in-Publication Data
Knowledge, belief, and strategic interaction / [edited by] Cristina
Bicchieri, Maria Luisa Dalla Chiara.
p. cm. – (Cambridge studies in probability, induction, and
decision theory)
Papers presented at the workshop knowledge, belief, and
strategic interaction which took place in Italy in June 1989.
ISBN 0-521-41674-4
1. Reasoning. 2. Belief and doubt. 3. Decision-making.
4. Rationalism. 5. Game theory. I. Bicchieri, Cristina.
II. Dalla Chiara, Maria Luisa. III. Series.
BC177.K58   1992
121–dc20   91–38415
                                        CIP

A catalog record for this book is available from the British Library.

ISBN 0-521-41674-4 hardback

Contents

Preface   *page* vii
List of contributors   xiv

1   Feasibility   1
    Isaac Levi

2   Elicitation for games   21
    Joseph B. Kadane, Isaac Levi, & Teddy Seidenfeld

*3   Equilibrium, common knowledge, and optimal sequential
    decisions*   27
    Joseph B. Kadane & Teddy Seidenfeld

4   Rational choice in the context of ideal games   47
    Edward F. McClennen

5   Hyperrational games: Concept and resolutions   61
    Jordan Howard Sobel

6   Equilibria and the dynamics of rational deliberation   93
    Brian Skyrms

*7   Tortuous labyrinth: Noncooperative normal-form games
    between hyperrational players*   107
    Wlodzimierz Rabinowicz

*8   On consistency properties of some strongly implementable
    social choice rules with endogenous agenda formation*   127
    Stefano Vannucci

9   Algorithmic knowledge and game theory   141
    Ken Binmore & Hyun Song Shin

10	*Possible worlds, counterfactuals, and epistemic operators* Maria Luisa Dalla Chiara	155
11	*Semantical aspects of quantified modal logic* Giovanna Corsi & Silvio Ghilardi	167
12	*Epistemic logic and game theory* Bernard Walliser	197
13	*Abstract notions of simultaneous equilibrium and their uses* Vittorioemanuele Ferrante	227
14	*Representing facts* Krister Segerberg	239
15	*Introduction to metamoral* Roberto Magari	257
16	*The logic of Ulam's games with lies* Daniele Mundici	275
17	*The acquisition of common knowledge* Michael Bacharach	285
18	*The electronic mail game: Strategic behavior under "almost common knowledge"* Ariel Rubinstein	317
19	*Knowledge-dependent games: Backward induction* Cristina Bicchieri	327
20	*Common knowledge and games with perfect information* Philip J. Reny	345
21	*Game solutions and the normal form* John C. Harsanyi	355
22	*The dynamics of belief systems: Foundations versus coherence theories* Peter Gärdenfors	377
23	*Counterfactuals and a theory of equilibrium in games* Hyun Song Shin	397

Preface

This book collects the papers presented at the workshop on "Knowledge, Belief, and Strategic Interaction," which took place in Castiglioncello, Italy, in June 1989. The workshop was a first attempt at an exchange between philosophers and game theorists, as in recent years game theorists have shown a growing interest in foundational problems and there are several areas in which their concerns overlap with those of philosophers.

Philosophers have a tradition of inquiring into topics such as rationality, learning, and knowledge, as well as the practical reasoning that results in deliberation and decision making. These issues have been mainly studied in an individual context: The knowing subject is depicted as facing a stationary natural environment of which he tries to explain and predict some features. Game theorists are equally involved in defining rationality, learning, and knowledge, although the environment faced by the individual is not a natural but a social one. In a game-theoretic context, one's knowledge or beliefs are about what other players plan to do, which is in turn determined by what these players know (or believe) about one's choices. This interdependency is captured by the concept of Nash equilibrium, which is the most common solution for noncooperative games. A traditional argument for the predictive significance of Nash equilibria asserts that a virtual process of reflection about what other rational players should expect, and accordingly should do, will converge to an equilibrium, so that rational agents who understand the game and think it through thoroughly before choosing their action should play in this way, even if the game is played only once. However, game theorists have come to acknowledge that, in many nontrivial cases, assuming that the only inputs needed for obtaining a solution are knowledge of the game being played and self-evident principles of rationality that make no use of any knowledge is not sufficient to guarantee that an equilibrium will be attained.

Many game theorists have questioned the adequacy of Bayesian decision theory as a foundation for a theory of strategic interaction, at least insofar as one sees the concept of Nash equilibrium as a central tenet of game theory. One way to overcome the problem is to add several assumptions about what players know about the game and about each other. For example, it is commonly assumed that the structure of the game, as well as their mutual rationality, are common knowledge among the players. Briefly stated, a fact f is said to be common knowledge among a group of people if for every finite n, the statement "everyone knows that everyone knows that . . . everyone knows f" is true, where "everyone knows" is repeated n times. Since the players in a game assign subjective probabilities to all uncertainty, including the actions of other players, and choose on the basis of these beliefs, common knowledge of players' beliefs is usually needed in order to attain a Nash equilibrium.

Some game theorists consider the assumption of common knowledge of beliefs as gratuitous, and their approach to bridging the gap between Bayesian decision theory and game theory typically does away with it. In a unified theory of rational decision, however, other decision makers should be regarded as part of nature, and rational decision will consist in maximizing one's expected payoff relative to one's uncertainty about the state of nature. The relevance of such a Bayesian viewpoint for the theory of games has become more widely appreciated as a result of fundamental papers by Harsanyi, Aumann, Pearce, Bernheim, and Kreps and Wilson. In all these approaches, common knowledge of rationality is preserved and – depending on what amount of a priori information about the setting is allowed – different solutions (such as rationalizability or correlated equilibrium) are appropriate. Recently philosophers have dealt with this issue, and there is a growing body of literature on the Bayesian foundations of game theory. The chapters by Kadane, Levi, and Seidenfeld; Levi; and Kadane and Seidenfeld exemplify this approach. Not all philosophers share this view, though, and Sobel's and McClennen's chapters take a different, critical approach to the possibility of bridging the gap between decision theory and game theory.

The framework of this literature, although Bayesian, is still static. The focus is on a solution that satisfies a set of conditions rather than on the procedures by which the players attempt to arrive at an optimal decision. The importance of the procedural aspect of rationality has long been emphasized by Herbert Simon. In accordance with this point of view, common knowledge of rationality in strategic situations is to be thought of as common knowledge of a rational deliberational procedure. In such a situation, computations must be conceived of as generating new information; as a result, probabilities can change as a result of pure thought. There are discussions of such a possibility under the name of "dynamic probability" in the writings of Good. The chapters by Skyrms and Rabin-owicz show that deliberation can be modeled as a dynamic process. Where deliberation generates new information relevant to the decision under consideration, a rational decision maker will feed back that information and reconsider. A firm decision is reached at a fixed point of this process – a deliberational equilibrium. A joint deliberational equilibrium on the part of the players is a Nash equilibrium of the game. Taking this point of view seriously leads to dynamic models of deliberation within which one can embed the theory of noncooperative games.

If one wants to maintain the centrality (and predictive value) of Nash equilibria, the first type of approach seems more adequate. It is, however, plagued with several problems. For one, the issue of belief formation is sidestepped. This deficiency should be remedied by supplementing the analysis with a model of how players learn. This represents a potentially fruitful and challenging application of theories of inductive inference, in that in a strategic environment players' beliefs may never converge to any single hypothesis, or even to any constant posterior distribution over the various hypotheses, and so agents may not converge upon any constant beliefs regarding the best action to take. Studies about how concensus is reached and how opinions can merge are directly relevant to this topic, and philosophers are giving important contributions to an understanding of the conditions under which conditional probability distributions ap-proach each other as available data increase, as is discussed in Vannucci's chapter and in the paper by Shervish and Seidenfeld.* Another problem is that modeling how players learn to predict another player's choice appears to lead to an infinite regress of "I learn that you learn that I learn. . . ." To avoid this problem it may help to introduce elements of bounded ra-tionality and computational complexity into the analysis, as is done in the chapters by Rubinstein and by Binmore and Shin.

A more general issue that encompasses both approaches is that of pro-viding formal theories of how players reason. In particular, it is impor-tant to formally model players' information about the game and about other players. Even if one rejects the assumption of common knowledge of beliefs, it is usually the case that common knowledge of mutual ration-ality and of the structure of the game are assumed. Following Lewis, Schiffer, and Aumann, common knowledge has been studied by relating it to concepts of knowledge and probability. In this context, the logical theory of epistemic operators plays a relevant role, and there have been

---

*The paper by Shervish and Seidenfeld was presented at the Castiglioncello workshop but does not appear in the present volume. See "A Fair Minimax Theorem for Two-person (zero-sum) Games Involving Finitely Additive Strategies," Technical Report No. 491, De-partment of Statistics, Carnegie Mellon University.

several attempts at constructing axiomatic theories of the game using epistemic logic. The recent developments in modal and epistemic logics are discussed in the chapters by Dalla Chiara and by Corsi and Ghilardi, and their application to game theory is presented in Walliser's and Ferrante's chapters. Corsi and Ghilardi investigate, among others, a basic problem of modal semantics, the "transworld identity" question: how to exist, at the same time, in different possible worlds? In spite of its metaphysical appearance, the notion of possible worlds represents a useful abstract tool that permits the modeling of different types of situations. As an example, let us think of the following statement referring to a particular chess game: "Had White moved the Queen, he would have won." A semantic analysis of statements of this kind gives rise to the question of how to represent the "alter ego" of White, living in a state of the world that does not actually occur. Corsi and Ghilardi propose an elegant formal analysis of this problem in the framework of a general categorical semantics. The basic intuitive idea of their approach goes back to Leibniz and to the theory of "counterparts" developed by David Lewis.

In spite of the many successful applications to different fields, possible-worlds semantics is not yet capable of providing an adequate analysis for the problems of epistemic logics. The reason for this failure is simple: Possible worlds are usually considered closed under logical consequence, in the sense that all the consequences of a logical truth in a given world must be true in that world. Hence any semantic analysis of the epistemic operators ("know," "believe," . . .) in terms of possible worlds can hardly avoid the unpleasant and unrealistic result – knowing the axioms of a given theory should be sufficient for knowing all the theorems – according to which agents should be logically omniscient. Dalla Chiara's chapter discusses an alternative approach to epistemic semantics, where epistemic operators are analyzed in terms of intensions, and where logical omniscience is avoided.

An even more basic formal analysis of games should include models of action. Segerberg's chapter proposes a general approach to dynamic logic in the framework of a standard set-theoretical setting. Taking as primitive the notion of state space, he analyzes the concepts of *state-of-affairs, event, process,* and *action.* The theory gives rise to a number of natural game-theoretic applications. One can try to characterize formally not only actions but also prescriptions of behavior or more generally moral principles. This aim is pursued in Magari's chapter, where an abstract characterization for the notion of Pascalian ethics is proposed. Roughly, a Pascalian ethics is identified with a function that permits one to maximize the probability of a given situation, determined by an action, multiplied by the value which one attributes to that situation.

Another area where logic can contribute to game theory is that of games in which there are communication stages. Mundici's chapter describes the logic of Ulam's game with lies: a logic where contradictory sentences do not lead to total ignorance. More precisely, an Ulam game where n lies are admitted corresponds to a Lucasiewicz $(n+2)$-valued logic. Lucasiewicz many-valued logics are examples of "paraconsistent" logics, where a certain degree of tolerance toward contradictions is admitted. For example, both the principle of noncontradiction (the negation of a contradiction is always true) and the Scotian principle ("ex absurdo sequitur quodlibet") are violated. From an intuitive viewpoint, these logics may represent a good framework for modeling epistemic situations in which contradictions do not play a totally destructive role, contrary to what happens in classical logic as well as in many alternative logics. Significantly enough, Lucasiewicz logics are also deeply related to AF C^*-algebras that are used in the description of quantum spin systems, as has been shown by Mundici in other papers. As a consequence, the logical interpretation of Ulam games might be useful in the abstract investigation of the physical systems that are studied in quantum mechanics.

Turning to the topic of common knowledge, one persistent problem is that it seems difficult to attain. Bacharach's and Rubinstein's chapters each give arguments why common knowledge cannot be attained in finite games. Moreover, it is not evident that common knowledge needs to be assumed in order to obtain a Nash equilibrium. For example, in finite, extensive-form games of perfect information, assuming the players to have only limited knowledge often provides a better approximation to real-life interactions and allows for solutions that are more plausible and intuitive, as the chapters by Bicchieri and Reny show.

Another important topic of relevance to game theorists and philosophers alike is a definition of what constitutes rationality of beliefs. After Nash, it was commonly assumed that once a strategy combination is a Nash equilibrium, it can be the solution for a noncooperative game. Selten pointed out that some Nash equilibria involve irrational moves and that such equilibria cannot serve as solutions for sequential games played by rational players. Equilibria may involve irrational moves because a player may have irrational beliefs about another player's choice. This happens because the beliefs that support a Nash equilibrium need only be internally consistent and self-fulfilling; no further restriction is imposed on their rationality. In extensive-form games, equilibria may involve irrational moves only at information sets that will never be reached if the players follow their equilibrium strategies. What happens out of equilibrium, however, affects what constitutes a best choice in a Nash equilibrium. In an extensive-form game, it makes sense to ask what another's reaction will be if one deviates from the prescribed equilibrium strategy. For suppose there are two equilibria, one of which involves some threat on the part of one of the players. Then it makes sense for the threatened party to ask whether the opponent, facing a deviation on his part, would fulfill the threat. If the answer is negative then the threat is not credible, and the equilibrium which is based on it is ruled out as not sensible. To rule out irrational equilibria, game theorists generally refer to the extensive form of the game, which conveys more information than the normal form. That normal and extensive form are not equivalent is not, however, an accepted tenet; Harsanyi's chapter is a defense of their basic asymmetry.

The vast literature on refinements of Nash equilibrium, while searching for criteria that restrict the number of possible equilibria, is mainly concerned with defining what it means to be rational at information sets off the equilibrium path. Within the class of refinements of Nash equilibrium, two different approaches can be identified. One solution aims at imposing restrictions on players' beliefs by explicitly allowing for the possibility of error on the part of the players. This approach underlies both Selten's notion of "perfect" equilibrium and Myerson's notion of "proper" equilibrium. The alternative solution is based instead upon an examination of rational beliefs rather than mistakes. The idea is that players form conjectures about other players' choices, and that a conjecture should not be maintained in the face of evidence that refutes it. This approach underlies the notion of "sequential" equilibrium proposed by Kreps and Wilson. All of these solutions aim at imposing restrictions on players' beliefs, so as to obtain a unique rational recommendation as to what to believe about the other players' behavior. This guarantees that rational players will select the unique equilibrium that is supported by these beliefs. Both approaches, however, fail to rule out some equilibria supported by beliefs that, although coherent, are still intuitively implausible. The limit of these approaches is that restrictions are imposed only on equilibrium beliefs, while out-of-equilibrium beliefs are unrestricted: A player will ask whether it is reasonable to believe the other player will play a given equilibrium strategy, but not whether the beliefs supporting the other player's choice are rational. To provide a satisfactory means of discriminating among equilibria, restrictions need to be imposed on all sorts of beliefs, even out-of-equilibrium ones.

A player, that is, must be able to rationally justify to himself every belief, and expect the other players to expect him to adopt such a rational justification. A possible solution lies in combining the heuristic method implicit in the "small mistakes" approach with the analysis of belief rationality characteristic of the sequential equilibrium notion. The "small mistakes" approach stresses the role of anticipated actions off the equilibrium path in sustaining the equilibrium. Because the reference point is an equilibrium of which one tests the stability against deviations, players are modeled as being involved in counterfactual arguments concerning what would happen if they were to deviate from their equilibrium strategy, or were themselves faced with other players' deviations. These arguments involve a change in the original set of beliefs, and for the process of belief change not to be arbitrary, rationality conditions must be imposed on it. Belief rationality, in this case, is a property of beliefs that are revised through a rational procedure. If there were a unique rational process of belief revision, then there would be a unique best theory of deviations that a rational player would be expected to adopt, and common knowledge of rationality would suffice to eliminate all equilibria that are robust only with respect to implausible deviations.

Philosophers have developed two main approaches to modeling epistemic states. One is the foundations theory, which holds that one needs to keep track of the justification for one's beliefs. The other is the coherence theory, which holds that one need not consider where one's beliefs come from. In coherence theory, the focus is on the logical structure of the beliefs; what matters is how a belief coheres with the other beliefs that are accepted in the present state. The foundations and the coherence theories have very different implications for what should count as rational changes of belief systems. According to the foundations theory, belief revision should consist, first, in giving up all beliefs that no longer have a satisfactory justification and, second, in adding new beliefs that have become justified. According to the coherence theory, the objectives are, first, to maintain consistency in the revised epistemic state and, second, to make minimal changes of the old state that guarantee sufficient overall coherence. A satisfactory model of belief change in games must encompass both approaches. Such dynamics of belief revision are extensively explored in Gärdenfors's chapter. Another approach to counterfactual reasoning in games is proposed by Shin, who applies the logic of conditionals developed by Stalnaker and Lewis.

We are greatly indebted to the Centro Fiorentino di Storia e Filosofia della Scienza and to the Centro Interuniversitario per la Teoria dei Giochi e le Applicazioni, which made the workshop possible. We are also grateful to Alessandro Pagnini, who organized a pleasant stay in Castiglioncello for all of us.

Pittsburgh and Firenze   *Cristina Bicchieri*
May 1991   *Maria Luisa Dalla Chiara*

Contributors

Michael Bacharach   *Christ Church, Oxford*
Cristina Bicchieri   *Department of Philosophy, Carnegie Mellon University*
Ken Binmore   *Department of Economics, University of Michigan*
Giovanna Corsi   *Dipartimento di Filosofia, Università di Firenze*
Maria Luisa Dalla Chiara   *Dipartimento di Filosofia, Università di Firenze*
Vittorioemanuele Ferrante   *Dipartimento di Economia, Università di Firenze*
Peter Gärdenfors   *Filosofiska Institutionen, Lund Universitet*
Silvio Ghilardi   *Dipartimento di Matematica, Università di Milano*
John C. Harsanyi   *School of Business Administration, University of California, Berkeley*
Joseph B. Kadane   *Department of Statistics, Carnegie Mellon University*
Isaac Levi   *Department of Philosophy, Columbia University*
Roberto Magari   *Dipartimento di Matematica, Università di Siena*
Edward F. McClennen   *Department of Philosophy, Bowling Green State University*
Daniele Mundici   *Dipartimento di Informatica, Università di Milano*
Wlodzimierz Rabinowicz   *Filosofiska Institutionen, Uppsala Universitet*
Philip J. Reny   *Department of Economics, University of Western Ontario*
Ariel Rubinstein   *Department of Economics, Tel Aviv University*
Krister Segerberg   *Filosofiska Institutionen, Uppsala Universitet*
Teddy Seidenfeld   *Department of Philosophy, Carnegie Mellon University*
Hyun Song Shin   *University College, Oxford*
Brian Skyrms   *Department of Philosophy, University of California, Irvine*
Jordan Howard Sobel   *Department of Philosophy, Scarborough College, University of Toronto*
Stefano Vannucci   *Dipartimento di Economia, Università di Siena*
Bernard Walliser   *CERAS, Ecole Nationale Des Ponts et Chaussees, Paris*

1

Feasibility

ISAAC LEVI

1. ACT, STATE, AND CONSEQUENCE

According to the procedure Savage (1954) and others have adopted as canonical for representing decision problems, three notions are deployed in the representation: the notion of an act, a state, and a consequence.[1] Many philosophers have followed the lead of Jeffrey (1965) in complaining about a wrong-headed ontology that insists on trinitarianism where monotheism should do. Jeffrey suggests that acts, states, and consequences are all events or propositions.

I do not want to quarrel with Jeffrey's suggestion. To me, something like it should turn out right. But I do not see why it should be supposed, as Jeffrey intimates, that Savage (or, for that matter, Ramsey) would disagree. Perhaps Ramsey and Savage may be convicted of what now seems like loose talk; but it is loose talk easily repaired without damage to the substance of their views. Instead of speaking of acts, states, and consequences, Savage could have spoken of act descriptions, state descriptions, and consequence descriptions, or of act propositions, state propositions, and consequence propositions.

There are, to be sure, important differences in a Savage framework between the attitudes the decision maker has toward act descriptions, state descriptions, and consequence descriptions. State descriptions are objects of personal or credal probability judgments; consequence descriptions are objects of utility judgment and act descriptions of *expected* utility judgment.

There is nothing in the Savage system to prevent assigning utilities to state descriptions or probabilities to consequence descriptions. Indeed, it seems clear that Savage intended the state descriptions to be evaluated with respect to utility in a certain way although, as is well known, his

Thanks are due to Teddy Seidenfeld for helpful suggestions.

axioms do not quite capture his intent. State descriptions cannot be assigned *unconditional* utility but may be assigned utility conditional on the act chosen. That is to say, given that act a_i is chosen and state s_j is true, the utility assigned to s_j conditional on a_i is equal to the unconditional utility assigned the consequence c_{ij}. Unless the consequences of all available options in a given state bear equal utility, the only way to derive an unconditional utility for the state is to compute the expectation of the conditional utilities of the state using unconditional probabilities for acts.[2]

Similarly credal probabilities are assignable to consequences, but these are conditional on the option chosen. Unless consequences are identical for all available options in a given state, the only way to compute unconditional probabilities for consequences is with the aid of unconditional probabilities for those options that yield them in some state or other.

Thus, unconditional utilities for states and unconditional probabilities for consequences are obtainable only if we can assign unconditional probabilities to acts. In the Savage formalism, however, unconditional credal probabilities are not assigned to acts – that is, to the agent's options. According to the Savage approach, one begins with a preference or value ranking of hypothetically available acts or options that satisfy the axioms proposed (and in which the preference ranking over the set of available options is embedded). The axioms are intended to ensure that the ranking evaluates the hypothetically available options with respect to expected utility. With this understood, one may derive a unique unconditional credal probability distribution over the states and an unconditional utility function over the consequences unique up to a positive affine transformation, where the utility of a consequence is independent of which state is true. Using this information, it is possible to derive a probability distribution over consequences conditional on acts and a utility function for states conditional on acts. However, the Savage theory fails to determine an unconditional credal probability over acts and, as a consequence, an unconditional utility function for states and an unconditional probability distribution over consequences.

Thus, the Savage trichotomy does not deny that acts, states, and consequences are propositions, sentences, or other truth-value–bearing entities. What it denies is that credal probability judgments and utility judgments are defined over the same set of propositions.

The Savage axioms, even when construed prescriptively as norms of rationality, are designed so that one might elicit an agent's probabilities and utilities from information about that agent's preferences among acts. However, information about preferences among acts unsupplemented by other data fails to yield conclusions about unconditional probabilities of acts.

2

Some might argue that this consideration justifies the conclusion that acts are not assigned credal probabilities. This conclusion may, perhaps, be too hasty. In any case, acts, states, and consequences could all be represented as propositions. The same sentence or proposition could qualify as a state description relative to one decision problem, and as a consequence description (or, indeed, even as an act description) relative to another. Given the propositional construal of acts, states, and consequences, this is not unexpected. It does not follow, however, that we must insist that probabilities and utilities be defined over all propositions without restriction in the setting of every specific decision problem. We may therefore endorse Jeffrey's insistence on regarding acts, states, and consequences propositionally without accepting his contention that the decision maker should assign "desirabilities" (utilities or expected utilities) and probabilities to propositions of these three kinds. The point under dispute is not an ontological one; rather, the concern is whether we should restrict the applicability of unconditional utility and probability judgments along Savage's lines or should follow Jeffrey in rejecting such restrictions.

In my opinion, Spohn (1977; 1978, pp. 72–5) is right in resisting Jeffrey on this issue. Decision makers should not assign credal probabilities to the acts available to them. If this is so, we should also reject the idea developed by Jeffrey of taking as fundamental a notion of preference among propositions for the purpose of deriving an agent's credal probability and utility judgments.

Spohn's chief argument for his view is that one need not assign unconditional probabilities to acts in order to determine expected utilities of options for the purpose of identifying optimal or admissible options. Spohn's point is well taken, but it does not entail a prohibition against the deliberating agent assigning credal probabilities to hypotheses predicting his decision. At best, it serves notice to those who insist that such probability assignments may be made (and, perhaps, ought to be made) that they should identify some function such probability assignments can serve other than that of guiding choice among the options under consideration. Spohn reminds us that assigning such probabilities is not crucial in applications of the prescription to maximize expected utility, or, for that matter, of my preferred recommendation that choice be restricted to E-admissible options (Levi 1974, 1980).

I think, however, that there is an additional argument that suggests good reasons why the deliberating agent ought to avoid assigning credal probabilities to predictions of what will be chosen from among the available options in the decision problem currently faced.[3] These considerations derive from reflection on the conditions that a proposition should satisfy, from the point of view of the deliberating agent deploying principles of

3

rational choice in deciding what to do, in order for it to be an act description representing a feasible option. The remainder of this chapter proposes an answer to this question and indicates some of its ramifications.

2. POSSIBILITY FOR AND POSSIBILITY THAT

To say that driving from New York to Boston in four hours on Monday is a feasible option for Sam is to say something about Sam's abilities. Sam is able to drive from New York to Boston in four hours on Monday as he chooses: It is *possible for* Sam to drive from New York to Boston in four hours on Monday. Such a claim is not to be confused with the claim that it is *possible that* Sam will drive from New York to Boston in four hours on Monday. The first modal claim predicates of Sam a certain ability. It is true or false of Sam depending on whether Sam satisfies the conditions for the ability attribution. The second modal claim differs in this respect. The agent X who makes the judgment (X may be Sam or some other party) is expressing a certain propositional attitude – an attitude indicating that nothing in what X takes for granted is inconsistent with Sam's making the trip on Monday in four hours. In making this judgment, X need not assume that Sam has the ability to make the trip. There may be some doubt as to whether Sam has that ability, yet X may still judge that it is possible that Sam will take the trip. Conversely, X may take for granted that Sam has the ability and yet judge it impossible that he takes the trip – because (say) X is certain that Sam will not exercise the ability.

Ability attributions are relative to what may be called initial or test conditions or experiments. For example, a person may be able to play the piano with training but lack the ability to play the piano by choice. The ability to play is relative to the test condition describable as undergoing training in piano playing. The inability is relative to the test condition describable as undertaking a deliberation.

The relativity, moreover, is not restricted to the abilities of agents to do things or have things done to them; it applies also to other objects or systems. A given coin may have the ability to land heads on a toss and also the ability to land tails on a toss. But it may lack the ability to land heads on a toss by Morgenbesser (although, if this were true, Morgenbesser himself would have some remarkable abilities). Less fancifully, the coin will lack the ability to land heads on a toss situating it in a mechanical state which, according to the laws of physics, destines it to land tails. Abilities are like dispositions in this respect; this is not surprising, because abilities are duals of dispositions. The ability to respond in manner R on a trial of kind T applies to x if and only if it is false that x

4

has the (sure-fire) disposition to fail to respond in manner R on a trial of kind T. And since dispositions are relative to test conditions, so are abilities.

Judgments of possibility (termed judgments of "serious possibility" in Levi 1980) are also relative. But the relativity is to the assumptions the inquiring agent takes for granted and hence judges with certainty to be true. To say that it is seriously possible that h, according to agent X at time t, is to say that h is consistent with X's full or settled beliefs at t.

Of course, whether X judges it seriously possible that h or not is true or false as the case may be. But X's judgment itself is neither true nor false any more than X's preferences, utility judgments, or judgments of credal probability are truth-valued. As a consequence, it makes no sense to say that X fully believes that it is seriously possible that h because the that-clause cannot represent a truth-value–bearing proposition. (One *can* say that X fully believes that it is seriously possible that h according to X.) On the other hand, it does make sense to say that X fully believes that Sam is able to take the trip to Boston on Monday at will. The claim that Sam has the ability is truth-value–bearing.

Although possibility for and possibility that are different notions, there is an important connection between them. Sometimes an agent X can ground or justify judgments of serious possibility in knowledge of ability.

Suppose then that X is certain that a coin has the ability to land heads on a toss. Suppose further that X is certain that coin will be tossed at time t. Do these assumptions warrant X's judging it to be possible (for all X knows) that the coin will land heads? This cannot be sufficient: X might also be certain that Morgenbesser will toss the coin at time t while being fully convinced that the coin lacks the ability to land heads on a toss by Morgenbesser (i.e., has the sure-fire disposition to fail to land heads on a toss). There is no inconsistency in believing that the coin has the ability to land heads on a toss, lacks the ability to land heads on a toss by Morgenbesser, and is tossed by Morgenbesser. If X has these convictions, it is incompatible with what he takes for granted that the coin will land heads. Of course, if in doubt as to whether the coin had the ability to land heads on a toss by Morgenbesser, X could judge it possible that it will land heads. But X could do that also if in doubt as to whether the coin had the ability to land heads on a toss. The important point is that X may know that a person or system has the ability to R on a trial of kind S and that a trial of kind S is being implemented. If X knows that the trial is also of kind S', then X may consistently continue to fully believe that the trial is of kind S, that the system has the ability to R on a trial of kind S and lack the ability to R on a trial of both kind S and kind S'. Under these circumstances, X should rule out as impossible the hypothesis that the system will respond in manner R, even though X is convinced that it is possible for the system to do so.[4]

These remarks have an important bearing on judgments of feasibility and how to understand act descriptions.

3. Passing judgment and deciding

If X supposes that the proposition that Sam will travel to Boston in four hours on Monday represents an act or option for Sam, then X presupposes that Sam has the ability to choose that he (Sam) will take the trip as the outcome of deliberation and that this choice will be efficacious. Moreover, X also assumes that Sam has or will engage in a process of deliberation by the appropriate time.

In general, if X takes the sentence "Sam behaves in manner R" to be an act description vis-à-vis a decision problem faced by Sam, then X is in a state of full belief that has the following contents:

(i) Sam has the ability to choose that Sam will R on a trial of kind S, where the trial of kind S is a process of deliberation eventuating in choice.[5] Let us call this the *ability* condition.

(ii) Sam is subject to a trial of kind S at time t; that is, Sam is deliberating at t. This is the *deliberation* condition.

Sam's state of full belief should also satisfy the following requirement:

(iii) Adding the claim that Sam chooses that he will R to X's current body of full beliefs entails that Sam will R – that is, travel to Boston. This is the *efficaciousness* condition.[6]

I shall say that X is committed to judging that Sam's doing R is a feasible option for Sam, for the purposes of passing judgment on the rationality of Sam's choice, if and only if X's state of full belief satisfies conditions (i)–(iii).

Suppose then that X judges that Sam is able to travel to Boston from New York on Monday in under four hours by his own choice. Suppose further that X is convinced that Sam will not choose to take the trip. X may be certain that (1) Sam will not even engage in deliberation resulting in choice or that (2) Sam will engage in such deliberation but that the deliberation will result in a decision not to make the trip.

In the first case, the "test condition" relative to which Sam is supposed to have the ability will not be realized (so X judges). If X also rules out Sam's being transported to Boston against his will, then X will judge it impossible that Sam travels to Boston on Monday. In this case, it should

6

be clear that traveling to Boston is not optional for Sam, so far as X is concerned, in any decision problem Sam is facing.

In the second case, X is convinced that Sam is facing a decision problem. X is also certain that the choice made will be not to travel to Boston. Again, if X also rules out Sam's taking the trip against his will, it is not a serious possibility (according to X) that Sam will take the trip. All of this holds even though X is convinced that conditions (i)–(iii) obtain.

For the purposes of passing judgment on the rationality of Sam's choice, counting Sam's taking the trip as an act description may seem quite acceptable. Even if X supposes that Sam will choose irrationally – perhaps, in an akratic manner – this seems to make good sense. But as we shall see shortly, appearances can be deceiving.

However, suppose X is not concerned to pass judgment on the rationality of Sam's choice but is rather engaged in advising Sam as to what his choice should be. In such a context, criteria of rational choice are being used not to pass judgment on what Sam did, does, or will do, but instead to decide what Sam should do. This use of rational choice principles is especially relevant when X is Sam himself (although it is not necessary that X be identical with Sam).

In this context, the fact that X's state of full belief satisfies (i)–(iii) is necessary but not sufficient for X's judging it feasible that Sam will R. To see that it is not sufficient, consider the situation where X is certain that Sam will choose not to R and that Sam's choice will be efficacious. From X's point of view, it is no longer a serious possibility that Sam will R. Although X might deplore this as a mark of Sam's irrationality, this is relevant only insofar as X is passing judgment on the rationality of Sam's decisions. If X is merely giving advice, it is pointless to advise Sam to do something X is sure Sam will not do. In this setting, advice concerning what Sam should rationally do should take into account only those propositions judged to represent feasible options in a stricter sense than the one captured by conditions (i)–(iii); a fourth condition must be satisfied by X's state of full belief:

(iv) Nothing in X's state of full belief is incompatible with Sam's choosing to travel to Boston on Monday. This is the *serious possibility* condition.

When and only when X's state of full belief satisfies conditions (i)–(iv) does X judge that doing R is feasible for Sam for the purpose of advising Sam concerning what to do.

I contend that the primary prescriptive function of principles of rational choice is to furnish criteria for determining what a rational agent ought to do, given the agent's goals and values. In particular, such criteria ought to be applicable when the deliberating agent is concerned with identifying optimal or admissible options from a given set of options judged to be feasible relative to the agent's state of full belief and values. (An admissible option is one that the agent is not forbidden to choose according to the principles of choice.)

From what has been said thus far, it appears that judgments of feasibility differ in the context of passing judgment on the rationality of choice versus the context of deciding what is to be done. But the difference can be exaggerated.

When passing judgment on the rationality of Sam's choice, there are two cases to consider. First, there is the question of whether Sam chose rationally relative to the information available to him. In this case, X should take the feasible options to be those that are judged to be feasible relative to Sam's state of full belief, when such judgments satisfy conditions (i)–(iv) applied to *Sam's* state of full belief. Even if X (who passes the judgment) is distinct from Sam, X should make the assessment on the assumption that Sam's state of full belief met these requirements prior to reaching a decision. X might actually disagree with Sam on some issue of feasibility; in particular, X can have views of his own as to what Sam will do. But when assessing the rationality of Sam's choice relative to the information available to him, feasibility should be as judged by Sam in the sense of meeting (iv) as well as the other requirements. The reason is that X is passing judgment on the rationality of Sam's judgment concerning what to do in the "thin" sense of rationality, a sense that covers minimal conditions of coherence and consistency of Sam's attitudes regardless of whether Sam's beliefs and values are sensible or accurate by X's lights or from any other point of view.

There is, to be sure, another approach to passing judgment on Sam's decision: X may evaluate the rationality of Sam's choice not relative to Sam's point of view prior to reaching a decision but relative to another body of information that (so X thinks) should have been available to Sam. Observe, however, that if the rationality of Sam's choice is at issue, then the body of information being used should meet the requirements for feasibility relevant to deciding what to do, so that – relative to that body of information – condition (iv) as well as conditions (i)–(iii) should be met.

To be sure, X might be in a state of belief of already knowing how Sam has chosen, so that condition (iv) cannot be satisfied for X's current judgments of feasibility. That is why I suggested that, in passing judgment on the rationality of Sam's choice, we may require only conditions (i)–(iii).

However, the importance of waiving condition (iv) can be exaggerated even in this context. If X judges that Sam's doing R represents a feasible

8

option for Sam in a given context where X is certain that Sam did not choose to do R, then X may still ask whether Sam should have chosen to do R relative to a system of beliefs like X's minus the information regarding what Sam did choose. By contracting X's state of full belief concerning Sam's decisions in a certain manner, the set of propositions judged to be feasible options relative to X's current state (and hence in the sense satisfying conditions (i)–(iii)) is converted into a set of propositions judged to be feasible relative to a contraction of X's current state meeting conditions (i)–(iv).

The point I mean to belabor is that passing judgment on the rationality of Sam's choices has little merit unless it gives advice as to how one should choose in predicaments similar to Sam's in relevant respects. In such situations, the evaluation of feasible options with respect to admissibility is undertaken on the assumption that conditions (i)–(iv) obtain.

Thus, even though prescriptions for rational choice can be used to pass a judgment on the rationality of a decision maker's choice, the fundamental type of application of such prescriptions is in determining what a rational agent ought to do. In particular, it becomes important to determine what a rational agent ought to do given the information available to that agent.

4. Foreknowledge of Rationality

The aim of a prescriptive theory of rational choice is to provide criteria for identifying a set of options that are optimal or, at least, "admissible" in the sense that they are not ruled out by the principles of choice, given the agent's beliefs and values. If the agent is to reach a stage in deliberation where such principles can be used to identify a set of admissible options, the agent must first identify a set A of feasible options or option descriptions from which the set $C(A)$ of admissible options is selected. No proposition can be admissible unless it is an option description – that is, represents a feasible option. And insofar as the deliberating agent is deciding what he or she is to do, and not passing judgment on what the deliberating agent or someone else has done or will do, judgments of feasibility must meet conditions (i)–(iii) as well as the serious possibility condition (iv).

In order for the agent to apply the choice criteria in determining what is to be done, the agent must not only be in a position to judge what is feasible (in the sense already explained) but must make other judgments as well. In particular, the agent must know enough about his own values (goals, preferences, utilities) and beliefs (both full beliefs and probability judgments), and have enough logical omniscience and computational

9

capacity, to use his principles of choice to determine the set $C(A)$ of optimal or admissible options. That is to say, the deliberating agent's state of belief must meet a *self-knowledge* condition and a *logical omniscience* condition. The agent does not need perfect self-knowledge or perfect logical omniscience – just enough so that, having identified a set A of feasible options, the agent's principles of choice can be used to identify the admissible set $C(A)$.

Suppose the agent meets conditions (i)–(iv), as well as the self-knowledge and logical omniscience conditions. In that case, the agent is certain what the elements of the admissible set $C(A)$ are. That is to say, the deliberating agent has identified which of the feasible options ought not to be chosen and which he is entitled to choose as a rational agent with given beliefs and goals.

Observe that nothing in the agent's beliefs need imply that the agent will choose rationally – that is, restrict the choice to an admissible option. The deliberating agent can determine what he ought rationally to do without assuming that he will actually succeed in doing it.

But suppose that the agent is confident of his or her own rationality and satisfies the *smugness* condition. The smugness condition states that the agent is certain that in the deliberation taking place at time t, X will choose an admissible option.

Assuming that logical omniscience enables the agent to identify elementary logical consequences of adding the smugness condition to (i)–(iv), to self-knowledge, and to logical omniscience, the agent must be certain of not choosing an inadmissible option. From conditions (iii) and (iv), it follows that no inadmissible option is feasible from the deliberating agent's point of view when deciding what to do; $C(A) = A$.

Though this result is not contradictory, it implies the vacuousness of principles of rational choice for the purpose of deciding what to do. To be nonvacuous, these principles should offer criteria for reducing a set of feasible options to a set of admissible options that is a proper subset of the feasible options. Sometimes the reduction fails and the two sets will coincide. If the reduction always fails, then the principles of rational choice are useless as criteria that deliberating agents might use to determine what they ought to do. If they are useless for this purpose then, by the argument of the previous section, they are useless for passing judgment on the rationality of choice as well. If principles of rational choice have (or are intended to have) a prescriptive application, we should reconsider the assumptions that lead to this result.

Notice that the argument invoked does not depend on endorsing any particular system of principles of rational choice. One can favor maximizing

10

expected utility, maximining, or any other principle and still reach the conclusion just obtained.[7]

It may, perhaps, be thought that vacuity can be avoided by appealing to the fact that deliberation takes time. Having identified a set of feasible options, it may be argued that they all remain feasible for the agent until deliberation is terminated by determining a set of admissible options. Once the set of admissible options is identified, the inadmissible options cease being feasible for the agent. But until that has happened, the options that have not yet (but will be) judged inadmissible remain feasible.[8]

The trouble with this suggestion is that, by applying the criteria of choice to identify a set of admissible options, the agent must have simultaneously identified the set of feasible options, his goals and values, and made the comparisons required of the feasible options. Once assured of choosing an admissible option, the agent cannot coherently regard any options as feasible other than admissible ones; hence, the only options to compare are those admissible ones. Emphasizing the temporal character of deliberation (in the manner briefly sketched in the previous paragraph) suggests that there is no stage in the deliberation where the agent can evaluate the feasible options with the aid of the principles of choice *unless* all feasible options are admissible. But that is precisely the threat of vacuousness.[9]

We cannot avoid the difficulty by rejecting the efficaciousness condition on the beliefs supporting judgments of feasibility. If the agent cannot by choice control whether A is true but only try to do A, then the agent's option is to try to make A true rather than making A true, and the efficaciousness condition applies to that option.[10]

Nor can we question the assumption that the agent has the logical omniscience and self-knowledge required to identify beliefs and values in sufficient detail to apply the principles of rational choice when determining which options are admissible. No doubt, decision makers often face problems too complex for them to address with their principles. Sometimes, however, these difficulties can be overcome by devising technologies that enable decision makers to make the requisite calculations or therapies that alleviate the psychological and social tensions that sometimes stand in the way of clear thinking. If decision makers never have the capacities to apply the principles of rational choice and cannot have their capacities improved by new technology and therapy, the principles are inapplicable. Inapplicability is no better a fate for principles of rational choice than vacuity.

There is a way to avoid both vacuity and inapplicability. In order to determine what he ought rationally to do, the deliberating agent should

11

be able to determine (via principles of choice) how to choose. But the agent need not assume he will choose rationally, that is, choose an admissible option. If the deliberating agent gives up the smugness condition, the admissible options need no longer coincide with the feasible ones.

The upshot of these considerations is that if principles of rational choice are to be nonvacuously applicable in deciding what to do, the decision maker should not be certain of choosing rationally, that is, of choosing an admissible option. More generally, the agent should be in a state of suspense as to which of the feasible options will be chosen.

5. Probability of choice

According to the argument developed thus far, if a proposition A is to qualify as an act proposition or act description from the point of view of a deliberating agent, then both A and its negation ought to be consistent with the agent's state of full belief K so that the truth of A is a serious possibility according to the agent at the time in question.

However, the argument does not yet prohibit the deliberating agent's assigning unconditional credal probabilities to act descriptions. I shall now argue that if such credal probability judgments are relevant in the evaluation of feasible options, then no such credal probabilities are assignable to act descriptions. If successful, this argument will sustain the approach of Savage et al. over the approach of Jeffrey.

Suppose decision maker X faces a choice between making proposition A true and making proposition B true. This assumption precludes X's having the ability to make both A and B true and constrains him to make at least one of these propositions true. Assuming that this is X's view of the matter, making B true is, from X's point of view, equivalent to making A false or making $\sim A$ true.

Suppose further that, prior to making a decision, X is offered a bet on the proposition that X will make A true. Given a fixed, positive, finite "stake" S (measured in units of utility) and a finite "price" P (measured in units of utility), X will receive $S - P$ units of utility by choosing A and will lose P units by choosing B. Suppose further that the value of the rewards and penalties is independent of the values of A and B.

X now has four options rather than two:

I: Choose to make A true and accept the bet.
II: Choose to make A true and refuse the bet.
III: Choose to make B true and accept the bet.
IV: Choose to make B true and refuse the bet.

12

Because of the efficaciousness condition, X must be certain of a better result from I than from II. The benefits of making A true are the same for both I and II; accepting the bet is better than refusing so long as $P < S$ but indifferent if $P = S$. If $P < S$ then I dominates II, but if $P = S$ then the two options are equivalued. Given the irrationality of being prepared to pay a value of P greater than S, we may assume that I is at least as good as II. Likewise, IV dominates III so long as P is positive and is equivalued with III if $P = 0$. As long as S is positive, it would be irrational for anyone to offer a bet with P negative. Thus we may concentrate on options I and IV.

Suppose X prefers A to B and recognizes this. Then I must be preferred to IV as long as P is nonnegative and less than S. Hence, as far as X is concerned, the fair price P^* for the bet ought to exactly equal S. This suggests not only that X's degree of belief that X will make A true should be 1 and the credal probability that X will make B true should be 0, but also that X regards it as not a serious possibility that X will make B true.[11]

This means, by our previous argument, that making B true is not feasible for X. If X can assign credal probabilities to hypotheses predicting his choice, then once more the criteria for rational choice become either vacuous or inapplicable for the purpose of deciding what X should choose.

To avoid this result, I suggest that one prohibit the decision maker from assigning unconditional probabilities to predictions as to how he will choose. This is, of course, the approach implicit in Savage's theory.

6. EXPLANATION OF BEHAVIOR

Schick (1979) is responsible for formulating the question of foreknowledge of one's choices in a way which shows that conditions (i)–(iv) on the decision maker's state of full belief – together with the conditions of self-knowledge, logical omniscience, and smugness – imply the coincidence of admissibility with feasibility. Schick takes the position that this result is entirely acceptable and simply means that the decision maker cannot predict which of the feasible (= admissible) options he will choose.

Schick is not distressed by the fact that principles of rational choice become empty as criteria that the decision maker might use in deciding what to do. Indeed, Schick displays little or no interest in the prescriptive function of principles of rational choice either in his 1979 paper or in his seminal 1984 book. His focus seems to be on principles of rational choice in explanation and prediction of human decisions.

But if one is to give up the smugness condition in order to allow principles of rational choice to serve their prescriptive function in guiding

13

decision makers, then the principles of rational choice cannot serve as covering laws in explanations of human decisions. Decision maker X must exempt the options he or she faces from the scope of the putative covering laws. What sort of law is it that agents claim cover every agent's decisions except their current decisions?

Notice that the objection is not to offering explanations of human behavior, but only to the claim that principles of rational choice can serve as true lawlike claims covering such behavior. It is, to be sure, of considerable (and indeed urgent) interest to ascertain the extent to which agents are and are not able to conform to the dictates of principles of rationality, as well as the extent to which diverse technologies and therapies can extend human capacities in this direction. Our preoccupation with these questions is itself testimony to our recognition that principles of rationality are not covering laws for the purpose of explanation, but rather prescriptions regulating what we ought to choose given our beliefs and goals.

Perhaps it will be suggested that we should abandon the prescriptive function of principles of rationality in order to save the explanatory power of such principles. In my judgment, we are far more likely to obtain good explanations of human behavior when we look beyond appeals to reasons and principles of rationality than when we restrict ourselves to such principles.

When we explain human behavior by appeal to reasons and principles of rationality, such explanations are best regarded either as stopgap explanations pending deeper psychological, biological, or physical explanations, or as determinations of the extent to which an agent has succeeded in living up to the requirements of rationality. We may not require that principles of rational choice be covering laws to proffer explanations in either of these senses. In that event, there is no reason to retain the smugness condition in order to preserve the explanatory role of principles of rationality from the point of view of the deliberating agent. And there is every reason to reject the smugness condition in order to secure the nonvacuous applicability of principles of rational choice as prescriptions concerning what to do.

7. COMMON KNOWLEDGE OF RATIONALITY

One casualty of the arguments of Sections 4 and 5 is the assumption that a plurality of agents engaged in a game can satisfy a condition of "common knowledge" of the rationality of the participants in the game.[12] Such a common-knowledge condition implies that each player is convinced of his or her rationality so that, for each player, criteria of rational choice are at best vacuously applicable.

14

The result is the same whether one wishes to make some sort of distinction between common knowledge and common belief. What the smugness condition requires, and what must be avoided, is the assumption that the decision maker is certain of choosing an admissible option. Whether the certainty is correct and justified, as would be required to qualify as knowledge according to received conceptions of knowledge (views I do not share), does not matter for the argument.

8. SEQUENTIAL CHOICE

Nothing in the argument offered implies, however, that the decision maker cannot assume that other agents will choose rationally or that he himself will choose rationally on some future occasion. The rejection of the smugness condition is relevant only to the deliberating agent's current set of options.

Consequently, in the context of sequential choice where the deliberating agent faces options whose choice results in further opportunities of choice at later "choice nodes," the agent may seek to predict what he will choose at subsequent choice nodes and may assume that he will choose rationally at such nodes without rendering vacuous the applicability of principles of choice to the current predicament. On the other hand, to the extent that he is focused on making predictions about subsequent actions, the agent is committed to regarding future choices as not currently subject to control, so that he cannot decide at present what will subsequently be chosen. This suggests that, even in individual sequential decision making, the extensive-form representation of the decision problem is not equivalent to a corresponding normal-form representation for the purpose of assessing admissibility. Once this equivalence is questioned, the cogency of money-pump arguments and appeals to diachronic consistency to sustain updating by Bayes theorem and standard choice consistency axioms is undermined (see Levi 1987, 1991).

9. CAUSAL DECISION THEORY

Like astrology, causal decision theory has a robustness enabling it to attract supporters in spite of the many decisive objections that have been leveled against it. (My own efforts are found in Levi 1975, 1982, 1983, 1985a,b.) Still, it is worth noting that the considerations introduced in this discussion suggest yet another difficulty.

Consider the Newcomb problem. We are told that a demon is a highly reliable forecaster of the decision maker's choices. Both ordinary language and technical discourse suggest that a reliable forecaster is one whose predictions are highly likely to be correct – that is, conditional on the forecaster predicting that h, the probability that h is high. Hence, the demon in Newcomb's problem is reliable in the sense that the probability is high that the decision maker picks the opaque box (or both boxes) conditional on the demon predicting that he will do so.[13]

In order to calculate expected utilities for the option of picking the opaque box and the option of picking both boxes, one needs to have the conditional probabilities of the demon predicting that the decision maker picks the opaque box (or both boxes) conditional on the decision maker doing so. However, these cannot be well defined if the conditional probabilities characterizing reliability are well defined, unless the unconditional probabilities of hypotheses as to what the decision maker will choose are also well defined. Unless both options are admissible for the agent, however, these alternatives cannot both bear positive probabilities – counter to what most advocates of causal decision theory seem to think.[14]

It will not do to turn to examples that rely on common causes or prisoner's-dilemma situations to motivate causal decision theory. The motivation cannot work without presupposing that act descriptions, whether admissible or not, bear positive unconditional probabilities.

10. CONCLUSION

The bulk of this paper has been devoted to explaining why the approach favored by Jeffrey, which insists that utilities and probabilities be defined over a common algebra of propositions, should be rejected in favor of views (like those of Savage) which prohibit assigning unconditional probabilities to act descriptions. The argument I have advanced pivots on a closer consideration of what the function of a prescriptive account of rational choice should be, and on how an account of this kind imposes constraints on what the deliberating agent may legitimately regard as a feasible option.

The later sections of the paper seek to indicate, in outline form, some of the important ramifications of the view of feasibility I am advocating. Whatever the merits of the conclusions I have sought to draw, I hope that enough has been said to suggest the philosophical importance of the issues raised.

NOTES

1. In constructing an axiomatization of principles of rational choice, Savage takes the conceptions of states (and events) and consequences as primitive and defines acts as functions from states to consequences. Other axiomatizations begin with acts and states as primitive. In this discussion, I am not concerned with

that issue. What is widely assumed, however, is that acts, states, and consequences are distinguishable components of representations of a decision problem regardless of how one might proceed to formalize the principles of rational choice.

2. By an "available" or a "feasible" option I mean the options judged to be available by the agent. In general, the number of options actually available may be quite small. For the purpose of eliciting or measuring probabilistic beliefs and utilities, however, Savage embedded the set of currently available options in a larger set of "acts" which are not, strictly speaking, available to the agent. As I understand him, Savage was inviting the decision maker to consider a counterfactual situation where a large number of options is available, eliciting preferences among options in such hypothetical situations. Having undertaken such elicitations, it is then assumed that the preference ranking for the hypothetically available options is preserved by the ranking of those of the hypothetically available options that are judged to be available.

I am not concerned in this discussion with problems of elicitation or measurement. Nor am I interested in deriving numerical probabilities and utilities from qualitative or comparative notions. I am not seeking to defend the "structural" axioms either explicitly or tacitly assumed by Savage. For example, I am not seeking to defend the claim that acts are functions from states to acts. If states are assigned unconditional probability but acts are not – which is, I take it, the Savage viewpoint – it follows that states must be probabilistically independent of acts. But it does not follow that acts are representable as functions from states to acts as Savage requires. In this chapter, I focus on the distinction between acts, states, and consequences that Savage shares in common with others who do not necessarily presuppose his other structural assumptions. Fishburn (1964, chaps. 2–3) clearly distinguishes the formulation of the decision problem from the questions of measuring probability and utility. Options or acts are called "prescriptive courses of action" (p. 25) and are contrasted with consequences. Each prescriptive course of action or option is then associated with a (finite) set of consequence descriptions. In order to evaluate an option one needs to compute the expected utility of the option conditional on its being implemented. To do this, appeal is made to the conditional probabilities of consequences given the implementation of the option and the unconditional utilities of the joint realizations of the option and each of the consequences. In chapter 3, Fishburn (1964) introduces states and situations where options are functions from states to consequences as a special case of his more general method of representing decision problems. In Levi (1980, p. 94) I suggest a general way of representing decision problems essentially like Fishburn's, taking note of other special cases including cases where acts are representable as functions from states to consequences (Levi 1980, chap. 4). For Fishburn and for me, distinctions like Savage's trichotomy remain important even though we do not assume that acts are functions from states to consequences.

3. Elements of this argument have been discussed for some time; see Jeffrey (1965, pp. 74–5, 167–8), Shackle (1966, p. 21), Goldman (1970, p. 194), Jeffrey (1977, pp. 136–8), Schick (1979, pp. 235–52), and Levi (1986, 58–67).

4. It may be thought that to say that x has the ability to R on a trial of kind T is to say that x is such that if x were subject to a trial of kind S it might R. (See Lewis 1973, pp. 36–43.) I think this mistaken. For one thing, the ability attribution is true or false as the case may be; conditionals, however, lack truth

17

values. (See Levi 1980, §11.9; 1988.) Suppose, however, that for the sake of the argument we grant that conditionals have truth values. The equation still fails. Assume that x (1) has the ability to R on a trial of kind T but (2) lacks the ability to R on a trial of kind T and T^*. Suppose further that (3) "x is subjected to a trial of kind T and T^* at the same time" is true. These three claims are clearly consistent. A coin constrained by its position in phase space to land heads may at the same time have the ability to land heads on a toss and at that very time be tossed. Take (1) to be equivalent to (1') "x is such that if it were subject to a trial of kind T it might R" and (2) to be equivalent to (2') "x is such that if it were subject to a trial of kind T and T^* it would fail to R." By virtue of (3), if we indulge in possible worlds then the actual world is one in which x is T and T^*. Hence the nearest world is, by the centering condition of Lewis (1973, p. 14), a world in which x is both T and T^*. So either (1') or (2') must be false by the criteria of Lewis. Efforts to construe attributions of abilities or possibility for in terms of *de re* conditionals understood as satisfying the centering requirement cannot be right. If we shift to the weak centering condition discussed by Lewis (1973, pp. 26–31), according to which the actual world is one of the possible worlds most similar to the actual world, we might try to allow (1'), (2'), and (3) all to be true by claiming that some worlds in which T holds but neither T^* nor R do are "closest" T-worlds to the actual world. We would have to say that the actual world in which T, T^*, and R all hold is just as similar to a world in which T holds but T^* and R do not as it is to itself. Advocates of possible worlds semantics for conditionals should be uncomfortable with the conception of similarity or of distance between possible worlds entailed by this approach.

5. Instead of saying that Sam has the ability to choose that Sam will R on a trial of kind S, one might say that Sam has the ability to R on a trial of kind S or that Sam has the ability to make it the case that Sam will R on a trial of kind S, where S is a process of deliberation eventuating in choice. Because Sam might terminate deliberation with a decision to R well before he actually implements the decision, one could allow for the logical possibility that Sam has the ability to choose that he will R while lacking the ability to *make it true* that he will R, where if it is made true that Sam will R then it is true that Sam will R. However, the efficaciousness condition (iii) ensures that unless X rules out that logical possibility as a serious possibility, X denies that R-ing is a feasible option for Sam. Conditions (i) and (iii) together ensure that the ability to choose to R on a trial of kind S is an ability to R on a trial of kind S and an ability to make it true that Sam will R on a trial of kind S.

6. If X's body of full belief does not satisfy the efficaciousness condition, then X is judging that Sam's traveling to Boston is not under Sam's control. Thus, if X has doubts as to whether the conditions for Sam being involved in an automobile accident or becoming enmeshed in a traffic jam will prevail on Monday, then X will not assume that it is optional to travel to Boston in four hours on Monday but perhaps only that it is open to him to try to do so (by setting off in his automobile, etc.). But then, relative to X's state of full belief, the proposition that Sam tries to drive to Boston does meet the efficaciousness condition.

7. The discussions in Jeffrey (1965, 1977) are tied closely to his own account of principles of rationality. Schick's (1979) version is more general.

8. This is the view of Jeffrey (1977).

18

9. Schick (1979, p. 240) complains that Jeffrey's (1977) proposal for avoiding the problem precludes the agent from satisfying conditions (i)–(iv), self-knowledge, logical omniscience, and smugness. That is to say, he complains that Jeffrey's proposal precludes the agent from applying the principles of rational choice to deciding what to do even while believing that he will conform to the dictates of these principles. Schick does not suppose that decision makers will always satisfy those conditions Jeffrey implies cannot be jointly satisfied, but insists that sometimes they will be.

10. This is the suggestion broached in Jeffrey (1965, pp. 74–5, 167–8). He seems to disavow it in Jeffrey (1977, p. 136), although he does not indicate why in 1977 he thought his 1965 view "inadequate."

11. There is no need to elaborate for those who insist that assigning hypotheses zero credal probability is equivalent to ruling them out as serious possibilities. I, for one, follow Jeffreys and de Finetti (among many others) in countenancing serious possibilities that carry zero probability (Levi 1989). Further elaboration is therefore in order.

   If X assigns zero probability to the hypothesis that X will make B true but still regards it as a serious possibility, option I remains better than option IV at the fair price even though the two options bear equal value when expected value, the stake, and the price are represented in terms of the standard reals. The fair price P^* would not be equal in value to the stake S even though its value as represented in the standard reals would be the same.

   According to the argument in the text, however, X ought to be prepared to pay any price for I in a pairwise choice with IV, up to and including S. The fair price P^* should be equal in value to the stake S not only in the standard representation, but also in a lexicographical ordering of the values or in a representation with the aid of the nonstandard reals. This precludes X's choosing B as a serious possibility.

12. There is a burgeoning literature on the topic of common knowledge and game theory. See in particular Aumann (1976, 1987), Bacharach (1985), and Binmore and Brandenburger (1988).

13. The reliability or accuracy of measuring instruments is understood somewhat differently. The measuring instrument's accuracy is determined by the conditional probability distribution over readings of the instrument given the true value of the quantity being measured. I emphasize presystematic understanding of notions like the reliability and accuracy of predictors and of measuring instruments because the examples invoked as intuitive support for causal decision theory appeal to such notions (or to kindred ones), and care must be exercised in determining what one may conclude about conditional probabilities from the information given in the examples. It is, of course, open to a causal decision theorist to advocate his or her point of view without appealing to such examples for intuitive support. I have argued elsewhere (e.g., in Levi 1975, 1982, 1983, 1985a,b) that there are reasons for rejecting causal decision theory on theoretical grounds. But in the present discussion, I focus exclusively on the appeal to examples to provide intuitive support.

14. In Levi (1975), I pointed out the need to assign unconditional probabilities to the options, but did not question the propriety of doing so. Instead I observed that, depending on what the unconditional probabilities happen to be, a one- or two-box solution will prove optimal according to strict Bayesian principles, and that if one allowed the unconditional probabilities to be indeterminate,

then both options would be admissible with respect to expected utility (E-admissible). Under such conditions, one might choose a maximin solution from among the E-admissible ones and this would lead to the two-box solution. I now consider this argument incoherent because the deliberating agent cannot have options if the agent is required to assign them credal probabilities.

REFERENCES

Aumann, R. (1976), "Agreeing to Disagree." *Annals of Statistics* 4: 1236–9.
Aumann, R. (1987), "Correlated Equilibrium as an Expression of Bayesian Rationality." *Econometrica* 55: 1–18.
Bacharach, M. (1985), "Some Extensions to a Claim of Aumann on an Axiomatic Model of Knowledge." *Journal of Economic Theory* 37: 167–90.
Binmore, K., and Brandenburger, A. (1988), "Common Knowledge and Game Theory." Discussion Paper #TE/88/167, Theoretical Economics Workshop, London School of Economics.
Fishburn, P. C. (1964), *Decision and Value Theory*. New York: Wiley.
Goldman, A. I. (1970), *A Theory of Human Action*. Englewood Cliffs, NJ: Prentice-Hall.
Jeffrey, R. C. (1965), *The Logic of Decision*. New York: McGraw-Hill.
Jeffrey, R. C. (1977), "A Note on the Kinematics of Preference." *Erkenntnis* 11: 135–41.
Lewis, D. (1973), *Counterfactuals*. Cambridge, MA: Harvard University Press.
Levi, I. (1974), "On Indeterminate Probabilities." *Journal of Philosophy* 71: 391–418.
Levi, I. (1975), "Newcomb's Many Problems." *Theory and Decision* 6: 161–75.
Levi, I. (1980), *The Enterprise of Knowledge*. Cambridge, MA: MIT Press.
Levi, I. (1982), "A Note on Newcombmania." *Journal of Philosophy* 79: 337–42.
Levi, I. (1983), "The Wrong Box." *Journal of Philosophy* 80: 534–42.
Levi, I. (1985a), "Epicycles." *Journal of Philosophy* 82: 104–6.
Levi, I. (1985b), "Common Causes, Smoking and Lung Cancer." In R. Campbell and L. Sowden (eds.), *Paradoxes of Rationality and Cooperation: Prisoner's Dilemma and Newcomb's Problem*. Vancouver: University of British Columbia Press, pp. 234–47.
Levi, I. (1986), *Hard Choices*. Cambridge: Cambridge University Press.
Levi, I. (1987), "The Demons of Decision." *The Monist* 70: 193–211.
Levi, I. (1988), "Iteration of Conditionals and the Ramsey Test." *Synthese* 76: 49–81.
Levi, I. (1989), "Possibility and Probability." *Erkenntnis* 31: 365–86.
Levi, I. (1991), "Consequentialism and Sequential Choice." In M. Bacharach and S. Hurley (eds.), *Foundations of Decision Theory*. London: Basil Blackwell, pp. 92–146.
Savage, L. J. (1954), *The Foundations of Statistics*. New York: Wiley.
Schick, F. (1979), "Self Knowledge, Uncertainty, and Choice." *British Journal for the Philosophy of Science* 30: 235–52.
Schick, F. (1984), *Having Reasons*. Princeton, NJ: Princeton University Press.
Shackle, G. L. S. (1966), *The Nature of Economic Thought*. Cambridge: Cambridge University Press.
Spohn, W. (1977), "Where Luce and Krantz Do Really Generalize Savage's Decision Model." *Erkenntnis* 11: 113–34.
Spohn, W. (1978), *Grundlagen der Entscheidungstheorie*. Scriptor.

2

Elicitation for games

JOSEPH B. KADANE, ISAAC LEVI, & TEDDY SEIDENFELD

1. INTRODUCTION

One of the questions often asked of subjective Bayesian game theorists is where the prior (on opponents' moves) "comes from." That is, how can such a probability distribution be elicited. This chapter addresses that question and shows that elicitation for games is exactly the same as elicitation for decision problems that involve uncertainty arising from other sources than the behavior of an opposing player. We review the elicitation of a single uncertain event, many events, dependence between action and state, and sequential decision problems. Each of these are shown to have application to game-theoretic contexts. The subjective Bayesian approach to game theory is thus shown to be operational.

We take standard Bayesian utility theory in the sense of Ramsey, Savage, and DeGroot as a standard of rational behavior. While we have studied various relaxations of these principles in other work, such as finite additivity: (Levi 1980, §§5.7, 5.9, 5.11, 12.16; Schervish, Seidenfeld, and Kadane 1984), state-dependent utility (Seidenfeld, Schervish, and Kadane 1990), and ordering (Levi 1974, 1980, 1986; Schervish, Seidenfeld, and Kadane 1990), the only relaxation of importance for this paper is possible dependence between act and state. To those principles we would add other principles or accept constraints on these principles only reluctantly. A general review of the consequences of this stance for game theory is given in Kadane and Larkey (1982, 1983) (see also Harsanyi 1982).

The issue we address here is how to find the prior opinions, with the intent of being operational. Our thesis is that the elicitation of prior opinions for games is no different in kind from elicitation in any other decision context. We first review elicitation of probabilities of events in Bayesian decision theory, and then consider the application of those ideas to games.

21

2. Elicitation of discrete probabilities

Suppose for the sake of simplicity that we concentrate on a single event, A, that it will rain tomorrow in Pittsburgh. Suppose you say that your personal probability for A is .37 and that your personal probability for A^c is .60. It is a theorem that at least one of these statements is in error. A theory of elicitation must be capable of dealing with such errors. A natural way for a (Bayesian) statistician S to think about such a situation is to imagine that you have personal probabilities that are reported perhaps with error. Then S has a joint probability distribution on your personal probabilities and the errors, or (equivalently) a joint distribution on your personal probabilities and your responses. By Bayes's theorem, the statistician S has a conditional distribution for your personal probabilities, given your responses, which (by Bayes's rule) leads to S's posterior distribution for your personal probabilities.

Some care must be given to the question of whose conditional probabilities these are. If you are S – that is, if it is the same person assigning the error probabilities and giving the responses – then the theory asks that person to have personal probabilities on his or her own personal probabilities. This is a very problematic situation, known as the Savage–Woodbury problem (Savage 1954, p. 58). Good (1983, p. 98) has addressed this problem, and Skyrms (1984, pp. 29–36) deals with it under the heading of "higher order" probability; see also Gaifman (1988). We believe this problem is well worth avoiding and, for this reason, we keep separate the views of the subject ("You"), whose probabilities about A and A^c are under investigation, from the observer (the "statistician"), who has a joint distribution over the subject's personal probabilities and responses. In this matter we follow the path indicated by Lindley, Tversky, and Brown (1979). Especially, we follow Lindley (1985) in *not* requiring that the subject's responses be coherent; that is, they need not form a coherent probability. (This additional complication is not pursued in detail in this chapter.)

We now return to the question of rain tomorrow. Suppose that You (as subject) have some probability of this event, unknown to statistical observer S. Then S does not know your log-odds for A, defined to be $y = \log(q/(1-q))$. Suppose S's prior on y is normal, with mean μ and variance σ^2.

Suppose now that You are asked about a probability for A and give probability p, or (equivalently) log-odds $x = \log(p/(1-p))$. What should S make of this information? S might suppose that X given y has a normal distribution with mean y and variance σ_2^2 (written $f(x \mid y) \sim N(y, \sigma_2^2)$). Some consequences of these beliefs are that the predictive distribution for X is

22

$$(1) \qquad f(x) \sim N(y, \sigma_1^2 + \sigma_2^2),$$

and that the conditional distribution for y given x is

$$(2) \qquad f(y|x) \sim N\left(\dfrac{\dfrac{\mu_1}{\sigma_1^2} + \dfrac{x}{\sigma_2^2}}{\dfrac{1}{\sigma_1^2} + \dfrac{1}{\sigma_2^2}}, \dfrac{1}{\dfrac{1}{\sigma_1^2} + \dfrac{1}{\sigma^2}} \right).$$

Formula (2) can be written in a neater form using the notation of precisions. Let $h_i = 1/\sigma_i^2$, $i = 1, 2$. Then (2) can be rewritten as

$$(3) \qquad f(y|x) \sim N\left(\dfrac{\mu_1 h_1 + x h_2}{h_1 + h_2}, \dfrac{1}{h_1 + h_2} \right).$$

In order to appreciate (3), we pause to examine some special cases. Suppose S is very unsure of what You think about A. This could be modeled by allowing σ_1^2 to become indefinitely large; thus $\sigma_1^2 \to \infty$ or $h_i \to 0$. In this case (3) takes the special form

$$(4) \qquad f(y|x) \sim N\left(x, \dfrac{1}{h_2} = \sigma_2^2 \right).$$

In this special case, S takes You at Your word and centers his opinion at x, where You say Your log-odds are, with precision h_2 or variance σ_2^2.

Now suppose S thinks that You are very poor at giving answers to such questions; for example, S thinks that You are drunk. Then S might take σ_2^2 to be very large. As $\sigma_2^2 \to \infty$, or $h_2 \to 0$, (3) simplifies to

$$(5) \qquad f(y|x) \sim N\left(\mu_1, \dfrac{1}{h_1} = \sigma_1^2 \right).$$

In this special case, S's opinions after hearing Your (drunken) log-odds X are no different than they were before. Hence both these special cases are behaving as one might hope they would. In the general case (3) the mean of S's posterior opinion about y is a weighted average of μ and x, with weights equal to the precisions.

Now suppose that, instead of a single event A whose probability is to be considered, we have a partition $A_1, \ldots, A_k$, mutually exclusive and exhaustive. Let $q_1, \ldots, q_k$ be Your probabilities on these events, and let

$$(6) \qquad \mathbf{y} = (\log(q_1/q_k), \log(q_2/q_k), \ldots, \log(q_{k-1}/q_k)).$$

As before, suppose that S's opinion about $\mathbf{y}$ has the multivariate normal distribution $\mathbf{y} \sim N(\boldsymbol{\mu}, \Sigma_1)$. Also suppose that You make available to S the elicited odds

$$(7) \qquad \mathbf{x} = (\log(p_1/p_k), \log(p_2/p_k), \ldots, \log(p_{k-1}/p_k)),$$

23

which are assumed by S to have the distribution, given $\mathbf{y}$,

$$(8) \qquad\qquad f(\mathbf{x}\,|\,\mathbf{y}) \sim N(\mathbf{y}, \Sigma_2).$$

Then the probability calculus yields the generalization of (3),

$$f(\mathbf{y}\,|\,\mathbf{x}) \sim N((\mathbf{x}H_2 + \mu_1 H_1)(H_1 + H_2)^{-1}, (H_1 + H_2)^{-1}),$$

where $H_i = \Sigma_i^{-1}$. The special cases just discussed generalize analogously.

A second extension occurs when You face a decision that may conceivably influence the outcome in question (moral hazard). A typical example would be whether an insurance company should insure a young executive against death in a parachute accident. The company, S, might reasonably think the executive's chances of dying in such an accident are higher if they grant the insurance than if they do not.

In this problem there are two events of interest: death if insured, A_1, and death if not insured, A_2. Suppose these events have probabilities to S of q_1 and q_2, and let

$$\mathbf{x} = (\log(p_1/(1-p_1)), \log(p_2/(1-p_2))).$$

Again the previous theory is applicable. Such a framework can model both independence and dependence among events, elicitation error, and so forth.

A third extension concerns sequential decision problems. These are handled by backward induction, with elicitation at each node. We need only consider the case where the information available at each node includes the hypothetical history that precedes it.

3. ELICITATION IN THE CONTEXT OF GAMES

Each of the four cases described in Section 2 has applications to games. The initial case of single dichotomy has an analog in the single-play game against an opponent who has two choices, in which there are no dependencies between my action and my opponent's. The multivariate case is analogous to the same situation where the opponent has k choices instead of two. The case of moral hazard may be compared to a model of the prisoner's dilemma against an opponent "just like me." If I choose to defect, my probability that my opponent will also is high; conversely, if I choose to cooperate, my probability that my opponent will so choose is also high. Although such a view may sound peculiar when applied to games, it is coherent. Finally, sequential decision problems find application also in the game context. The Harsanyi–Selten example discussed by Harper (1989) is one example where a prior reasonably changes according to the history before a particular node is reached. See also DeGroot and

Kadane (1983) for another example of a sequential game with more than one decision maker.

It might be claimed that the Bayesian theory sketched here does not give guidance about what a person *should* think in the context of a game. It does not. On the other hand, Bayesian theory does not purport to tell a person what to think about the weather or any other kind of uncertain event either. Rather, its general advice – to make a probabilistic model, to condition on all available information, etc. – applies to both game and nongame problems.

Perhaps it is also worth remarking that concepts of common knowledge, Nash equilibrium (or any other equilibrium concept), minimax, and so on have not been mentioned, and seem to play no particularly natural role. For those of us who wish to reject McClennen's (1992) advice and keep subjective expected utility maximization in a central role in noncooperative game theory, this approach suggests that we can do so directly, without adding more normative concepts.

REFERENCES

DeGroot, M. H., and Kadane, J. B. (1983), "Optimal Sequential Decisions in Problems Involving More than One Decision Maker." In H. Rizvi, J. S. Rustagi, and D. Siegmund (eds.), *Recent Advances in Statistics – Papers Submitted in Honor of Herman Chernoff's Sixtieth Birthday.* New York: Academic Press, pp. 197–210.

Gaifman, H. (1988), "A Theory of Higher Order Probabilities." In W. L. Harper and B. Skyrms (eds.), *Causation in Decision, Belief Change and Statistics,* vol. II. Dordrecht: Kluwer.

Good, I. J. (1983), *Good Thinking.* Minneapolis: University of Minnesota Press.

Harper, W. (1989), Unpublished paper presented at the workshop on "Knowledge, Belief, and Strategic Interaction" (June 1989), Castiglioncello, Italy.

Harsanyi, J. (1982), "Subjective Probability and the Theory of Games: Comments on Kadane and Larkey's Paper." *Management Science* 28: 121-3.

Kadane, J. B., and Larkey, P. D. (1982), "Subjective Probability and the Theory of Games." *Management Science* 28: 113–20.

Kadane, J. B., and Larkey, P. D. (1983), "The Confusion of Is and Ought in Game Theoretic Contexts." *Management Science* 29: 1365–79.

Levi, I. (1974), "On Indeterminate Probabilities." *Journal of Philosophy* 71: 391–418.

Levi, I. (1980), *Enterprise of Knowledge.* Cambridge, MA: MIT Press.

Levi, I. (1986), *Hard Choices.* Cambridge: Cambridge University Press.

Lindley, D. V., Tversky, A., and Brown, R. (1979), "On the Reconciliation of Probability Assessments." *Journal of the Royal Statistical Society* A 142: 146–80 (with discussion).

Lindley, D. V. (1985), "Reconciliation of Discrete Probability Distributions." In J. M. Bernardo, M. H. DeGroot, D. V. Lindley, and A. F. M. Smith (eds.), *Bayesian Statistics,* vol. 2. Amsterdam: North Holland, pp. 375–90.

McClennen, E. (1992), "Rational Choice in the Context of Ideal Games." In C. Bicchieri and M. L. Dalla Chiara (eds.), *Knowledge, Belief, and Strategic Interaction.* Cambridge: Cambridge University Press.

Savage, L. J. (1954), *Foundations of Statistics.* New York: Wiley.

Schervish, M., Seidenfeld, T., and Kadane, J. B. (1984), "The Extent of Non-Conglomerability in Finitely Additive Probabilities." *Zeitschrift für Wahrscheinlictkeitstheorie und verwandte Gebiete* 66: 205–26.

Schervish, M. J., Seidenfeld, T., and Kadane, J. B. (1990), "State Dependent Utilities." *Journal of the American Statistical Association* 85: 840–7.

Seidenfeld, T., Schervish, M., and Kadane, J. B. (1990), "Decisions without Ordering." In W. Seig (ed.), *Acting and Reflecting: The Interdisciplinary Turn in Philosophy.* Dordrecht: Kluwer, pp. 143–70.

Skyrms, B. (1984), *Pragmatics and Empiricism.* New Haven, CT: Yale University Press.

3

Equilibrium, common knowledge, and optimal sequential decisions

JOSEPH B. KADANE & TEDDY SEIDENFELD

1. INTRODUCTION

In a paper described as an initial exploration of what two Bayesians need to know in order to play a sequential game against each other, DeGroot and Kadane (1983) argue that optimal sequential decisions need not conform to much of what traditional game theory requires of rational play. Specifically:

(1) The players' optimal strategies need not form a Nash equilibrium.
(2) Nor do the players need to know (or even believe that they know) the optimal choices of their opponent; there is no requirement of "common knowledge," in that sense.

Nonetheless, these authors propose that

(3) Reasoning by backward induction succeeds in locating optimal play.

Each of these three claims is a point of active dispute. For example, regarding (1), in an extended defense of a refined equilibrium concept, Harsanyi and Selten (1988) argue that because of common knowledge (of mutual rationality) the agents *ought to* settle on an equilibrium solution – but a refined equilibrium (see also Harsanyi 1989). Aumann (1987), like Harsanyi and Selten, seeks to reconcile Bayesian and game-theoretic rationality but is led to a theory of correlated equilibrium based on an assumption of a "common prior" for different players. Relating to (2), Binmore and Brandeburger (1988), wary of common-knowledge assumptions, seek other grounds for justifying equilibrium solutions. Concerning (3), Bicchieri (1989) questions the validity of backward induction in cases where the agents have too much or too little common knowledge.

We dedicate this paper to the memory of our dear friend and colleague, *Morris H. DeGroot*.
    Research for this work was supported, in part, by NSF Grants DMS-8705646, DMS-8701770, SES-8900025, by ONR Contract N00014-89-J-1851, and by the Buhl Foundation.

The position we take in this chapter is based on the initial exploration of DeGroot and Kadane (1983). We extend the central example of that work to include the Harsanyi–Selten "trembling hand" model of choices. We argue that, in the extended example as in the original one, equilibrium is not a norm for rational play. (This is a position already announced by Kadane and Larkey 1982.) Based on our views about what may serve as states of uncertainty in a common prior across players, we argue that rational play need not result in a correlated equilibrium. And we argue that there is no problem of too much common knowledge. That is, with respect to the standards of expected utility, backward induction is a valid method for arriving at optimal play. In short, we respectfully disagree with each of the authors mentioned above!

The analysis we offer in this chapter supports the condition of "rationalizability" for strategies, a view intelligently defended in papers by Bernheim (1984) and Pearce (1984). In addition, a version of the DeGroot-Kadane game that introduces common priors leads us to a conclusion similar to that reported by Rubinstein (1989); namely, approximating common knowledge does not yield strategies that approximate optimal play under common knowledge. Thus, we endorse the attitude expressed by many who emphasize a careful assessment of *what* players in a game know of each other's beliefs and, especially for sequential games, *when* they come to know it.

2. The DeGroot–Kadane game

The DeGroot-Kadane game is played between two agents by successive moves of a visible pointer located on the real line. Following their (1983) presentation, suppose the game has three moves: First player 1 moves the pointer, then player 2, and last player 1 moves it again. The payoff (a loss given in utiles) to each player is a function of two components: how far the player has moved the pointer and, after the final move, how far the pointer is from that player's designated target value (x for player 1 and y for player 2). Put formally, let s_0 be the initial location of the pointer and let s_i be its location after the three moves ($i = 1, 2, 3$). Thus, $u = s_1 - s_0$ is player 1's first move, $v = s_2 - s_1$ is player 2's move, and $w = s_3 - s_2$ is player 1's second (and final) move. The payoff (loss) to player 1 is

$$(2.1) \qquad L_1 = q(s_3 - x)^2 + u^2 + w^2$$

and the payoff to player 2 is

$$(2.2) \qquad L_2 = r(s_3 - y)^2 + v^2.$$

DeGroot and Kadane examine two version of this game. In both versions, s_0, q, and r are quantities known to both players. However, in the first version each player knows both targets (so that x and y are common knowledge), while in the second version of the game each player knows only his or her respective target and is uncertain about the opponent's target. In both versions of the game the players are (expected) utility maximizers, and each player models the opponent in that way too. Thus, in the language of Pearce (1984) and Bernheim (1984), the players construct rationalizable strategies.

In the simple version, where the targets are common knowledge, De-Groot and Kadane show that optimal play, as identified by backward induction, yields the following strategies:

$$(2.3) \qquad\qquad w = \frac{q(x-s_2)}{q+1},$$

$$(2.4) \qquad\qquad v = (1-k)(m-s_1),$$

and

$$(2.5) \qquad\qquad u = \frac{(q+1)(x-s_0)+r(x-y)}{(q+1)[1+(q+1)/qk^2]},$$

where

$$m = (q+1)y - qx \quad \text{and} \quad k = \frac{(q+1)^2}{r+(q+1)^2}.$$

Evidently, these strategies are in equilibrium; that is, each constitutes the best reply if the opponent's play is as specified above. Because the solutions are unique, these strategies make for a "strong equilibrium" in Harsanyi's (1977, p. 104) sense. Moreover, the game has no maximin value for either player. That is, given a proposed value $V < 0$, there are "silly" moves the opponent is permitted to make that force the player's loss to exceed V.

Because the players cannot cooperate in this game (there are no binding agreements), it is not surprising that optimal play does not yield Pareto efficiency. For example, as noted by DeGroot and Kadane (1983, Thm. 2-ii, p. 202), if $s_0 < x < y$ then player 1's first move u is negative – *player 1 moves the pointer away from both targets* – if and only if $x - s_0/(y-x) < r/(1+q)$. Such a move is clearly Pareto inefficient, as it leads to increased losses for each player compared to what they can achieve subject to binding agreements.

3. An objection by Bicchieri

In response to a challenge raised by Bicchieri (1989) concerning the legitimacy of backward induction, we note that these strategies are optimal under the assumptions of the model for the game. As we understand Bicchieri's worry, it is that hypothetical reasoning used with backward induction does not accurately reflect how players would react were the hypothetical conditions realized. In terms of the simple DeGroot–Kadane game, we believe her objection takes the following form.

The optimality of player 2's move, given by (2.4), depends upon the assumption that player 1 makes the final move according to (2.3). (That choice of w is determined, according to backward induction, by the fact that – regardless of what has happened on the first two moves – player 1 minimizes L_1 by adhering to (2.3).) Then, regardless of player 1's initial move, player 2 does best by conforming to (2.4). However, according to (our interpretation of) Bicchieri's objection, were player 2 to observe that the initial move by player 1 fails to satisfy (2.5), then the assumption that player 1 will satisfy (2.3) also becomes questionable; hence the backward induction reasoning leading to (2.4) is undone by player 2's observation of the initial move by player 1. To conclude (our account of) Bicchieri's analysis: The hypothetical reasoning that (2.4) is best for player 2, *regardless* of what player 1 chooses as an initial move, is not correct; hence, backward induction is not valid.

Our response to this objection is that the backward induction reasoning, as illustrated by the argument that player 2's move should agree with (2.4), does not indicate what player 2 should do if the model fails, as would be indicated by player 1 failing to satisfy (2.5). Rather, backward induction is used by DeGroot and Kadane merely as an algorithm for determining what is optimal *under the model* for their game. Backward induction employs hypothetical reasoning of the form: "If player 1 has moved to s_1 and this accords with the model (i.e., if that is best for player 1), then what is player 2's best move?" Of course, we discover that player 1's final move accords with the model if and only if it satisfies (2.3), player 2's move accords with the model if and only if it satisfies (2.4), and player 1's first move accords with the model if and only if it satisfies (2.5). But we can use backward induction (and hypothetical reasoning) to discover this without requiring the model be consistent with each (physically) possible move that the players are capable of making.

What is "the model" for the game? In the DeGroot–Kadane game it is the combined assumptions that the players know the initial position s_0, targets (x, y), and the permissible moves, they know the loss functions (2.1) and (2.2), and they accept that each player is concerned to minimize his or her loss and knows this of the other. Under the DeGroot–Kadane model, the solution (2.3)–(2.5) is optimal; that is, it maximizes each player's utility and backward induction correctly identifies the solution. Reasoning by backward induction to arrive at the DeGroot–Kadane solution is not to be confused with reasoning how (for example) player 2 would play when the model fails.

Let us illustrate the difference. Suppose, contrary to the DeGroot–Kadane model, that by some action (internal or external to the game) player 1 can cause player 2 to believe that the final move might not conform to (2.3). Under this alternative model, optimal play for player 1 can differ from that prescribed by (2.5).

Consider the game with $q = r = 1$, $s_0 = 0$, $x = 1$, and $y = 2$. Optimal play (under the DeGroot–Kadane model) yields choices $u = 4/33$, $v = 19/33$, and $w = 5/33$, with a payoff (loss) to player 1 of $L_1 = 2/33$ (≈ 0.06) and to player 2 of $L_2 \approx 1.66$. Suppose, in fact, that player 2 adopts the DeGroot–Kadane model for the game and that player 1 knows it. What would result if the first player were to depart from the strategy (2.5), $u = 4/33$, and instead make the surprising move $s_1 = s_0$? What would happen if player 1 chose not to move the pointer on the first round ($u = 0$)?

With (2.5) failed, it would establish conclusively for player 2 that the hypothesized DeGroot–Kadane model is false. How would player 2 react? Might not player 2 come to doubt that player 1 will maximize in choosing w? Might not player 2 come to think player 1 will again refuse to move the pointer and choose $s_3 = s_2$ (corresponding to $w = 0$)? If that is player 2's reaction to the surprise of $u = 0$ (i.e., if then player 2 predicts the choice $w = 0$), the very best player 2 can do is to make the move $v = 1$ and earn the loss $L_2 = 2$. But that yields player 1 the best possible score, $L_1 = 0$, by choosing $w = 0$ – ironically, just as player 2 predicts. Can player 1 anticipate player 2's reaction to the initial move ($u = 0$) and fool player 2 into this false model for the choice of w (which coincidentally happens to yield a correct prediction for the third move if player 2 chooses $v = 1$)? Such a reaction by player 2 improves player 1's payoff over the strategy (2.3)–(2.5). Does this hypothetical reasoning refute the backward induction argument leading to the strategies (2.3)–(2.5)?

We do not think the question of how player 2 would respond to a failure of his model for the game is to be answered by the logic of decisions. How player 2 would react to the move $u = 0$ is instead an empirical matter. Our point here is simple: A correct interpretation of the backward induction argument is to see it as reasoning used to identify optimal play *under the conditions of a model for the game*. Backward induction does not include (or require) the counterfactual reasoning that is needed when a player's model of the game is falsified, so the strategy (2.3)–(2.5) is optimal under the conditions of the DeGroot–Kadane model. In Section 4 we illustrate how a small change in the DeGroot–Kadane game leads to a model of rational play consistent with all possible observations.

4. Trembling hands in the simple DeGroot–Kadane game

John Harsanyi and Reinhard Selten question the adequacy of Nash's equilibrium concept when applied to the normal-form version of an extensive-form game. They deny the equivalence of normal and extensive game forms. Instead, they advocate a refined equilibrium concept for extensive-form games, based on a "trembling-hand" model of choice. (But even such refined equilibria are subject to criticism, as illustrated by Pearce 1984, p. 1044.) An equilibrium for extensive forms is acceptable, according to their account, provided it is robust over small perturbations in choice. Specifically, in order to avoid "imperfect equilibria," they alter the basic moves in a game so that an agent selects one from a set of distributions (on pure options); a player chooses a mixed strategy rather than a pure option.

One of their examples beautifully illustrates the difference between the two kinds of equilibria. Figures 1 and 2 report (respectively) the extensive and normal forms of their game. Figure 3 gives the normal form for the perturbed game, where players may choose one of two mixed strategies in a perturbed extensive-form game (not pictured).

In their game, each player has two pure strategies. In the extensive form, strategy $\{a, b\}$ for player 1 and – provided 1's information set is reached (provided player 1 chooses a) – strategy $\{c, d\}$ for player 2; in the corresponding normal form, $\{A, B\}$ for player 1 and $\{C, D\}$ for player 2. (Note that the normal form fails to distinguish between the extensive form of Figure 1 and a different game where both play simultaneously, i.e., where player 2's information set does not reflect whether player 1 chooses a or b.) In the perturbed game, the normal-form options given in Figure 3 arise by using a two-point distribution, with probabilities $(1 - \epsilon)$ and ϵ assigned to each pure option in the corresponding perturbed extensive form.

Observe that, corresponding to the normal-form Figure 2, there are two equilibria: the pairs $\{A, C\}$ and $\{B, D\}$. However, the latter is "imperfect" in the extensive form of Figure 1, as that requires player 2 to (threaten to) play option d in case choice node **2** is reached. Of course, at node **2**, player 2 maximizes by playing option c instead, and player 1 knows this fact.

In the perturbed versions of the game, this difference between the two solution pairs (which are in equilibrium in the game form of Figure 2) is made evident. In the normal form of Figure 3, only the pair $\{A^*, C^*\}$ is in equilibrium. The $\{B^*, D^*\}$ pair is not in equilibrium because, when player 1 chooses B^*, player 2 improves 2's (expected) payoff by shifting from D^* to C^*; that is, D^* is not player 2's best response to B^*.

32

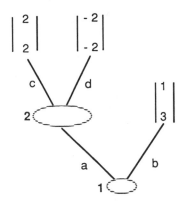

Figure 1

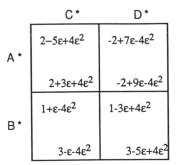

	C	D
A	2	- 2
	2	- 2
B	1	1
	3	3

Figure 2

	C*	D*
A*	$2-5\varepsilon+4\varepsilon^2$	$-2+7\varepsilon-4\varepsilon^2$
	$2+3\varepsilon+4\varepsilon^2$	$-2+9\varepsilon-4\varepsilon^2$
B*	$1+\varepsilon-4\varepsilon^2$	$1-3\varepsilon+4\varepsilon^2$
	$3-\varepsilon-4\varepsilon^2$	$3-5\varepsilon+4\varepsilon^2$

Figure 3

33

The Harsanyi–Selten idea is that imperfect equilibria are deficient because, in extensive game forms, they require a player to choose an outcome that fails to maximize utility. Nonetheless, the suspect choice is justified by Nash's criterion of equilibrium in the corresponding normal form – and it can be viewed as "threat" in the extensive form.

We agree with the Harsanyi–Selten objection to such imperfect equilibria. In the extensive form of their game, player 2 does not maximize utility by choosing option d (if node 2 arises) – d is an idle threat. That move is inconsistent with the assumption that the players are utility maximizers. In contrast, the trembling-hand model of choice eliminates the imperfect equilibrium from the normal form of the game. When the choices are mixed options, the imperfect equilibria fail to be Nash equilibria.

We may incorporate trembling hands in the DeGroot–Kadane game by limiting players to distributions for the location of the pointer, rather than supposing that player moves fix the pointer location exactly. Suppose a player moves by determining the mean of the distribution for the pointer, and suppose that distribution has a fixed and finite variance. Thus, a player may aim as follows:

(4.1) the player may fix $E_k(s) = k$ by aiming the pointer at location k; and

(4.2) the distribution has known, finite variance, $E_k[(s-k)^2] = c$ ($0 \leq c < \infty$), where c does not depend upon k but may reflect the stage of the game and past locations of the pointer.

The DeGroot–Kadane game (of Section 2) is a special case, where $c = 0$. The loss functions for the modified game are again given by (2.1) and (2.2), as formulated in terms of the successive, observed pointer locations. We do this in order to preserve complete information in the game. (We understand the sequential game to have *complete information* if the payoffs to each player are a known function of the public outcomes, outcomes that both players observe. This condition does not require that the players' *choices* be known to both, as the following example illustrates.) The feature of complete information would be lost if, instead, losses were defined through the unobserved aiming points.

Aside. The class of distributions specified by (4.1) and (4.2) is more general than Harsanyi–Selten's version of trembling hands. In our version, by contrast, (i) we do not require either that there be a point mass concentrated on a single pure strategy – there does not have to be probability mass assigned to point k; (ii) nor do we suppose that errors are symmetrically distributed across the alternative pure strategies.

34

Under this modification of the simple version of the DeGroot–Kadane game, backward induction leads to the same choice of aiming points as in the degenerate case, where $c = 0$.

Theorem. *For the (modified) simple version of the DeGroot–Kadane game, with moves specified by (4.1) and (4.2), the optimal aiming points are given by (2.3)–(2.5), which is the same solution as in the original game where $c = 0$.*

Proof. The reason for the theorem is the well-known fact (about point estimates) that the mean of a distribution minimizes expected squared error.

We illustrate the calculation for player 1's final move, on the third round of the game, given the observations of s_0, s_1, and s_2 (respectively, the initial location of the pointer, its location after player 1's first move, and after player 2's move). The prospective loss to player 1 after the final move, given by (2.1), is $L_1 = q(s_3 - x)^2 + u^2 + w^2$. Player 1's selection of moves is the choice of where to aim s_3 $(= s_2 + w)$, where $E_k[s_3] = s_2 + E_k[w] = k$. Thus, the expected loss in choosing $E_k(s_3) = k$ is

$$
\begin{aligned}
E_k[L_1] &= E_k[q(s_2 + w - x)^2 + u^2 + w^2] \\
&= q(s_2 - x)^2 + 2q(s_2 - x)E_k(w) + (q+1)E_k[w^2] + u^2 \\
&= q(s_2 - x)^2 + 2q(s_2 - x)(k - s_2) + (q+1)(c + [k - s_2]^2) + u^2.
\end{aligned}
$$

Solving the equation $0 = dE_k[L_1]/dk$ (to minimize expected loss) yields

$$
w = (k - s_2) = \frac{q(x - s_2)}{q+1},
$$

as required by (2.3). Equations (2.4) and (2.5) follow in a similar fashion. $\square$

Thus, the DeGroot–Kadane solution (2.3)–(2.5) is also the refined, perfect equilibrium solution advocated by the Harsanyi–Selten theory. (This result obtains because (2.3)–(2.5) are, trivially, limit points of the solutions generated by letting $c \to 0$.) There is, nonetheless, an interesting difference between the two forms of the (perfect information) DeGroot–Kadane game. In the original version, with $c = 0$, the model for the game is consistent with exactly one line of play, as dictated by (2.3)–(2.5). But, in the modified form of the simple game, if the error distribution has full support on the real line (e.g., using a normally distributed shot aimed at its mean), then each (logically) possible combination of locations for s_1, s_2, and s_3 is consistent with the model that both players are utility maximizers. No matter what player 2 sees for the location s_1 – that is, no matter where the pointer stops after player 1's first move – that outcome is consistent with the hypothesis that player 1 chose optimally in accord with (2.5).

Hence, in the trembling-hands version of this simple game ($c > 0$), with suitable error distributions, backward induction reasoning may be used to answer hypothetical questions of the form, "What would player 2 do if u is observed equal to u_1?" In other words, with such trembling-hand moves, Bicchieri's concern is satisfied: Within the model that players are utility maximizers and know this of each other, backward induction accommodates all possible moves. No possible outcome leads to a counterfactual situation.

Of course, some outcomes will be surprising though consistent under the model. Even so, we do not require of the players that they retain their belief in the model, regardless of the observed outcomes. Again, we advocate the strategies (2.3)–(2.5) as optimal under the model. Our view on this matter is no different from our view regarding statistical models generally. Sometimes observations force reevaluation of the statistical model; other times, as with "outliers," it is reasonable to do so even though the data are (formally) consistent with the model.

5. Version 2 of the DeGroot–Kadane game: Targets are not common knowledge

In the first versions of the DeGroot–Kadane game (with and without "trembles"), optimal play according to expected utility theory leads to strategies that are in perfect equilibrium. That consequence does not obtain when the game is modified and a target point is known *only* to the player for whom it is the target – where player 1 knows x but not y and player 2 knows y but not x, and this information difference itself is common knowledge. Let us rehearse the DeGroot–Kadane solutions to the second game (where $c = 0$) to see why optimal play will not form a Nash equilibrium.

According to backward induction, at the final move with u and v given, in order to minimize loss L_1, player 1 takes no interest in player 2's target y. Hence player 1's last move w is determined once again by (2.3):

$$(5.1) \qquad w = \frac{q(x - s_2)}{q + 1}.$$

Player 2 does not know x, but player 2 knows that player 1 will choose w to minimize L_1. Thus the argument leading to (2.4) does not apply directly. However, as an expected utility maximizer, player 2 has a personal probability for x (given the datum u) with mean $E_2(x \mid u)$. In light of the squared-

36

error form of the loss function (2.2) and knowing that player 1 will choose w according to (5.1), player 2 minimizes expected loss by choosing

$$(5.2) \qquad\qquad v = (1-k)[M(u) - s_1],$$

where $M(u) = (q+1)y - qE_2(x|u)$.

How shall player 1 determine the initial move u? Player 1 knows neither y nor the quantity $E_2(x|u)$. But, knowing that player 2 solves (5.2) to find v, player 1 establishes an optimal choice for u in terms of the personal joint probability for these two quantities: y and $E_2(x|u)$. The resulting optimization is given by solving the equation

$$(5.3) \qquad\qquad 0 = \frac{1}{2}\frac{d}{du}E_1\{K^2(u)\} + \frac{u(q+1)}{q},$$

where $K(u) = (s_0 + u)k - x + (1-k)M(u)$.

Although these moves are the best responses to what a player believes about the opponent's moves – that is, they maximize the subjective expected utilities of each player (under the common model that they are subjective utility maximizers) – these strategies do not form a Nash equilibrium. The strategies are not in equilibrium for the simple reason that the model does not include the targets as common knowledge. The model does not result in players knowing what the other will do, even under optimal play.

For example, if at stage **2** of the game player 1's rule for choosing w (formula (5.3)) were made known to player 2 by revealing the target x, this would alter player 2's belief set – unless player 2 thinks u reveals where x is or that x is already known, in which case $\mathrm{Var}_2[x|u] = 0$ – with the result that player 2's best move would change to agree with (2.4). Likewise, if player 1 learns both 2's target y and that player 2 learns x prior to 2's move at stage **2** of the game, then w is selected according to (2.5), not according to (5.3).

In short, exposing details of the opponent's strategy, as the Nash condition requires for ascertaining whether a replay is also a "best response," radically changes the epistemic conditions of the game. The common-knowledge assumptions leading to strategies (5.1)–(5.3) are not consistent with players verifying that theirs is a best response. The epistemic change required for satisfying the Nash condition is inconsistent with the model for the second version of the game.

This feature of our analysis is not affected by the use of trembling-hand moves. That is, if (as in Section 4) a player moves by choosing an aiming point rather than by fixing the quantity u, v, or w for certain, then optimal play does not form a Nash equilibrium, just as optimal play according to (5.1)–(5.3) does not result in Nash equilibrium. This is shown by the following.

37

Theorem. *For the (modified) second version of the DeGroot–Kadane game – where target points are not common knowledge – with trembling-hand moves specified by (4.1) and (4.2), the optimal aiming points are again given by (5.1)–(5.3).*

Proof. As before, the solution arises because the mean squared error of an estimate is minimized at the mean. In particular, player 1's final move (the choice of where to aim s_3, given u and v) is optimized by aiming w so that $0 = dE_k[L_1]/dk$; hence w satisfies (2.3). Likewise, at stage **2**, player 2 knows how the first player will aim the last shot, though player 2 may remain uncertain of player 1's target x. Nonetheless, player 2 has a personal probability for x, given u, whose mean $E_2(x|u)$ enters the optimization just as in the previous version of the game (the version without trembling hands). Because the mean minimizes the expected loss, player 2 chooses the aiming point for v according to (5.2). Equation (5.3), governing the first aiming point, is obtained in the same fashion.   □

To repeat the point of this exercise, under a model for the DeGroot–Kadane game where players have common knowledge that they are expected utility maximizers but where they lack common knowledge of their targets, and where moves are subject to trembles, optimal play does not result in a Nash equilibrium.

6. AUMANN'S CORRELATED EQUILIBRIUM AND THE DEGROOT–KADANE GAME

Aumann (1987) proposes an original unification of the game-theoretic and Bayesian decision-theoretic viewpoints. He identifies the game-theoretic perspective with a generalized account of (Nash) equilibrium, leading to what Aumann terms *correlated equilibrium*. These are best-response strategies that may rely on correlated (rather than independent) joint distributions to form mixed options. That is, the distribution used by player 1 to create a mixed strategy can be correlated with the distribution used by player 2. Aumann's account of Bayesian rationality in games leads to the result that Bayes-rational players will adopt strategies that are in correlated equilibrium. Moreover, each correlated equilibrium can be a model (with specific informational constraints on the individual players) for Bayes-rational play. Hence, there is a reconciliation of the two viewpoints.

We agree with Aumann that Bayesian decision theory should apply to games; the logic of choice is the same whether our uncertainty is about "Nature" or an opponent's moves (see Kadane and Larkey 1982). However, we take issue with (what we understand to be) Aumann's formulation of Bayes rationality in games. He requires a very rich space Ω of states of the world:

The term "state of the world" implies a definite specification of all parameters that may be the object of uncertainty on the part of any player of [the game] G. In particular, each ω includes a specification of which action is chosen by each player of G at that state ω. Conditional on a given ω, everybody knows everything; but in general, nobody knows which is really the true ω. (1987, p. 6)

Though agents are permitted private information about Ω, Aumann requires that (each) player i's personal probability (over Ω), here denoted by $p_i(\Omega \mid D_i)$, is a conditional probability that arises from a common prior: $p_i(\Omega) = p(\Omega)$ given i's (perhaps) private data $d_i \in D_i$; however, the prior is the same for each player i. That is, apart from private evidence, the players are required to have the same opinions about the set of states Ω. Because (by the severe assumption that) each state ω specifies "all parameters that may be the object of uncertainty on the part of any player," Aumann argues that the information sets D_i (though not the private information d_i) also are common knowledge to all players.

We object to Aumann's condition that there be a common prior (across players) in games. He recognizes this challenge in Section 5 of his paper. Concerning ordinary decisions, we believe there is no basis within (say) Savage's decision theory for that assumption, regardless of the detail with which states (of Nature) are defined. Savage's opposition to what he called "necessary" Bayesian theory (Savage 1954, p. 61; 1962, p. 102; 1967) leads us to think he rejected a common-prior requirement even in the structured setting of parametric statistical inference, where likelihoods are specified, a fortiori in less structured game settings where likelihoods are not so determined.

Our second concern is with consequences of demanding that Ω be as detailed as Aumann proposes. In particular, we are uncomfortable with the prospect that agents are required to hold (nontrivial) probabilities over their own current choices. (Again, we observe that Savage's theory is carefully formulated to distinguish between acts and states; states but not acts are assigned personal probability.) We do not see a problem when an agent assigns personal probabilities (more accurately, personal conditional probability) now to future choices, because the agent cannot now make those states true or false. Nor do we find a conceptual problem in assigning a personal probability to past choices, since the agent may have forgotten those past choices. The difficulty with personal probability over one's current choices is that such probabilities do not support the familiar betting-odds interpretation. (See Spohn 1977, Kadane 1985, and Levi 1989 for related discussions.)

The second version of the DeGroot–Kadane game serves to illustrate our position on this issue. Recall that in the second version, players know their respective targets but are uncertain of the other's target, and this informational structure is itself common knowledge. Recall also that it is the uncertainty about the opponent's target that alone differentiates the two versions of the game. In the first version of the game, when the informational structure of the game includes common knowledge of the targets, the optimal strategies (2.3)–(2.5) are common knowledge, too; there is no uncertainty for either player about what he or she will do.

We propose, therefore, to analyze the second version of the game (without trembles) using pairs of targets for the states $\Omega' = \{(x, y): x$ is player 1's target, y is player 2's target$\}$. We introduce a common prior $p(\cdot)$ over these states to allow a comparison with Aumann's theory. As we make clear shortly, the set Ω' is not Aumann's set of states Ω for this game. (Also, to agree with Aumann's presentation, we are prepared to use the game's normal form. That is, we see the selection of "states" as the relevant issue here, not the collapse of extensive to normal form.)

Suppose the two players begin their analysis with a common prior over Ω'; that is, they do not yet know their targets, yet they share the following background information: It is given that both players are utility maximizers, that their respective loss functions are L_1 and L_2, that the initial pointer location is s_0, and that all this is common knowledge. For simplicity, before learning their targets, assume the players have a (common) joint distribution $p(x, y)$ which is bivariate normal (μ, Σ), with known means $\mu = (x_0, y_0)$, with known and equal variances $(\sigma_x = \sigma_y = \sigma)$, and with (x, y) independent $(\rho_{xy} = 0)$.

Then, after player 1 learns x, 1's probability for player 2's target, $p_1(y \mid x) = p_1(y)$, is normal $N(y_0, \sigma^2)$, since x and y are independent. Likewise, after learning y, player 2 has uncertainty about x, denoted by $p_2(x \mid y) = p_2(x)$, which is normal $N(x_0, \sigma^2)$. These distributions are common knowledge. In particular, prior to any moves, player 1 knows that player 2's expected value for x is x_0; that is, player 1 knows $E_2(x \mid y) = E_2(x) = x_0$ and player 2 knows $E_1(y \mid x) = E_1(y) = y_0$.

Despite the common prior, this common knowledge does not induce a correlated equilibrium with respect to Ω'. That the addition of a common prior for Ω', even one that makes (x, y) independent variables, fails to yield a correlated equilibrium is explained by tracing the impact of the common prior on the solutions (5.1)–(5.3). With respect to player 1's choice of a final move w, the prior $p(x, y)$ is irrelevant because x is known and y plays no role in minimizing L_1 through the choice of s_3. At the second move, when player 2 is contemplating the choice of v, what is relevant is the quantity $E_2(x \mid u, y)$. But the common prior $p(x, y)$ does not determine this expectation! It fails to do so since it leaves open what might be player 2's beliefs about player 1's choice of u. That is, all of $p_2(u)$, $p_2(u\,|\,y)$, and $p_2(u\,|\,x,y)$ are underdetermined by the common prior on Ω'.

For instance, both players know that

$$p_2(x\,|\,u,y) = p_2(u\,|\,x,y) \cdot p_2(x\,|\,y)/p_2(u\,|\,y).$$

Also, it is common knowledge that $p_2(x\,|\,y) = p_2(x)$, where x is a normal $N(x_0, \sigma^2)$ distribution. But the common prior in (x, y) does not fix the ratio $p_2(u\,|\,x,y)/p_2(u\,|\,y)$, which is known to player 2 only. The terms $p_2(u\,|\,x,y)$ and $p_2(u\,|\,y)$ cannot be derived using Bayes's theorem merely by giving player 2 privileged information about (x, y). Specifically, the probability $p_2(u\,|\,x,y)$ should not be confused with the (point mass) solution for u, given by (2.5) from the first version of the DeGroot–Kadane game (where both targets are common knowledge and player 2 knows u for certain). It is important to correctly interpret the compound conditioning event in $p_2(u\,|\,x,y)$. That conditioning event specifies both targets, but it leaves x known to player 1 only. It is important to distinguish two cases:

(1) conditioning on the event (x, y) when these are common knowledge, as in the first version of the DeGroot–Kadane game, leading to (2.3)–(2.5); and

(2) conditioning on the event (x, y), when target x is known to player 1 alone and y is known to player 2 alone.

When (x, y) are not common knowledge, as in the second version of the game, it is the second of these two cases that the players face when evaluating the term $p_2(u\,|\,x,y)$. For some discussion on the range of values $p_2(u\,|\,x,y)$ can take (all of which are unknown to player 1), see Corollaries 1 and 2 in DeGroot and Kadane (1983, p. 206).

Thus, we see the impact of Aumann's selection of fine-grained states Ω on his result equating Bayes rationality (subject to a common prior) with correlated equilibria. For Aumann's theorem to apply, the agents must include player 1's choice of u, as well as the targets x and y, in the states of uncertainty. Then, with a common prior over the refined states $\Omega = \{(u, x, y)\}$, the problematic term $E_2(x\,|\,u,y)$ becomes common knowledge. However, to demand a common prior over Ω mandates two conditions that we find unwarranted for rational play in this game. Aumann's analysis mandates:

(i) that player 2's beliefs about player 1's choice u are transparent to player 1; and

(ii) that player 1 holds nontrivial probabilities about 1's own actions.

41

What is the basis for demanding condition (i)? What is the interpretation, from player 1's perspective, of assigning (nontrivial) probabilities to the choice u?

7. ON RUBINSTEIN'S THESIS:
ALMOST COMMON KNOWLEDGE IS NOT GOOD ENOUGH

The preceding section explored a consequence of imposing a common prior distribution $p(x,y)$ on the set of target states Ω' for the second version of the DeGroot–Kadane game. The upshot of that analysis is that a common prior on Ω' is insufficient for defining player 2's choice of move v, since it leaves open player 2's conditional distribution for player 1's move u, given targets x and y. Thus, the common prior on Ω' also leaves open player 1's first move u, since that depends upon player 1's expectation of player 2's expectation of u, and so forth.

This argument is valid for each value $\sigma^2 > 0$. (Recall that σ^2 is the common variance for the targets.) However, the first version of the game, with targets (x_0, y_0) common knowledge, corresponds to the limiting distribution $\sigma^2 = 0$. Thus, the first version of the DeGroot–Kadane game is not necessarily the limit of the second-version games with common priors, where $\sigma^2 \to 0$. That is, the optimal strategy (2.3)–(2.5) for the first game (where targets are common knowledge) may not equal the limit (as $\sigma^2 \to 0$) of optimal strategies for the second version of the game, constrained by a common prior.

Let us illustrate the point. To simplify the formulas, take $q = r = 1$ and $s_0 = 0$. With the variance $\sigma^2 > 0$ given, denote with subscripts u_σ and v_σ the choices for the first two moves. And, with some slight abuse of notation, use the subscripted u_0 and v_0 to denote the limit of these moves as $\sigma \to 0$. Suppose player 2 reasons as follows.

In the first version of the DeGroot–Kadane game, with targets common knowledge, according to (2.5) my opponent's first move u is linear in the targets x and y. That is, were our targets known, player 1 would choose

(7.1) $$u = (3x - y)/8.25.$$

So, I'll take my expectation for x to be linear in u_σ and y:

(7.2) $$E_2(x \mid u_\sigma, y) = a_\sigma + b_\sigma u_\sigma + c_\sigma y.$$

Then my move v_σ satisfies

(7.3) $$v_\sigma = 0.2[2y - a_\sigma - (1 + b_\sigma)u_\sigma - c_\sigma y].$$

42

The move v_σ, (7.3), contrasts with player 2's choice (from 2.4) of

(7.4) $$v = 0.2[2y - u - x]$$

for the case where targets are common knowledge. Recall that x and y are uncorrelated. Therefore, in order to make (7.3) equal (7.4) as $\sigma \to 0$ (i.e., for $v_0 = v$), it is necessary and also sufficient that $a_\sigma \to 0$, $b_\sigma \to 1$, $c_\sigma \to 0$, and $u_\sigma \to x$.

Now, in case player 1 knows that player 2 has the linear expectation (7.2) (without in general knowing the coefficients a_σ, b_σ, and c_σ), DeGroot and Kadane (1983, p. 206) have shown that player 1's optimal choice of u_σ, given x, satisfies

(7.5) $$u_\sigma = \frac{E_1\{(0.2b_\sigma - 0.8) \cdot (0.2[(2 - c_\sigma)y - a_\sigma] - x)\}}{2 + E_1\{(0.2b_\sigma - 0.8)^2\}}.$$

For the limiting values of the coefficients necessary to make (7.3) and (7.4) agree, this yields

(7.6) $$u_0 = E_1\{(0.6x - 0.24y)/2.36\}.$$

However, $u_0 \neq u$, (7.6) does not agree with (7.1) (which is player 1's choice for u when targets are common knowledge), and neither does $u_0 = x$, as is necessary for (7.3) to agree with (7.4).

In short, the limit of optimal play (with $\sigma \to 0$ here does not correspond to the optimal play at $\sigma = 0$. The singularity (at $\sigma^2 = 0$) occurs because merely shrinking the variance ($\sigma^2 \to 0$) of the prior distributions for the targets does not suffice also for shrinking player 2's conditional probability $p_2(u \mid x, y)$ to the point mass for u concentrated at the solution (2.5). It fails to do so because, in part, the correct interpretation of this conditional probability in the second version of our game is not to be confused with the first-version interpretation, which corresponds to common knowledge of the targets.

Though the limit ($\sigma^2 \to 0$) of the common priors is common knowledge of the targets, the limit of the optimal strategies based on these common priors need not be the optimal strategy based on common knowledge of the targets. Rubinstein is correct: Almost common knowledge is not good enough!

Remark. By supplying the two players with additional, common evidence about the targets (x, y), we can implement the dynamics of a common "posterior" with shrinking variance. For example, if both players observe n pairs (x_i', y_i') ($i = 1, \ldots, n$) of i.i.d. bivariate normal variates, with (unknown) means ($\mu_{x'} = x$, $\mu_{y'} = y$), known (equal) variances σ'^2, and zero correlation ($\rho_{x'y'} = 0$), then their common posterior distribution for the targets will be as independent bivariate normal variates with a (common) variance that shrinks to 0 as the sample size n grows without bound.

8. Conclusion

We have used a relatively simple sequential game between two utility maximizers to emphasize that optimal play among Bayesians (who model each other as such) does not put their strategies into equilibrium. Even with a common prior over the uncertain components of the game (which itself is common knowledge), that is, even with a common prior over the target points, optimal play does not require a correlated equilibrium. The optimal extensive-form strategies are rationalizable in the sense of Pearce (1984), as the reasons for (5.1)–(5.3) make clear. That is, those strategies are derived by backward induction using the common knowledge that the opponents are utility maximizers.

It is right to develop Bayesian game theory. A decision against an opponent, rather than against Nature, does not require novel principles. However, especially in sequential games, the challenge of doing Bayesian game theory against a Bayesian opponent is considerable. It has many facets: Not only must players represent their uncertainties about ordinary events (which, in the DeGroot–Kadane game, corresponds to the players' beliefs about each other's target), but each player must be prepared to formalize how his own actions will affect the other's subsequent choices. In order to do that while respecting the model of common knowledge (where each player is an expected utility maximizer), each must think about how the other models himself. That is, I must ponder what the opponent believes about my beliefs, and so on. The complexity of this thought, the depth to which each player must evaluate iterated expectations of beliefs about the other in order to apply backward induction, depends upon the number of turns in the game. Already, in the simple three-move game of this chapter (without common knowledge of the targets), that task is not trivial for player 2.

The subtleties that attend the difference between "common knowledge" and "almost common knowledge" hint at the number of different faces of Bayesian game theory. In this chapter we have focused on one game where it is common knowledge that players are rational. Not all games have that form, even when the players are, in fact, all rational. Perhaps this is a direction to look in to gain a better understanding of such tactics as bluffs and feints. We trust the challenges of Bayesian game theory will be met: some through analysis and some through empirical enquiry.

REFERENCES

Aumann, R. J. (1987), "Correlated Equilibrium as an Expression of Bayesian Rationality." *Econometrica* 55: 1-18.

Bernheim, B. D. (1984), "Rationalizable Strategic Behavior." *Econometrica* 52: 1007-28.

Bicchieri, C. (1989), "Backward Induction without Common Knowledge." In A. Fine and J. Leplin (eds.), *PSA 1988,* vol. 2. East Lansing, MI: Philosophy of Science Association. Presented at the workshop on "Knowledge, Belief, and Strategic Interaction" (June 1989), Castiglioncello, Italy.

Binmore, K., and Brandeburger, A. (1988), "Common Knowledge and Game Theory." Technical Report 89-06, Department of Economics, University of Michigan, Ann Arbor.

DeGroot, M. H., and Kadane, J. B. (1983), "Optimal Sequential Decisions in Problems Involving More Than One Decision Maker." In Rizvi, Rustagi, and Siegmund (eds.), *Recent Advances in Statistics.* New York: Academic Press, pp. 197-210.

Harsanyi, J. C. (1977), *Rational Behavior and Bargaining Equilibrium in Games and Social Situations.* Cambridge: Cambridge University Press.

Harsanyi, J. C. (1989), "Game Solutions and the Normal Form." Presented at the workshop on "Knowledge, Belief, and Strategic Interaction" (June 1989), Castiglioncello, Italy.

Harsanyi, J. C., and Selten, R. (1988), *A General Theory of Equilibrium Selection in Games.* Cambridge, MA: MIT Press.

Kadane, J. B. (1985), "Opposition of Interest in Subjective Bayesian Theory." *Management Science* 31: 1586-8.

Kadane, J. B., and Larkey, P. D. (1982), "Subjective Probability and the Theory of Games." *Management Science* 28: 113-20.

Levi, I. (1989), "Feasibility." Department of Philosophy, Columbia University, New York.

Pearce, D. G. (1984), "Rationalizable Strategic Behavior and the Problem of Perfection." *Econometrica* 52: 1029-50.

Rubinstein, A. (1989), "The Electronic Mail Game: Strategic Behavior Under 'Almost Common Knowledge'." *American Economic Review* 79: 385-91.

Savage, L. J. (1954), *The Foundations of Statistics.* New York: Wiley.

Savage, L. J. (1962), *The Foundations of Statistical Inference: A Discussion.* London: Methuen.

Savage, L. J. (1967), "Implications of Personal Probability for Induction." *Journal of Philosophy* 64: 593-607.

Spohn, W. (1977), "Where Luce and Krantz Do Really Generalize Savage's Decision Model." *Erkenntnis* 11: 113-34.

4

Rational choice in the context of ideal games

EDWARD F. MCCLENNEN

1. INTRODUCTION

Traditionally, the problem for the theory of two-person games has been to establish the solution to an ideal type of interdependent choice situation characterized by the following background condition.

(1) **Common knowledge.** There is full common knowledge of (a) the rationality of both players (whatever that turns out to mean), and (b) the strategy structure of the game for all players, and the preferences that each has with respect to outcomes.

The force of this condition is that if a player i knows something that is relevant to a rational resolution of i's decision problem, then any other player j knows that player i has that knowledge. This is typically taken to imply (among other things) that one player cannot have a conclusive reason, to which no other player has access, for choosing in a certain manner. That is, there are not hidden arguments for playing one way as opposed to another.

In addition, one invariably finds that the analysis proceeds by appeal to the following (at least partial) characterization of rational behavior for the individual participant.

(2) **Utility maximization.** Each player's preference ordering over the abstractly conceived space of outcomes and probability distributions over the events that condition such outcomes can be represented by a utility function, unique up to positive affine transformations, that satisfies the expected-utility principle.

(3) **Consequentialism.** Choice among available strategies is strictly a function of the preferences the agent has with respect to the outcomes (or disjunctive set of outcomes) associated with each strategy.

Following Hammond (1988), condition (3) can be taken to imply that strategies are nothing more than neutral access routes to outcomes (or disjunctions of outcomes); the latter are what preferentially count for the agent. In particular, then, if two strategies yield exactly the same probabilities of the same outcomes occurring, then the agent will be indifferent between those strategies.

2. The conceptual problem and its traditional solution

At the very outset of the formal study of games, one finds von Neumann and Morgenstern (1953) suggesting that one faces a conceptual as distinct from a technical difficulty when moving from the study of the isolated individual (the proverbial Robinson Crusoe) who faces an "ordinary maximum problem" to the study of interacting persons. With regard to the former:

Crusoe is given certain physical data (wants and commodities) and his task is to combine and apply them in such a fashion as to obtain a maximum resulting satisfaction. There can be no doubt that he controls exclusively all the variables upon which this result depends – say the allotting of resources, the determination of the uses of the same commodity for different wants, etc. (1953, p. 10)

To be sure, outcomes may be conditioned by such "uncontrollable" factors as weather; but these, they go on to indicate, can be registered by appeal to statistical assumptions and mathematical expectations.

What happens when one shifts to the case of a participant in a social exchange economy? In this case, the authors argue,

the result for each will depend in general not merely upon his own action but on those of the others as well. Thus each participant attempts to maximize a function . . . of which he does not control all the variables. (1953, p. 11)

This is a problem, they suggest, that cannot be overcome simply by appeal to probabilities and expectations:

Every participant can determine the variables which describe his own actions but not those of the others. Nevertheless these "alien" variables cannot, from his point of view, be described by statistical assumptions. This is because the others are guided, just as himself, by rational principles – whatever that may mean – and no *modus procedendi* can be correct which does not attempt to understand those principles and the interactions of the conflicting interests of all participants. (1953, p. 11)

Kaysen, in a very early review of von Neumann and Morgenstern's *Theory of Games and Economic Behavior,* takes this to mean that the theory of

. . . games of strategy deals precisely with the actions of several agents, in a situation in which all actions are interdependent, and where, in general, there is no possibility of what we have called parametrization that would enable each agent (player) to behave as if the actions of the others were given. In fact, *it is this very lack of parametrization which is the essence of a game.* (1946-7, p. 2; emphasis added)

What is interesting about this, of course, is that starting with von Neumann and Morgenstern and continuing on in more or less unbroken fashion ever since, the basic strategy has been to solve the game problem by showing that, contrary to Kaysen's suggestion, parametrization of one sort or other is possible. In very general terms, the notion is that if (1)-(3) govern the behavior of both players then this (together with certain additional assumptions) will enable each player to frame expectations about the behavior of the other player, expectations that are sufficiently determinate to enable each to treat the decision problem as a single maximization problem.

Within this sort of framework, the typical first move has been to defend the following as a necessary condition of rational choice.

Rejection of dominated strategies. An agent should never choose a strategy whose associated set of possible outcomes is strictly dominated by the set of outcomes associated with some other strategy.

The rationale for the dominance condition is thought to be clear enough. To say that one strategy *strictly dominates* another is to say that, regardless of one's expectation concerning how the other player will choose, the utility of the expected outcome of the former is strictly greater than the utility of the corresponding expected outcome of the latter. But then, (2) and (3) together would seem to require that one reject the latter in favor of the former.

Dominance, even when invoked iteratively, does not get one very far. For most game theorists, however, conditions (1)-(3) have been thought to be sufficiently strong to provide a grounding for Nash's (1951) equilibrium concept. Luce and Raiffa (1957) is the *locus classicus* here. The argument is indirect in form, and provides a model of how to show that parametrization will be possible within the framework of assumptions (1)-(3). Luce and Raiffa begin by assuming that there exists a theory of rational interdependent choice sufficiently determinate that, under conditions of common knowledge, each player will be able to predict what each other rational player will do; they then proceed to explore what conclusions can be drawn about the content of that theory:

It seems plausible that, if a theory offers A_{i_0} and B_{j_0} as suitable strategies, the mere knowledge of the theory should not cause either of the players to change his choice: just because the theory suggests B_{j_0} to player 2 should not be grounds

49

for player 1 to choose a strategy different from A_{i_0}; similarly, the theoretical prescription of A_{i_0} should not lead player 2 to select a strategy different from B_{j_0}. Put in terms of outcomes, if the theory singles out (A_{i_0}, B_{j_0}), then:

(i) No outcome O_{ij_0} [i.e., one that 1 could realize, given that 2 plays B_{j_0}, by playing some strategy other than one picked out by the theory] should be more preferred by 1 to $O_{i_0 j_0}$.

(ii) No outcome $O_{i_0 j}$ [i.e., one that 2 could realize, given that 1 plays A_{i_0}, by playing some strategy other than one picked out by the theory] should be more preferred by 2 to $O_{i_0 j_0}$.

And A_{i_0} and B_{j_0} satisfying conditions (i) and (ii) are said to be in *equilibrium,* and the *a priori* demand made on the theory is that the pairs of strategies it singles out shall be in equilibrium. (1957, p. 63)

One can mark in this an implicit appeal to all three of the conditions delineated. In accordance with (1), not only is the strategy and payoff structure of the game presumed to be common knowledge, it is also presumed that each player is aware of what the (postulated) theory of rational choice instructs each other player to do. But this information, when coupled with (2) and (3), provides that player with a basis for choice.

To be sure, the equilibrium condition has more recently come under a certain amount of attack, as evidenced in Bernheim (1984, 1986), Pearce (1984), and Aumann (1987). Bernheim (1986) nicely sorts out a crucial issue – namely, whether it is possible to develop a theory determinate enough for each player to be able to predict the specific choice that each other player will make. Given such predictability, the equilibrium condition is a necessary condition on the solution to any ideal game: that is, the solution to any game will have to be a refinement of the set of Nash equilibria. If predictability does not hold then it appears that one must retreat to some weaker condition, such as rationalizability. Bernheim (1986) also provides a useful exploration of what assumptions, in addition to (1)–(3), suffice to characterize the various alternative solution concepts that have emerged. For my purposes here, however, what is significant is that all of these revisionist moves have in common with the original equilibrium perspective both a commitment to the principle of iterated dominance and all three of the conditions listed in Section 1. My concern is with the implications of any theory of this type, so it will usually suffice, when a representative example is needed, to refer to the equilibrium theory.

3. PROBLEMS WITH CONCEPTUALIZING INTERACTIVE CHOICE AS A SPECIES OF PARAMETRIZED CHOICE

What characterizes the theories just discussed is that each rational player is presumed to face the task of framing some sort of "reasonable" estimate as to how the other players will choose, and then, per conditions (2) and (3), responding to that estimate by choosing among personal strategies so as to maximize his or her (subjectively defined) expected utility over associated outcomes. However, it is a matter of considerable interest, from both an analytic and a historical point of view, that von Neumann and Morgenstern adopted a quite distinct way of conceptualizing the task facing the individual player in the zero-sum (i.e., perfectly competitive) game.

Von Neumann and Morgenstern begin by explicitly appealing to an indirect argument, parallel to but quite distinct from the one employed by Luce and Raiffa:

Let us now imagine that there exists a complete theory of the zero-sum two-person game which tells each player what to do, and which is absolutely convincing. If the players knew such a theory then each player would have to assume that his strategy has been "found out" by his opponent. The opponent knows the theory, and he knows that a player would be unwise not to follow it. Thus the hypothesis of the existence of a satisfactory theory legitimatizes our investigation of the situation when a player's strategy is "found out" by his opponent. (1953, pp. 147–8)

What is the implication of a player expecting that his choice will be found out? In the case of the zero-sum, two-person game, that the other player will correctly anticipate the given player's strategy choice, and maximize from that player's own perspective, implies that a given player must expect to end up receiving the minimum utility associated with whatever strategy is chosen. In the light of that expectation, (2) and (3) imply that the player should choose a strategy whose associated minimum-valued outcome takes on a maximum value; that is, the player should *maximin*. In this context, then, their indirect argument directly ratifies not the equilibrium requirement but rather the principle that each player should employ a maximin strategy.

Of course, within the context of zero-sum, two-person games, it is easy to establish that pairs of maximin strategies are also equilibrium pairs and vice versa. However, von Neumann and Morgenstern's indirect argument is quite general; nothing in its formulation limits its application to zero-sum games. In a more general setting, this indirect argument implies that a player must still expect that the other player will choose a utility-maximizing response to what the first player chooses. But this in turn implies that the first player should choose so as to maximize expected utility, computed on the presupposition that whatever choice is made, the other player will maximize (from that player's own perspective) in response. Let us designate any strategy that satisfies this condition a *maxilor* strategy. Correspondingly, the expected return from playing a given strategy, on the assumption that the other player responds in a utility-maximizing fashion, can be characterized as the *maxilor return* for that strategy. Given these distinctions, one can now formulate the following condition on a rational solution.

Maxilor condition. If a theory prescribes that player 1 select strategy A_{i_0}, then A_{i_0} should maximize expected utility for player 1 in the light of a maximizing response on the part of player 2; similarly, if a theory prescribes that player 2 select strategy B_{j_0}, then B_{j_0} should maximize expected utility for player 2 in the light of a maximizing response on the part of player 1.

Here, then, is a quite distinct way to arrive at the notion that each player is in a position to take the behavior of the other player as a given. To be sure, the choice behavior of the other player is in this instance a dependent variable. The net effect, however, is the same: for each available strategy, the agent need only specify the expected value of the consequence of selecting that option; thus, once again, the agent faces an ordinary maximization problem.

Note that both the equilibrium argument and the maxilor argument turn on the assumption that rational players satisfy (2) and (3). The former argument invites one to think of oneself as an outcome maximizer against an estimated choice on the part of the other player; the latter invites one to think about the requirements of outcome maximizing when one must contend with a counterpart player who is maximizing against oneself, under conditions of common knowledge. The former line of reasoning proceeds, in effect, to trace out the implications for a utility maximizer of being actively able to find out what the other player will do; the latter traces out the implications of passively anticipating that the chosen strategy will be found out by the other player. Both implications are seemingly forced upon us by the consideration that, under ideal conditions, each will be able to anticipate what the other will do.

It turns out that for the zero-sum, two-person game, a solution will satisfy the equilibrium condition if and only if it also satisfies the maxilor condition. In this context, the existence of rival conceptions of parametrizing the choice situation poses no problem. But in the non–zero-sum case, the two indirect arguments will typically yield conflicting requirements. In the much-discussed game of "chicken," for example, the maxilor solution has the two players crash head-on; equilibrium solutions exclude that case and include the two cases where one player swerves. If both arguments can be sustained, then one has an impossibility result with respect to the rational solution for non–zero-sum, two-person games.

This conflict can be resolved, of course, by dropping the maxilor condition. But there remains a problem, and it is one that arises even within the context of zero-sum games. As I sought to argue in McClennen (1972), because equilibrium strategies are not necessarily unique best replies to the choice of an equilibrium strategy by the other player, the equilibrium solution concept cannot be squared with the implications of consequentialism and expected-utility theory, specifically with respect to strategies with the same expected utility.

It should also be noted that von Neumann and Morgenstern's maxilor model at the very least serves as a striking reminder that even the strict dominance condition cannot be directly derived from the framework of conditions (1)–(3). Most game theorists have taken strict dominance as bedrock, but in the presence of a belief that the other player's choice is probabilistically dependent upon one's own choice of a strategy – precisely the sort of belief that characterizes the maxilor perspective – regimentation to the principle of strict dominance is not necessarily rational. Yet even in the recent revisionist work of Bernheim (1984, 1986), Pearce (1984), and Aumann (1987), this issue seems to be begged: certain logically possible forms of probabilistic dependence apparently have been ruled out of court.

Another (and much more remarked upon) problem that arises in connection with the standard approach for non–zero-sum games is that "rational" solutions will typically be suboptimal. The problem of suboptimality is usually introduced in connection with what is known as the prisoners' dilemma game. But the dilemma is, of course, endemic to the whole class of non–zero-sum, n-person games. If rational choice for interactive situations under the ideal conditions specified in (1) must satisfy conditions (2) and (3), as those have usually been interpreted, then one is forced to the unhappy conclusion that fully rational players who have common knowledge of each other's rationality, and common knowledge of the strategy and payoff structure of the game, must each nonetheless deliberately choose in a manner that leaves both with a less preferred outcome than they could have achieved if only they had coordinated their choices. It is also clear that there is nothing to be gained in this respect by shifting from an equilibrium to a maxilor perspective.

The response has typically been to hold fast to some version of the standard theory and insist that the suboptimality of solutions to most games is simply an unavoidable anomaly of an adequate theory of rational choice. Once this point is made, one usually finds a remark to the effect that the problem can be circumvented if provision is made for binding agreements. The suggestion is that it will be rational to ensure that agreements are binding, since both parties stand to gain thereby. But all of this carries with it the rather curious implication that rational agents will be willing to expend resources on restructuring their environment to ensure that agreements will be binding (and thereby increase their return), and yet be unwilling to agree to and act on a self-policing approach – despite the fact that typically, in virtue of the costs of making agreements binding, the latter approach would result in even greater returns to each.

Consider the following very simple game, a demi-version of the standard prisoners' dilemma:

		Player 2	
		B_1	B_2
	A_1	3, 4	1, 3
Player 1	A_2	4, 1	2, 2

Player 1 has a dominant strategy, A_2, whereas player 2 does not. Player 2 would prefer to choose B_1 and thereby cooperate so as to realize the outcome (3, 4), but only if player 2 were convinced that player 1 would be cooperative. On the standard way of reasoning, however, that player 1 has a dominant strategy suffices to determine the rational outcome of this game. The rational choice for player 1 is to play this dominant strategy, and in turn player 2, given conditions of full information and common knowledge of the rationality of each, will expect player 1 to behave just so. However, in light of this expectation, player 2's best response will be B_2. On the usual account, then, barring some way to make binding agreements with one another, rational agents who know each other to be such must settle for the outcome (2, 2), despite the fact that each would prefer the outcome (3, 4).

What characterizes the standard way of thinking about interactive rationality is the manner in which it anchors the choice of a strategy. As suggested in Section 1, there is invariably (if often only implicitly) an appeal to a consequentialist perspective, according to which strategies are merely neutral access routes to consequences. Within the framework of this assumption, any preference with respect to strategies must be accounted for by reference back to preferences for expected consequences.

Yet consequentialism so interpreted is worrisome. In particular, it is precisely player 1's (allegedly rational, and consequentialist based) disposition to choose the dominant over the dominated strategy that precludes coordination with player 2 to jointly implement a plan the consequences of which are preferred to the consequences of choosing A_2 and player 2 responding with B_2. That is, by hypothesis, player 1 prefers the outcome of coordination to the outcome of what is alleged to be rational interaction. Notice, moreover, that player 1 cannot rationalize having to settle for a utility payoff of 2 rather than 3 by reference to the dispositions of agent 2. Agent 2 would be quite willing to coordinate on a plan that will realize the joint payoff $(3, 4)$ – once assured that player 1 will really cooperate – so what stands between agent 1 and the larger payoff is just agent 1's own disposition.

This suggests that perhaps more is involved in the standard arguments than simply an appeal to conditions (2) and (3), utility maximization with respect to abstractly considered outcomes, and consequentialism. Adapting an argument offered in McClennen (1988, 1990) with respect to dynamic foundations for expected-utility reasoning, one can suggest that there is, in addition, an implicit appeal to a separability assumption. In the agent's deliberation concerning the choice of neutral means to preferred outcomes, it is invariably presupposed that the agent can separate out the piece of the problem that pertains just to the choice to be made – that is, consider it *in abstraction from the context of the interactive problem* itself – and consider how to evaluate just those options if it were the case that, instead of interacting with another rational player who is also deliberating about what choice to make, the agent needed only to take into account some parameter (about whose value the agent may, of course, be uncertain). The choice made under these transformed conditions is then the one that should be made in the context of the interactive situation (the game) itself. This may be stated somewhat more formally as follows.

Separability. Let G be any game, and let D be the problem that a given player in G would face, were the outcomes of the available strategies in G conditioned not by the choices of another player but rather by some "natural" turn of events in the world, so that the player faces (in effect) a classic problem of individual decision making under conditions of risk or uncertainty. Suppose further that the player's expectation with regard to the conditioning events corresponds to the expectations held with regard to the choice that the other player will make in G. Then the first player's preference ordering over the options in G must correspond to that player's preference ordering over the options in D.

For the decision situation D_{dpd}, corresponding to the demi–prisoners' dilemma game G_{dpd}, consequentialism appears to unproblematically imply that the player's preference ordering of outcomes in D_{dpd} determines the ordering of the alternatives in D_{dpd}. In particular, on the assumption that the turn of events in D_{dpd} is not linked (causally or probabilistically) with choices made, agent 1's preferences for the corresponding outcomes – together with the dominance principle – imply that agent 1 must choose

A_2. The point is that if it is certain that B_1 would occur then agent 1 would obviously choose A_2, and if it is certain that B_2 would occur then agent 1 would also choose A_2; so A_2 is the best choice, regardless of the turn of events in the world. In this case, then, no matter what the expectations with regard to conditioning events, there is a clear choice. But separability, in turn, requires that player 1's ordering of the alternatives in D_{dpd} determines the ordering of the alternatives faced in G_{dpd}. Thus, these two principles taken together yield the standard conclusion.

Notice, more generally, that expected-utility maximization, consequentialism, and separability together imply that if a given agent can anticipate the choice to be made by the other agent (or at least assign a probability distribution over an exclusive disjunction of possible choices), then the first agent should maximize (expected) return, given this anticipation. Correspondingly, they also imply that if the agent believes that the other player will correctly anticipate whatever choice is made, and maximizes in response, then the agent should choose a maxilor strategy. In this case, however, the appropriate model is one of decision making against nature under conditions where states of nature are causally dependent upon the agent's choice of an action. The separability assumption, then, plays a key role in both of the ways (equilibrium and maxilor) in which, contrary to Kaysen's suggestion, ideal forms of interactive choice can be treated as presenting each player with a parametrized choice problem.

Note the logic of the evaluation of choices in any such separable framework. An agent who is committed to separability will choose so as to maximize with respect to preferences for expected consequences, given the agent's expectations as to how the other agent will choose. This has the further (and most important) implication that the evaluation of any proposed coordination plan proceeds from the evaluation of what the plan calls upon a given agent to choose, holding all other features of the plan fixed; that is, it proceeds from the evaluation of each segment of that plan to the whole plan. A plan must be judged as not acceptable if it calls for some agent to make a particular choice that the agent would not be disposed to make if the decision problem were viewed as separable in the sense just introduced. Returning to game G_{dpd}, the plan calling for agent 1 to choose A_1 and agent 2 to choose A_2 must, from this perspective, be rejected: It calls upon agent 1 to make a choice that would not be made were that same set of outcomes presented in abstraction from the interactive setting of G_{dpd}.

Thus, separability places substantial restrictions on the capacity of an agent to coordinate choices with others. Indeed, separability in this context precludes coordination in any meaningful sense of that term. What is left to the agent who is committed to such a separability principle is not coordination but strategic interaction: The agent's task is to estimate how the other agent will choose and then to make unilateral adjustments in choice so as to maximize expected return.

5. The case for nonseparability in the non–zero-sum context

What recommends separability as a necessary condition for non–zero-sum rational interactive choice? As already remarked, it might well seem that, within the framework of conditions (2) and (3), it is one and the same whether the outcome of choosing an action is conditioned by choices that another agent makes or by natural events. That is, the rational agent is to conceive the problem here as no different from that faced in the case of statistical decision making against nature. This problem is one that calls for the agent to make independent adjustments in choice, against independently or dependently fixed values of the other variables, so as to achieve (by means of such an adjustment) a maximum expected return.

But this is *consequentially* costly. In the case of our demi–prisoners' dilemma, adoption of such a separable perspective precludes agent 1 from being able to agree to a cooperative scheme with agent 2 and then adhere to it. Effective cooperation between two rational agents in this situation requires agent 1 to be willing to refrain from an independent readjustment of choice (i.e., switching to A_2), given the expectation that agent 2 will choose B_1. But this is precisely what an agent who is committed to separability cannot do. To be sure, agent 1's commitment to viewing things from a separable perspective supplies a motive for persuading agent 2 to believe that agent 1 will cooperate, but it is equally certain that cooperation is not rational for agent 1. Under conditions of common knowledge, agent 2 must then expect that agent 1 will not cooperate, and so on. Thus, the agent who views rationality from such a separable perspective must forego the gains that coordinated choice would make possible.

This not only serves to undercut the plausibility of separability as a criterion of rational choice, but it does so, interestingly enough, by reference to a consequentialist consideration. In very general terms, the notion is that a condition C cannot be taken as a criterion of a consequentially oriented theory of rational choice for a given class of games if acceptance of C works to the agent's own disadvantage. For the case in question, an agent committed to a separable perspective ends up with an outcome (consequence) less preferred to one that could have been realized had that agent been disposed to coordinate choices with the other agent. For this class of cases, then, the consequence of a commitment to separability is that the agent does less well in interaction with other rational agents than it would be possible to do. But this renders suspect the claim that separability is a necessary condition of rational choice.

One can attempt to ground the separability condition in some other way, by appeal (say) to some intuitive notion of "consistency." But granting that, what requires us to take any such intuitive basis as overriding when it comes into conflict with the revised version of consequentialism? That is, what sense can we make of a consistency requirement the imposition of which implies that rational persons under conditions of common knowledge must settle for less than they could otherwise obtain?

I do not expect, of course, that all will be converted by this argument. But it does seem at the very least that there is a need to sort out more carefully the presuppositions that characterize the modern theory of rationality. Ever since the prisoners' dilemma pattern was first identified, theorists have persisted in treating a consequentialist perspective as requiring mutual defection, even though the consequence is that both players do less well.

It is interesting to note that von Neumann and Morgenstern partially abandon the separable perspective when they move to the theory of n-person (rather than two-person) zero-sum games. In the case of a game between three or more players, there can be a parallelism of interests that makes cooperation desirable; this will, in at least some cases, lead to an agreement between some of the players involved. If the game is zero-sum then of course it cannot be in the interests of *all* the players to join in a grand coalition, but smaller coalitions may still form. When this happens, von Neumann and Morgenstern imagine that the coalition will coordinate to secure the maximum payoff possible for members of that group, thereby ensuring that between the coalition and those who remain outside there will be a strict opposition of interest (1953, Section 25). This, then, provides a real place for full cooperation within their theory of n-person, zero-sum games.

Von Neumann and Morgenstern also sketch a theory of non–zero-sum games that retains the presupposition that rational agents will be disposed to coordinate when there are gains to be secured thereby. In particular, they suppose that any non–strictly competitive game involving n agents can be embedded in a strictly competitive game in which there is one additional "fictional" player – might not one think of this as nature? – whose payoff is simply the negative of the payoff that the n players can achieve if they form a coalition of all n players. The suggestion is that in a game such as G_{dpd}, the two agents can think of themselves as jointly playing a strictly competitive game against "nature," where their best strategy is to fully cooperate with one another and thereby force the maximum joint payoff possible from nature (1953, Section 56).

58

6. Nonseparable interactive rationality

I am not at all sure what a full theory of rational interactive choice would look like within a nonseparable framework. I suggest, however, that a theory that is prepared to reject the separability condition for non–strictly competitive games would take the familiar principle of collective rationality – the Pareto optimality condition – as a necessary condition of rational interaction under conditions of common knowledge. I have sought to say something about how one might motivate this view in McClennen (1985). On such an alternative conception of rational interaction, rational agents – who are able to communicate with one another (or who can tacitly bargain) and who have common knowledge of each other's rationality, and so forth – will not face the classical prisoners' dilemma problem. Such agents will be able to reach an agreement on, and then implement, a plan that satisfies the Pareto optimality condition. In a corresponding manner, models of suboptimal equilibrium outcomes are best understood as models of interaction under nonideal conditions, that is, those of imperfect rationality or imperfect information.

What, in addition to the Pareto optimality condition, is likely to figure in a theory of interaction between rational agents under circumstances of common knowledge? Recent work in the theory of bargaining and negotiation is clearly relevant here. Unfortunately, since the prevailing view has been that interactive situations in general are best understood in terms of models of strategic (noncooperative) rather than cooperative choice, bargaining theory is somewhat less than fully developed. However, important contributions to such a theory are to be found, for example, in Nash (1953), Kalai and Smorodinsky (1975), and Gauthier (1986, chap. V). One might hope, moreover, that appreciating the unsatisfactory implications of the standard approaches just surveyed will lead theorists to consider reintroducing the concept of coordination into an area from which it has been systematically banished – namely, the theory of the non–zero-sum game between ideally rational players – and that this will, in turn, spur increased interest in the subject of bargaining theory.

REFERENCES

Aumann, R. J. (1987), "Correlated Equilibrium as an Expression of Bayesian Rationality." *Econometrica* 55: 1–18.
Bernheim, B. D. (1984), "Rationalizable Strategic Behavior." *Econometrica* 52: 1007–28.
Bernheim, B. D. (1986), "Axiomatic Characterizations of Rational Choice in Strategic Environments." *Scandinavian Journal of Economics* 88: 473–88.
Gauthier, D. (1986), *Morals by Agreement*. Oxford: Clarendon Press.
Hammond, P. (1988), "Consequentialist Foundations for Expected Utility." *Theory and Decision* 25: 25–78.

Kalai, E., and Smorodinsky, M. (1975), "Other Solutions to Nash's Bargaining Problem." *Econometrica* 43: 513–18.

Kaysen, K. (1946–7), "A Revolution in Economic Theory?" *Review of Economic Studies* 14(1): 1–15.

Luce, R. D., and Raiffa, H. (1957), *Games and Decisions.* New York: Wiley.

McClennen, E. F. (1972), "An Incompleteness Problem in Harsanyi's General Theory of Games and Certain Related Theories of Non-Cooperative Games." *Theory and Decision* 2: 314–41.

McClennen, E. F. (1985), "Prisoners' Dilemma and Resolute Choice." In R. Campbell and L. Sowden (eds.), *Paradoxes of Rationality and Cooperation.* Vancouver: University of British Columbia Press.

McClennen, E. F. (1988), "Dynamic Choice and Rationality." In B. R. Munier (ed.), *Risk Decision and Rationality.* Dordrecht: Reidel, pp. 517–36.

McClennen, E. F. (1990), *Rationality and Dynamic Choice: Foundational Explorations.* Cambridge: Cambridge University Press.

Nash, J. (1951), "Non-Cooperative Games." *Annals of Mathematics* 54: 286–95.

Nash, J. (1953), "Two-Person Cooperative Games." *Econometrica* 21: 128–40.

Pearce, D. G. (1984), "Rationalizable Strategic Behavior and the Problem of Perfection." *Econometrica* 52: 1029–50.

Von Neumann, J., and Morgenstern, O. (1953), *Theory of Games and Economic Behavior,* 3rd ed. New York: Wiley.

5

Hyperrational games: Concept and resolutions

JORDAN HOWARD SOBEL

This chapter studies normal form games of a special kind. It studies games played by very knowledgeable causal expected-value maximizers. These games are defined, and theorems concerning their resolutions are developed.

Hyperrational games are in several dimensions highly idealized objects that, while approached, are probably never realized. I am interested in the theory of these objects, but not for the light it promises to cast on actual games, or as a source of prescriptions for actual games. Rather, I think that this theory, and especially the part to do with problems of hyperrational games,[1] can contribute to explanations and understandings of real agents and cultures, and can contribute justifications of such aspects of culture as coercive institutions. Also, though this is no part of my motivation, some scholars may take an interest in the theory because of the grist it can seem to provide for criticisms of Bayesian rationality.

THE CONCEPT OF A HYPERRATIONAL NORMAL-FORM GAME

1

I begin with conditions for normal-form games in a certain strict or ideal sense, and then proceed to conditions specific to hyperrational games.

Axiom 1. In a *pure strategy game,* each of finitely many players has as an option exactly the members of a finite set of strategies.

Axiom 2. Each player in a game has an expected value for each possible interaction of strategies.

I am particularly indebted to Willa Freeman-Sobel and Wlodek Rabinowicz for help with this work.

Expected values of propositions are taken in the sense of World Bayesianism: They are probability-weighted averages of values of ways propositions might work out, or worlds in which they might take place:

$$\text{EV}(p) = \sum_{P(w \text{ given } p) > 0} [P(w \text{ given } p) \cdot V(w)]$$

(Sobel 1989b). Sets of strategies and expected values for interactions determine normal-form structures of games. For example, in the two-person pure strategy normal-form game

Game 1

	C1	C2	C3
R1	3, 1	0, 2	2, 0
R2	0, 1	1, 0	4, 3

Row's options are R1 and R2, and Column's are C1, C2, and C3. First numbers are Row's expected values for interactions or conjunctions of strategies, and second numbers are Column's. Thus, $\text{EV}_r(\text{R1 \& C1}) = 3$ and $\text{EV}_c(\text{R2 \& C3}) = 3$.

Axiom 3. Strategies of players in games are *causally independent* of one another: What a given player does cannot influence what any other player does.

Axiom 4. Players in a game are rational in choices of strategies.

Axiom 5. Each player in a game knows its normal-form structure, and knows that Axioms 3 and 4 are satisfied.

Axiom 6. It is "common knowledge" among players in a game that Axiom 5 is satisfied: Each player knows that Axiom 5 is satisfied, knows that it is known by each player that Axiom 5 is satisfied, knows that it is known by each player that it is known by each player that Axiom 5 is satisfied, and so on ad infinitum.

Rationality and knowledge conditions for hyperrational extensive-form games would need to be "subjunctified" and more complicated.

2

For mixed-strategy games, or mixed extensions of pure-strategy games, it is assumed that each player has as an option every chance-mix of a finite set of pure strategies. To define mixed-strategy games, Axiom 1 is replaced by the following.

Axiom 1m. In a *mixed-strategy game,* each of finitely many players has as an option every chance-mix based on some finite set of pure strategies.

To "have as an option a chance-mix" is to have the wherewithal somehow to commit irrevocably to chance, with chances fixed as in the mix, which of the strategies mixed is eventually enacted. A *genuine* mixed strategy accords positive chances to at least two distinct pure strategies. A non-genuine or *degenerate* mixed strategy makes some one pure strategy certain; in a mixed-strategy game it is the surrogate for that pure strategy. (The pure strategy is not *per se* an option in the game, though to avoid circumlocutions I will write as if it were an option and identical with the degenerate mixed strategy.)

I stipulate as a basic assumption for mixed-strategy games that players' expected values for interactions of mixed strategies can be computed, using the formula of utility theory for computing utilities for lotteries given utilities for their prizes. The basic assumption is that, with respect to lotteries that would have as outcomes interactions of pure strategies, players are "risk neutral" in that their preferences for these lotteries are represented by standard utility functions. For example, in the mixed extension of game 1, the interaction in which Row uses $(\frac{1}{3}R1, \frac{2}{3}R2)$, the mixed strategy that fixes the chance for R1 at $\frac{1}{3}$ and that for R2 at $\frac{2}{3}$, and in which Column uses $(\frac{1}{6}C1, \frac{1}{3}C2, \frac{1}{2}C3)$, has for Row and Column (respectively) the expected values $\frac{3}{2}$ and $\frac{15}{18}$:

Game 1

	C1	C2	C3	(1/6 C1, 1/3 C2, 1/2 C3)
R1	3, 1	0, 2	2, 0	
R2	0, 1	1, 0	4, 3	
(1/3 R1, 2/3 R2)				3/2, 25/18

Our basic assumption licenses the following calculation for Row's expected value for this interaction of mixed strategies:

$$\text{EV}_r[(\tfrac{1}{3}R1, \tfrac{2}{3}R2) \,\&\, (\tfrac{1}{6}C1, \tfrac{1}{3}C2, \tfrac{1}{2}C3)]$$

$$= (\tfrac{1}{3}\cdot\tfrac{1}{6}\cdot 3) + (\tfrac{1}{3}\cdot\tfrac{1}{3}\cdot 0) + (\tfrac{1}{3}\cdot\tfrac{1}{2}\cdot 2) + (\tfrac{2}{3}\cdot\tfrac{1}{6}\cdot 0) + (\tfrac{2}{3}\cdot\tfrac{1}{3}\cdot 1) + (\tfrac{2}{3}\cdot\tfrac{1}{2}\cdot 4)$$

$$= \tfrac{1}{6} + 0 + \tfrac{1}{3} + 0 + \tfrac{2}{9} + \tfrac{4}{3} = \tfrac{27}{18} = \tfrac{3}{2}$$

63

It is a theorem of World Bayesianism – both of Richard Jeffrey's evidential version (in which the terms $P(w$ given $p)$ in the definition of $EV(p)$ are cast as conditional probability terms $P(w/p)$), and of the causal theory assumed for the present study (in which the terms $P(w$ given $p)$ are cast as probability terms for causal conditionals $P(p \square\rightarrow w)$) – that, for any lottery L in which the known chance for outcome O_i is c_i,

$$EV(L) = \Sigma_i[c_i \cdot EV(L \& O_i)].$$

However, in contrast with standard utility theory (e.g., that of Luce and Raiffa 1957), it is *not* a theorem that

$$EV(L) = \Sigma_i \, c_i \cdot EV(O_i).$$

World Bayesianisms are roomier than standard utility theories; for example, they accommodate Allais and Ellsberg preferences. Thus an assumption is needed to license standard calculations for expected values of mixed strategy interactions (Sobel 1989b).

3

For *hyperrational* games, Axioms 1, 1m, and 4 are elaborated; two additional epistemic conditions are inserted between Axioms 4 and 5; and Axiom 5 is readdressed to the expanded and elaborated set of prior conditions (Axioms 1–4).

3.1. Bayesian decision theories that would have agents maximize expected value need to be explicit about "proper" partitions of agents' options, and about what alternatives are relevant for purposes of maximizing comparisons. Such theories, to be complete and exact, need to say precisely which expected values are relevant to the assessments of options.

It might seem sufficient to say that an action maximizes in a manner significant for choice if and only if its expected value is not exceeded by that of any of its alternatives. But this approach encounters problems with relatively indeterminate, disjunctive actions; for expected values of disjunctions (even of exclusive disjunctions, on my causal theory) are not necessarily bounded by expected values of their disjuncts. Given this fact, although action "*A* or *B*" is open whenever *A* is, one would want to count it as a relevant alternative to *A* and to *B* only when it is open in a distinctively disjunctive manner: only when one can choose it without choosing either of them. One reaction to such problems might be to make primary the assessments of "most specific" actions, but this can seem wrong in some cases where the agent can choose to leave open certain specifications, especially specifications pertaining to future stages of (courses of) actions. Another idea might be to make primary the assessments of "most specific minimal" actions that settle nothing for the future that can be left open. But this can seem wrong for cases in which the agent can choose that certain possible future options not be left open.

The problem of proper partitions of options, as it arises for the causal theory of this chapter, is addressed in Sobel (1983). In view of difficulties with approaches that would identify agents' options with most specific or most specific minimal actions, I propose a theory in which agent options are identified with certain *choices* for actions.

I assume that hyperrational games are built on proper partitions of options, and for definiteness spell out this assumption in terms of my theory of proper partitions (though its details, and hence the correctness of these details, are not presently important). Axioms 1 and 1m are transformed, for hyperrational games, as follows.

Axiom 1h. In a pure-strategy hyperrational game, each of finitely many players has a finite set of strategies that are all and only the "precise choices" open to him.

Axiom 1mh. In a mixed-strategy game, each of finitely many players has as an option every chance-mix based on some finite set of pure strategies, and these are all and only his possible "certain precise choices."

To flesh out the terminology (without fully explaining it), I offer these statements:

[A] *precise choice* of an action x [is] a choice of this action that is not accompanied by a choice of any action y such that y entails but is not entailed by x. (Sobel 1983, p. 179)

[A mixed] choice is . . . a *certain precise choice* if its agent is . . . sure he can make it without making any refined version of it. (Sobel 1983, p. 182)

An important dividend of including propriety conditions on strategy sets is that such conditions should ensure that there can be for a given actual game only one hyperrational idealization, so far as the identities and numbers of agents' options are concerned. This consequence neutralizes certain objections to the application (in Axiom 4.1) of the "principle of insufficient reason" (Rabinowicz 1986, p. 215) in hyperrational games when it is known that a player is confronted with a tie. For example, cloning a maximizing strategy S in a hyperrational game G leads not to an equivalent hyperrational game G' but rather to another game, one wherein either in place of the option S a player has two versions of S, or in addition to S he has the option of a version of S. The player in G' would then have somewhat greater control over his actions than he has in G. (S could be getting one over the plate. Versions of S would then include throwing a curve for a strike, and throwing a hard fast one. Pop!)

3.2. The definition of a strict or ideal normal-form game leaves open the character of rationality in games. For hyperrational games this issue is settled in favor of causal expected-value maximization. Players in these games are causal expected-value maximizers – nothing more and nothing less.

Axiom 4h. Each player in a hyperrational game is rational in the choice of strategies; that is, each player:
  (i) is a self-conscious and deliberate causal maximizer;
 (ii) does only things he knows are rational in causal maximizing terms; and
(iii) when confronted with several tied maximizing strategies, is indifferent between them and makes an indifferent (unprincipled) choice of one of them.

That players are causal maximizers in hyperrational games raises a problem. For although an option's *evidential* expected value can be analyzed in terms of its evidential expected values in conjunction with the members of any partition of propositions or possible circumstances, not every partition is in this way adequate for the analysis of an option's causal expected value. It turns out, however, that according to our theory of causal expected values, possible actions of others in a game always do make a partition of circumstances that is adequate in this way, so that any strategy's expected value is a weighted average of the expected values of interactions in which it might participate, where weights are its agent's probabilities for remainders of these interactions. Though not every partition is adequate for an action, every partition *is* adequate that satisfies the following condition: it is certain that the action is not open in conjunction with both members of any pair of circumstances in it. (This is proved in Sobel 1989a.) Games always satisfy this condition. For example, in a prisoner's dilemma it will be clear to me that I cannot both confess alone and confess along with you; one or the other is possible (depending on what you will do), but of course not both. So your possible actions make an adequate partition of circumstances for mine, and, for example (using subscript m's to stress that *my* expected values and probabilities are at issue),

$EV_m($ I confess $)$

$\quad = [P_m($ I confess $\square\!\!\rightarrow$ you confess $) \cdot EV_m($ I confess $\&$ you confess $)]$

$\quad\quad + [P_m($ I confess $\square\!\!\rightarrow$ you do not confess $)$

$\quad\quad\quad \cdot EV_m($ I confess $\&$ you do not confess $)].$

Given that I know that your actions are causally independent of mine, this equation reduces to

EV_m(I confess)

$= [P_m(\text{you confess}) \cdot EV_m(\text{I confess & you confess})]$

$+ [P_m(\text{you do not confess}) \cdot EV_m(\text{I confess & you do not confess})].$

3.3. One special epistemic condition for hyperrational games is about players' expectations concerning the strategies of others whom they know to be confronted with ties.

Axiom 4.1. When a player i in a hyperrational game considers the behavior of another player j who i knows is confronted with a tie, then player i judges j's several tied pure strategies to be *equally probable;* their probabilities sum to 1.

Axiom 4.1 is plausible for mixed-strategy as well as pure-strategy games, given the dense and uniform character of sets of tied maximizing strategies in mixed-strategy games. It is a consequence of the basic assumption for mixed-strategy games that, if MAX is the set of a player's maximizing strategies for a mixed strategy game, then MAX contains all and only the mixed strategies that are based on only pure strategies in MAX. (I say that a mixed strategy M is based on certain pure strategies if and only if each of these, and no others, have positive chances in M.)[2]

As Rabinowicz has noted (1986, p. 228, n. 3), Axiom 4.1 would be "troublesome" if it required equal probabilities for all tied strategies and not only for all pure strategies in a tie. That trouble might be dealt with by recourse to equal nonstandard infinitesimal probabilities or, as Rabinowicz suggests, to density functions; I deal with it by not raising it. Although interesting, deductions of Axiom 4.1 from a condition that for mixed-strategy games stipulated something like equal probabilities for all tied strategies would not contribute significantly to the plausibility of this condition on hyperrational games.

3.4. Here is our second epistemic condition for hyperrational games.

Axiom 4.2. There is no "private practical knowledge" in a hyperrational game; if any player knows at his time of action that some strategy is uniquely maximizing, so that he ought (according to causal maximizing principles) to do it, or knows that some strategies are tied and all are maximizing so that he ought to choose from among them, then these things are known to everyone in the game at their times of action.

I believe that Axiom 4.2 is a consequence of other stipulations, for I think that hyperrational players could know nothing of relevance beyond

67

what we (as students of their games) can know, given that we know the normal-form structures of their games and that their games satisfy all conditions for hyperrational games except for Axiom 4.2. And I think that hyperrational players would know everything of relevance that we students of their games can know.

According to Axiom 4.2, hyperrational players choose with knowledge of and precisely *not* "in ignorance of others' choices" (Bernheim 1984, p. 1009). They do not merely rationalize their choices in terms of internally consistent conjectures concerning others' choices, others' conjectures of others' choices, and so forth. Hyperrational players in the end know not only what maximizes for themselves, but also what maximizes for others.

4

4.1. It is a further (hardly resistible) stipulation for hyperrational games that strategies of players are not only causally independent but also evidentially independent, at least when it is time to act. Players in hyperrational games must, when they act, know what is maximizing for themselves, and know what was, is, or will be maximizing for others at their times of action. But then, last-minute news to a player – shocking news – that he will not do what he knows to be the rational maximizing thing should not be evidence that others also will not do the maximizing things for them. Amazing news that one player will make a mistake should not be evidence that mistakes will be rampant – it should not be evidence that mistakes will extend at all beyond that player's own case.

Consider, for example, the following hyperrational game:

Game 2

You

		C1	C2
Me	R1	2, 2	1, 0
	R2	1, 0	0, 1

Consider this game "at the moment of choice or action."[3] Presumably I know what I should do, for R1 dominates R2. And I should know what you should do, because setting aside R2 (as you should be in a position to do) reduces the game to a 1×2 problem where C1 dominates C2. So for me $P(C1)$ should be either 1 or nearly 1, [1]. At the time of action I should not still have best-response conditional probabilities for your actions, so that not only $P_m(C1/R1) = [1]$ but also $P_m(C2/R2) = [1]$. For though both of my probabilities, $P_m(R2 \& C2)$ and $P_m(R2 \& C1)$, are nearly 0, unless both are exactly 0 the first should be much lower than the second. My probability for not only my doing what I now see to be a very dumb thing, but for your, coincidentally, joining me and also succumbing to the sillies, should be *much* lower than my probability for my "going off" on my own. If $P_m(R2) > 0$ still, at the moment of action, then $P_m(R2 \& C2)/P_m[(R2 \& C1) \vee (R2 \& C2)]$ should then be [0]. And in any case $P_m(C2, R2)$ should then be [0], notwithstanding that C2 would be your best response to R2, and what I am confident you would do were you to know that I was doing R2. (I assume here nonstandard conditional probabilities defined for all possibilities, and not only for all positively probable ones as in Sobel 1987.)

4.2. So I think that, in hyperrational games, strategies that are causally independent are also evidentially independent, at least by times for action. Presumably, however, there can be kinds of evidential dependence of actions "early on," supposing that decisions in even hyperrational games can take time to make, and that time for decisions is allowed. One interesting hypothesis for early-on conditional probabilities in two-person games would be that each player's initial conditional probabilities are such that news that he was going to do a certain action would mean that the other player was going to employ a best response to it. Some motivation for this hypothesis can be gathered from Harper (1988, p. 30):[4]

Suppose you assume there is a unique rational choice ... and that you will end up committing yourself to it, but you haven't yet figured out what [your] choice will be. When you consider the hypothetical news provided by your assumption that you choose strategy *A* you hypothetically assume that *A* is the rational act. You keep fixed your assumption that what you will end up committing yourself to will be the rational choice, and assume hypothetically that reasoning legislated choosing *A*. . . . [Y]ou assume your opponent . . . will have been able to reconstruct the reasoning that leads you to choose *A* and will have predicted your choice. Thus, you assume that she or he will choose some best response to *A* when you assume that you will choose *A*.

But these grounds for best-response conditional probabilities early on, which are specific to two-person games and not easily extendable even to three-person ones, fail to justify these probabilities even for two-person games. The trouble, as Harper himself comes to observe, is with the player's initial assumption of a unique rational choice. There may not be such a choice; even when there is, what seems called for – if early-on best-response conditional probabilities are to be justified – is not its assumption but its discovery. Harper himself comes to quite undercut his motivation for "best-response conditional priors":

When I hypothetically assumed I would end up choosing strategy A, perhaps all I should have assumed is that A is some strategy in the solution set [i.e., not *the* rational act, but only *a* rational act]. But, this will not allow me to hypothetically assume that an opponent who completely understands the game and the demands of rationality will have predicted my choice. My ground for using a best response prior would seem to be undercut . . . . (p. 38; cf. Rabinowicz 1986, pp. 215–16, and n. 1, p. 227, from whence Harper got this idea)

What suggests itself is not that the idea of best-response priors in two-person games be abandoned, but that it be spelled out somewhat differently. Rather than begin with conditional probabilities that exclude all but best responses to choices of particular strategies, one might more reasonably begin with conditional probabilities that exclude all but best responses to possible indifferent choices from various sets of one's strategies. Given Axiom 4.1, such best responses will be well defined in hyperrational games.

I do not, however, think that initial best-response conditional probabilities, no matter how they are articulated, are especially appropriate for hyperrational games. Indeed, I doubt that anything completely general can be said about appropriate conditional probabilities early on in hyperrational games. Even so, there may be some interest in discussing particular games under one or another hypothesis concerning early-on conditional probabilities; for example, kinds of best-response conditional probabilities and (for some games) kinds of like-response conditional probabilities. I do not here explore these avenues, and in what follows make absolutely no assumptions concerning prior, initial, or early-on conditional probabilities.

Turning from possible early-on conditional probabilities to early-on unconditional ones, it is plausible that when players have no clear ideas what they will do – as one might rule should always be the case very early on – they should then judge all pure strategies in each given player's strategy set to be equally probable, or judge feasible pure strategies equally probable. It may be interesting to examine particular hyperrational games under this early-on equal-probabilities hypothesis (see Rabinowicz 1989, where the considerable possible interest of such examinations is demonstrated). However, I myself favor the idea that hyperrational players are, very early on, quite devoid of determinate expectations or probabilities for one another's strategies, though nothing that follows depends on this idea.

The first theorems about hyperrational games are for *necessary* conditions on resolutions. These theorems say that hyperrational games can resolve only in something like equilibria of sorts. The second group of theorems set out *sufficient* conditions for resolutions. These theorems say that hyperrational games can resolve by various processes of elimination, and would resolve in the choices to which these processes would lead. The two sets of theorems are related somewhat as the "indirect" and "direct" arguments of von Neumann and Morgenstern concerning a satisfactory theory for zero-sum, two-person games are related. They saw their indirect argument as "[narrowing] down the possibilities to one," but felt that it was "still necessary to show that the one remaining possibility [was] satisfactory" (1953, p. 148, n. 5). A major difference between my project and theirs is that they sought, and to their satisfaction found, a complete theory of solutions for all two-person, zero-sum games. In contrast, I do not seek a complete theory of resolutions for all would-be hyperrational games. Indeed, I think the main interest of the theory of hyperrational games lies precisely in its necessary incompleteness.

NECESSARY CONDITIONS FOR, OR LIMITATIONS ON,
RESOLUTIONS OF HYPERRATIONAL GAMES

5

Random strategies are mixed strategies in which certain pure strategies have equal chances; these chances sum to 1. Random strategies, while not strictly identical to indifferent choices from the pure strategies on which they are based, are in a certain way equivalent to such choices in hyperrational games. In a hyperrational game, knowledge that a player will make an indifferent choice, and knowledge that he will employ the random strategy that corresponds to that choice, would give rise to the same expectations and probabilities for pure strategies involved.

The *random extension* of a pure-strategy game is the part of its mixed extension reached by adding to players' options only and all random mixed strategies. Random extensions of pure-strategy games are finite. For example, the random extension of a 2×2 game is a 3×3 game, and the random extension of a 3×3 game is a 7×7 one, for there are exactly four distinct random strategies based on a set of three pure strategies.

The *random contraction* of a mixed-strategy game is the random extension of the pure-strategy game of which this mixed-strategy game is the mixed extension. The basic assumption for mixed-strategy games, according to which expected values of interactions of mixed strategies are the usual weighted averages of expected values of interactions of pure strategies, covers random extensions as well as mixed extensions and random contractions. This assumption is always in force herein, though not all results depend on it.

71

The *randomization* of a pure-strategy game is its random extension; of a mixed-strategy game, its random contraction. To illustrate, here is the randomization of game 1:

	C1	C2	C3	[C1, C2]	[C1, C3]	[C2, C3]	[C1, C2, C3]
R1	3, 1	0, 2	2, 0	3/2, 3/2	5/2, 1/2	1, 1	5/2, 3/2
R2	0, 1	1, 0	4, 3	1/2, 1/2	2, 2	5/2, 3/2	5/2, 2
[R1, R2]	3/2, 1	1/2, 1	3, 3/2	1, 1	9/4, 5/4	7/4, 5/4	5/3, 7/6

Here [R1, R2] abbreviates $(\frac{1}{2}R1, \frac{1}{2}R2)$; similarly, [C1, C2, C3] abbreviates $[\frac{1}{3}C1, \frac{1}{3}C2, \frac{1}{3}C3]$.

5.1. I proceed to my first theorem concerning hyperrational games. It states that these games can resolve only in equilibria of sorts.

Theorem 1. *A hyperrational game G can resolve only in indifferent choices from (possibly singleton) strategy sets such that the interaction of random strategies that correspond to these choices is an equilibrium in the randomization of G.*

For substantiation, suppose that a two-person hyperrational game resolves in indifferent choices from singleton sets: Suppose that in the game Row does R_i and Column does C_j. Column knows that he ought to do C_j, which is his maximizing strategy (Axiom 4h). Row also knows this (Axiom 4.2). Row is thus sure that Column will do C_j. So R_i is Row's unique best response to C_j: if not, then Row would not be choosing indifferently from the singleton set $\{R_i\}$. Similarly, C_j must be Column's unique best response to R_i. So (R_i, C_j) is an equilibrium in the game, and thus in its random extension.

This argument can be generalized to hyperrational games of all sizes that resolve in indifferent choices from sets of all sizes. Its generalization to mixed-strategy games depends mainly on results already noted concerning the dense and uniform character of sets of maximizing mixed strategies, and the consequent character of a player's expectations for another's pure strategies when it is known that this other will make a certain indifferent choice. In any hyperrational game, pure or mixed, it will be as if each player is, at the time of action, certain that others have used (or will use) random strategies corresponding to indifferent choices from their maximizing strategies, and thus to indifferent choices from their maximizing pure strategies (see Axiom 4.1 and comments thereon). So the player's own indifferent choice needs to correspond to a random strategy that is a best response to those random strategies.

5.2. Theorem 1 says that hyperrational games resolve only in equilibria of sorts, roughly, only in equilibria in random extensions or random contractions. It is important that it is possible to be more restrictive, in terms of an idea (due to Rabinowicz) of a kind of equilibrium of "intermediate strength." Let an interaction $(s_1, \ldots, s_n)$ in a randomization be an *equilibrium** if and only if it is an equilibrium and

each [player i] would lose in utility if he, instead of s_i, played some *pure* strategy not belonging to [the set of pure strategies on which s_i is based] when other players play their strategies in $(s_1, \ldots, s_n)$. (Rabinowicz 1989, n. 6)

Every strong equilibrium in a randomization is an equilibrium*. In contrast, while some weak equilibria of randomizations are equilibria*, others are not. For example, in the randomization of a game discussed in Pearce (1984, p. 1035),

Game 3

	C1	C2	[C1, C2]
R1	0.5	−1, 3	−1/2, 4
R2	0.0	−1, 3	−1/2, 3/2
[R1, R2]	0, 5/2	−1, 3	−1/2, 11/4

neither (weak) equilibrium (R1, C1) nor (weak) equilibrium (R2, C2) is an equilibrium*: Against (R1, C1) I note that, though R2 is not in {R1}, Row does not lose by playing R2 when Column plays his strategy in (R1, C1); against (R2, C2) I note that Row does not lose if she defects to R1. In contrast, (weak) equilibrium ([R1, R2], C2) of this randomization is an equilibrium*: There are no strategies outside {R1, R2} to which Row can defect; and Column loses if he defects to C1, the only strategy to which he can defect.

For another example, in the randomization of

Game 4

	C1	C2	C3
R1	0, 0	1, 1	1, 1
R2	0, 0	1, 1	1, 1
R3	0, 0	1, 1	1, 1

73

mixed (weak) equilibrium ([R1, R2], [C2, C3]) is not an equilibrium*. This is so because R3, although not in {R1, R2}, is still a best response to [C2, C3], as the following display of a part of this randomization makes plain:

	C1	C2	C3	[C2, C3]
R1	0, 0	1, 1	1, 1	1, 1
R2	0, 0	1, 1	1, 1	1, 1
R3	0, 0	1, 1	1, 1	1, 1
[R1, R2]	0, 0	1, 1	1, 1	1, 1

Row could defect from ([R1, R2], [C2, C3]) without loss to a strategy outside the base of [R1, R2]. For similar reasons, not one of the six pure-strategy (weak) equilibria is an equilibrium*, and not one of the five displayed pure/mixed-strategy (weak) equilibria is an equilibrium*.

Pure-strategy equilibria are either strong or weak, and a pure-strategy equilibrium in a randomization is strong if and only if it is an equilibrium*. All mixed and pure/mixed equilibria (i.e., all equilibria that do not involve only pure strategies) are weak, but, as we have seen, some but not all are equilibria*.

The arguments for Theorem 1, when allowed to follow their natural course, lead as well to the following.

Theorem 1*. *A hyperrational game G can resolve only in indifferent choices from (possibly singleton) strategy sets such that the interaction of random strategies that correspond to these indifferent choices is an equilibrium* in the randomization of G.*

If a hyperrational game resolves, then each player in it makes an indifferent choice from all and only those strategies that are best responses to the indifferent choices each player knows that the others are making or have made. In a mixed-strategy game, a player's best responses, the maximizing strategies, constitute a set that is closed in a certain manner under mixing: It contains precisely the mixed strategies that are based on pure strategies it contains. And in any game that resolves, pure or mixed, it will be as if its players see themselves as engaged not in the game itself but rather in its random extension or contraction, and as if each player is, at the time of action, certain that others have used (or will use) random strategies corresponding to indifferent choices from their maximizing pure strategies. These random strategies must be best responses each to the others, and must constitute not only an equilibrium but also an equilibrium* in the game's randomization.

5.3. Theorems 1 and 1* may be compared with views of Rabinowicz regarding two-person mixed strategy games, views that generalize naturally to all n-person games:

[A] proper demand on any member α of R_a, the set of a's rational strategies, is . . . that it should be a best response to R_b. That is, any α in R_a should maximize expected utility on the assumption that b is going to choose one of the strategies in R_b, where all the strategies in R_b are taken to be equiprobable. The analogous demand applies to R_b. (1986, pp. 215–16)

If these demands are satisfied, Rabinowicz will say that R_a and R_b make a "probabilistic equilibrium." Rabinowicz endorses as a part of any adequate theory of games the additional demand that "R_a and R_b should, if possible, constitute a *complete* probabilistic equilibrium," where this condition is met if and only if every strategy α $[\beta]$ that "is a best response to R_b $[R_a]$ (in the above-explained sense)" is in R_a $[R_b]$ (p. 216). Theorem 1* corresponds to this stronger demand.

I proceed in terms of equilibria* and randomizations. However, computing randomizations and scanning for equilibria* is tedious and error-prone. It is therefore useful that, as Rabinowicz has observed in conversation, equilibria* of randomizations can be quickly and securely found by exhaustive searches for complete probabilistic equilibria. Using for illustration

Game 1

	C1	C2	C3
R1	3, 1	0, 2	2, 0
R2	0, 1	1, 0	4, 3

the following table

Sets of Row's pure strategies	Column's best pure responses to these sets	Row's best pure responses to these responses
{R1}	{C2}	{R2}
{R2}	**{C3}**	**{R2}**
{R1, R2}	{C3}	{R2}

establishes that this game has a unique complete probabilistic equilibrium, which is ({R2}, {C3}). It follows that ([R2], [C3]) – or, for short,

75

(R2, C3) – is the unique equilibrium* in this game's randomization, as inspection confirms:

	C1	C2	C3	[C1, C2]	[C1, C3]	[C2, C3]	[C1, C2, C3]
R1	3, 1	0, 2	2, 0	3/2, 3/2	5/2, 1/2	1, 1	5/2, 3/2
R2	0, 1	1, 0	4, 3	1/2, 1/2	2, 2	5/2, 3/2	5/2, 2
[R1, R2]	3/2, 1	1/2, 1	3, 3/2	1, 1	9/4, 5/4	7/4, 5/4	5/3, 7/6

For another illustration, we revisit game 3 and observe that the table

Sets of Row's pure strategies	Column's best pure responses to these sets	Row's best pure responses to these responses
{R1}	{C1}	{R1, R2}
{R2}	{C2}	{R1, R2}
{R1, R2}	**{C2}**	**{R1, R2}**

establishes that this game has the unique complete probabilistic equilibrium ({R1, R2}, {C2}). It follows that ([R1, R2], C2) is the unique equilibrium* in this game's randomization, as was recently established by inspecting its randomization.

5.4. We proceed to several corollaries of Theorems 1 and 1* that concern cases involving unique equilibria.

Theorem 2. *If the randomization of a hyperrational game has a unique equilibrium, then this game can resolve only in indifferent choices that correspond to the random strategies in that equilibrium.*

Theorem 3. *A hyperrational mixed strategy game G can resolve only in indifferent choices that correspond to an equilibrium* in its random contraction that is an equilibrium in G itself.*

Theorem 2 follows directly from Theorem 1. Theorem 3 follows from Theorem 1*, given that an equilibrium* in the random contraction of a mixed strategy game must be an equilibrium, and indeed an equilibrium*, in the game itself.

Theorem 3 is, for the games it addresses, no more restrictive than Theorem 1*. It might seem, however, that Theorem 3 can be strengthened as follows.

¿Theorem 3'? *A hyperrational game G, mixed or pure, can resolve only in indifferent choices that correspond to an equilibrium* in its randomization that is an equilibrium in G itself.*

This principle is more restrictive for pure-strategy games than Theorem 1* itself, since not every equilibrium in the random extension of a pure-strategy game need correspond to an equilibrium in the game itself. But for this reason it is very doubtful that the principle is valid. Consider the following pure-strategy game, displayed here with its random extension:

Game 5 An Appointment in Samarra

	C1	C2	[C1, C2]
R1	0, 1	1, 0	1/2, 1/2
R2	1, 0	0, 1	1/2, 1/2
[R1, R2]	1/2, 1/2	1/2, 1/2	1/2, 1/2

It seems that this game might well resolve for hyperrational players in indifferent choices corresponding to the equilibrium* in its random extension. I shall imply that it *would* resolve for hyperrational players in such choices: See Theorem 9 (cf. Rabinowicz 1986, pp. 218–19, game G6).[5]

Theorem 4. *If a hyperrational mixed strategy game has a unique equilibrium, then it can resolve only in that equilibrium.*

This theorem follows from Theorem 3.
    It might seem that Theorem 4 can be strengthened, as follows.

¿Theorem 4'? *If a hyperrational game, mixed or pure, has a unique equilibrium, then it can resolve only in that equilibrium.*

Against this conjecture I offer the following pure strategy game:

Game 6

	C1	C2	C3
R1	2, 2	0, 2	0, 0
R2	2, 0	2, 1	1, 2
R3	0, 0	1, 2	2, 1

This game has a unique equilibrium, (R1, C1). Its random extension has two equilibria, specifically, (R1, C1) and ([R2, R3], [C2, C3]). But of these only the second is an equilibrium*.

	C1	C2	C3	[C1, C2]	[C1, C3]	[C2, C3]	[C1, C2, C3]
R1	2, 2	0, 2	0, 0	1, 2	1, 1	0, 1	2/3, 4/3
R2	2, 0	2, 1	1, 2	2, 1/2	3/2, 1	3/2, 3/2	5/3, 1
R3	0, 0	1, 2	2, 1	1/2, 1	1, 1/2	3/2, 3/2	1, 1
[R1, R2]	2, 1	1, 3/2	1/2, 1	3/2, 5/4	5/4, 1	3/4, 5/4	7/6, 7/6
[R1, R3]	1, 1	1/2, 2	1, 1/2	3/4, 3/2	1, 3/4	3/4, 5/4	5/6, 7/6
[R2, R3]	1, 0	3/2, 3/2	3/2, 3/2	5/4, 3/4	5/4, 3/4	3/2, 3/2	4/3, 1
[R1, R2, R3]	4/3, 2/3	1, 5/3	1, 1	7/6, 7/6	7/6, 5/6	1, 4/3	10/9, 10/9

By Theorem 1* game 6 *cannot* resolve for hyperrational players in its sole equilibrium, for that (weak) equilibrium does not correspond to an equilibrium* in its random extension. This game can resolve for hyperrational players only in indifferent choices corresponding to the sole (mixed) equilibrium* in its random extension. In Theorem 9 I imply that it would resolve in this manner.

Game 6 creates a problem because its sole equilibrium is weak. Perhaps then the following enhancement of Theorem 4 is valid.

¿**Theorem 4″**? *If a hyperrational mixed strategy game has a unique equilibrium, then it can resolve only in it; and if a hyperrational pure strategy game has a unique strong equilibrium, then it can resolve only in it.*

Without deciding the status of this principle, I note first that it is not an easy corollary of Theorem 1*. For even if a pure-strategy hyperrational game does have a unique strong equilibrium, its random extension can harbor additional equilibria*. Here is a game illustrative of this point:

Game 7

	C1	C2	C3
R1	5, 5	4, 4	0, 0
R2	4, 4	2, 2	3, 3
R3	0, 0	3, 3	2, 2

78

The random extension of this game has two equilibria, specifically, (R1, C1) and ([R2, R3], [C2, C3]), both of which are equilibria*. (I note that every strong equilibrium is an equilibrium*. The distinction between equilibria and equilibria* is significant only for weak equilibria.)

	C1	C2	C3	[C1, C2]	[C1, C3]	[C2, C3]	[C1, C2, C3]
R1	5, 5	4, 4	0, 0	9/2, 9/2	5/2, 5/2	2, 2	3, 3
R2	4, 4	2, 2	3, 3	3, 3	7/2, 7/2	5/2, 5/2	3, 3
R3	0, 0	3, 3	2, 2	3/2, 3/2	1, 1	5/2, 5/2	5/3, 5/3
[R1, R2]	9/2, 9/2	3, 3	3/2, 3/2	15/4, 15/4	3, 3	9/4, 9/4	3, 3
[R1, R3]	5/2, 5/2	7/2, 7/2	1, 1	3, 3	7/2, 7/2	9/4, 9/4	7/3, 7/3
[R2, R3]	2, 2	5/2, 5/2	5/2, 5/2	9/4, 9/4	9/4, 9/4	5/2, 5/2	7/3, 7/3
[R1, R2, R3]	3, 3	3, 3	5/3, 5/3	3, 3	7/3, 7/3	7/3, 7/3	23/9, 23/9

It is not a corollary of Theorem 1* that game 7 can resolve only in its unique equilibrium (R1, C1). Theorem 1* leaves open that game 7 can resolve instead in indifferent choices corresponding to the other equilibrium* in its random extension, ([R2, R3], [C2, C3]).

I leave as an open question whether ¿Theorem 4″? – though not a corollary of Theorem 1* – is a valid limitation on resolutions of hyperrational games in its own right. Considerations presented in the long version of this chapter (see note 1) to show that game 7 would not resolve for hyperrational players, and that it is not a possible hyperrational game, provide reasons for thinking that ¿Theorem 4″? is a valid principle in its own right.

SUFFICIENT CONDITIONS FOR AND WAYS TO RESOLUTIONS
OF HYPERRATIONAL GAMES

6

Resolutions of hyperrational games require that players have settled expectations and probabilities for one another's strategies. For only then are maximizing options defined; to act, players in hyperrational games must know which of their options to maximize. In this section, ways are considered for hyperrational players to settle their expectations and escape from the labyrinthine courses of deliberation in ordinary situations where, in order to maximize, a player seeks to determine what others are likely to do while realizing that they may well be engaged in like efforts and even trying to take into account his awareness of that possibility. Compare:

It is characteristic of social interaction that I cannot maximize my own utility without "taking into account" (in some vague sense which we understand only darkly) what *you* are up to. I must have some theories or intuitions about how you are likely to behave, how you will respond to my actions, and the like. Worse than that, I must be aware that you are likely doing the same thing in regard to me, and I have to take *that* possibility into account as well. But of course you may know that I am aware of this possibility, and you [may] adjust your behaviour accordingly. And so it goes – we are both involved in a tortuous labyrinth of relations, and though we act in this way quite easily and freely, it is better than even money that neither of us could even *begin* to given an explicit account of how we do it. (Moore and Anderson 1962, p. 413)

This section is about ways in which hyperrational players, declining to begin regressions of deliberation or to enter into endless labyrinths of higher- and higher-order reflections, can in some cases make up their minds what to expect of one another.

Theorems to follow depend on the idea that, if a game *can* resolve in a certain way for hyperrational players – that is, if there is a process by which their expectations can be settled sufficiently to resolve the game in some pattern of choices – then the game *would* resolve for them in that way; speedy hyperrational players would find that way, or another to the same end. Let a *potentially resolvable game* be a game whose randomization contains an equilibrium*; let a potentially resolvable game that can, and so would, resolve for hyperrational players be a *possible hyperrational game*. This terminology leaves open whether or not every potentially resolvable game is a possible hyperrational game.

6.1. Here is a first simple principle, of very limited coverage.

Theorem 5. *A two-person game resolves for hyperrational players if at least one player has a strongly dominant strategy.*

Consider, for example, the following game with its randomization:

Game 8

	C1	C2	[C1, C2]
R1	3, 2	2, 3	5/2, 5/2
R2	2, 2	1, 1	3/2, 3/2
[R1, R2]	5/2, 2	3/2, 2	2, 2

80

Row and Column can settle by elimination their probabilities for each other's choices, and, having settled these probabilities, settle the expected values of their own options, and so settle or make their own choices. Taking Column first, he can be sure that whatever Row's probabilities at the time of action, Row will not employ R2. For Column can see that R1 is strongly dominant, and knows that Row realizes that Column's actions are causally independent of his own. So Column can set aside R2. When this is done, C2 emerges as uniquely maximizing. Row in turn can, by indirect reasoning, eliminate the possibility that Column will employ C1. She can reason that if Column did employ C1, he would know that Row, expecting C1, would employ R1. That knowledge would lead to C2's uniquely maximizing. On C1's elimination, R1 emerges as uniquely maximizing. Game 8 can in this way resolve into the interaction (R1, C2), and so it does.

It is not important to the resolution of game 8 that C2 is uniquely maximizing for Column. Suppose the game were slightly changed, as follows:

Game 8'

	C1	C2
R1	3, 2	2, 2
R2	2, 2	1, 1

Column can reach an indifferent choice from his strategies in the way described. When R2 is set aside, there is nothing in the choice between C1 and C2 (or between any mixed strategies based on them). And Row can, by indirect reasoning, eliminate the possibility that there will be (at the time of action) something in Column's choice, thus settling her own probabilities at $P(C1) = P(C2) = \frac{1}{2}$.

Here is a second simple principle that is closely related to Theorem 5.

Theorem 6. *A two-person game in which each player has a strongly dominant strategy resolves for hyperrational players in the interaction of these strategies.*

It follows from Theorem 1 that such a game can resolve for hyperrational players only in the indicated interaction, for that interaction will be the only equilibrium in the game's randomization. The argument for Theorem 5, especially the part concerning the settling of Column's probabilities and expected values, indicates just how such a game can resolve for hyperrational players in that interaction. Each player can begin with thoughts about the availability to the other player of a strongly dominant option.

6.2. I confess to a certain unease with the arguments just given. Column, for example, is allowed to settle by elimination that Row will employ R1. Column is not required to wonder how Row can come to employ R1; nor is he required to explain how Row can settle her expectations for Column, and her expected utilities, so that she can see that R1 uniquely maximizes. Capitalizing on his knowledge of the game's form and hyperrationality, Column finds out by elimination what Row *must* do, and stops there. Similarly for Row.

Each comes first to know that the other will make some choice, to know by elimination that the other player will somehow settle his or her expectations (and thus his or her expected utilities) in certain ways, and make some particular choice. This knowledge of the other's choice settles a player's own expectations and expected utilities, and terminates his or her prechoice cogitations. My players refrain from asking a question that must naturally arise for them, the question whose answer would be "by elimination." Upon determining that their fellows must settle their expectations in certain ways, my players refrain from asking how their fellows will manage to do that, and from asking how these fellows think *their* fellows will manage to settle their expectations concerning *their* fellows, and so on. My players exercise restraint in their deliberative ratiocinations. (Whether this restraint is admirable in the circumstances is debatable.) It's as if my players have learned that the best answers to some questions are: "Don't ask! You don't want to know." All arguments in Section 6 stand under this cloud, to which I return in Section 6.6.

6.3. Let a strategy x be strictly dominated if and only if its player has a strategy x' such that, for each combination of strategies for all other players, x' is better than is x. Let a game G "reduce by a strictly dominated discard" to G' if G' comes from G by deletion of a strictly dominated strategy. For example,

Game 9

	C1	C2	C3
R1	2, 4	1, 0	0, 0
R2	3, 1	2, 2	3, 3
R3	4, 0	1, 1	2, 0

reduces by a strictly dominated discard to

	C1	C2	C3
R2	3, 1	2, 2	3, 3
R3	4, 0	1, 1	2, 0

Let a game G "reduce by a sequence of strictly dominated discards" to G' if there is a sequence of games beginning with G and ending with G' such that each comes from its immediate predecessor by a strictly dominated discard. For example, game 9 reduces by such a sequence, first to the 2×3 game above, and then by discards in turn of C1 and R3 all the way to

	C3
R2	3, 3

Here is a principle built on the idea of strictly dominated discards.

Theorem 7. *If a game G reduces by a sequence of strictly dominated discards to G', and game G' resolves for hyperrational players in a certain way, then game G resolves for hyperrational players in this way.*

If a strategy is strictly dominated in a hyperrational game, then in calculating expected values of strategies all players can assign zero probability to this strategy. Hyperrational players can know that such a strategy must eventually be set aside by its agent and thus by everyone. Hence, each player can see that eventually everyone's deliberations will be as they would be from the beginning in the simpler reduced game; similarly, in turn, for that game, if a strategy is strictly dominated in it; and so on until no further deletions of this kind are possible. Applications of Theorem 7 to mixed-strategy games can concentrate on pure strategies, since if a pure strategy is strictly dominated then so is every mixed strategy in which this pure strategy is assigned a positive chance. It is not difficult to see that an interaction, if it can be reached by a sequence of strictly dominated discards from a game, corresponds to the sole equilibrium* in the game's randomization. Compare:

The outcome is determined by the fact that everybody ignores dominated actions, everybody expects everybody else to ignore dominated actions, and so on. (Lewis 1969, p. 19)

But there are two differences between Lewis's idea and mine. First, while he seems to have in mind that everybody ignores his *own* dominated options, etc., my idea is that everybody ignores dominated options of *others,* etc. I cannot comfortably rationalize a hyperrational player's ignoring his own dominated options before he has settled his expectations for others (and thus expected values for all his options), so that he can compare and see which of them maximize expected value. (The hyperrational player will of course then ignore dominated ones or set them aside, seeing that they are submaximal.) But it makes sense that before this player has settled his expectations for others, and indeed in order to do so, he should set aside *their* dominated options as things he is sure they will not do.

The other difference between our ideas is that Lewis's definition of "strictly dominated" is nonstandard:

A *strictly dominated* choice is one such that, no matter how the others choose, you could have made some other choice that would have been better. (p. 17; brackets have been deleted)

Would-be resolutions by discards of Lewis-dominated options are discussed in a here-omitted appendix; see note 1.

6.4. Not every game that can resolve for hyperrational players is covered by Theorem 7. Consider, for example,

Game 10

	C1: $1.10	C2: $1.05	C3: $1.00
R1: $1.10	4, 4	0, 5	0, 3
R2: $1.05	5, 0	2, 2	0, 3
R3: $1.00	3, 0	3, 0	1, 1

No strategy is strictly dominated, yet C1 could be discarded by hyperrational players. As both players could realize, Row could reason as follows:

> If Column is *sure* that I will use R3, then his best response is C3 and he will not use C1. If Column is *not* sure that I will use R3, then – whatever his probabilities for my possible actions R1, R2, and R3 – his expected value for C2 exceeds his expected value for C1, and he will not use C1.

They can thus set C1 aside. The effect of that reduction is a game in which R3 is strictly dominant, and in which C3 is best against R3. Column can be sure that Row will employ R3, and Row can be sure that column will employ C3. The game can in this way resolve in the interaction (R3, C3).

84

For a principle that covers such resolutions, let a probability function P_i be *admissible* for a player i if and only if, for each other player j, either $P_i(s_j) = 1$ for some one strategy s_j of player j, or P_i makes several of that player's strategies equally probable, their probabilities summing to 1. Let a strategy be *eligible* if and only if it maximizes under a probability function that is admissible for its agent. Here is the promised principle:

Theorem 8. *If a game G reduces by a sequence of discards of ineligible strategies to G', and game G' resolves for hyperrational players in a certain way, then game G resolves for hyperrational players in this way.*

If a pure strategy is ineligible, then so is every mixed strategy in which this pure strategy is assigned a positive chance. Thus applications of Theorem 8 can be made by inspections of relevant randomizations. For example, inspection of

	C1	C2	C3	[C1, C2]	[C1, C3]	[C2, C3]	[C1, C2, C3]
R1	4, 4	0, 5	0, 3	2, 9/2	2, 7/2	0, 4	4/3, 4
R2	5, 0	2, 2	0, 3	7/2, 1	5/2, 3/2	1, 5/2	7/3, 5/3
R3	3, 0	3, 0	1, 1	3, 0	2, 1/2	2, 1/2	7/3, 1/3
[R1, R2]	9/2, 2	1, 1/2	0, 3	11/4, 11/4	9/4, 5/2	1/2, 13/4	11/6, 17/6
[R1, R3]	4, 2	3/2, 5/2	1/2, 2	5/2, 9/4	2, 2	2, 9/4	11/6, 13/6
[R2, R3]	2, 0	5/2, 1	1/2, 2	13/4, 1/2	9/4, 1	3/2, 3/2	7/3, 1
[R1, R2, R3]	4, 4/3	5/3, 7/3	1/3, 7/3	17/6, 11/6	13/6, 11/6	1, 7/3	2, 2

confirms that C1 is ineligible: It is not maximum on any row in this randomization of game 10. Setting C1 aside yields a game in which the resolution (R3, C3) can be reached by the sequence (R1, R2, C2) of dominated discards.

6.5. Presumably there are potentially resolvable games not covered by Theorem 8 that would resolve for hyperrational players. One supposes that games with unique equilibria* in their randomizations should resolve for hyperrational players, but there are hyperrational games that have unique equilibria* in their randomizations, where every pure strategy is eligible. Here is such a game with its randomization (I owe this modification of game 7 to Rabinowicz):

Game 11

	C1	C2	C3	[C1, C2]	[C1, C3]	[C2, C3]	[C1, 2, C3]
R1	5, 5	4, 4	1, 1	9/2, 9/2	3, 3	5/2, 5/2	10/3, 10/3
R2	4, 4	2, 2	3, 3	3, 3	7/2, 7/2	5/2, 5/2	3, 3
R3	1, 1	3, 3	2, 2	2, 2	3/2, 3/2	5/2, 5/2	2, 2
[R1, R2]	9/2, 9/2	3, 3	2, 2	15/4, 15/4	13/4, 13/4	5/2, 5/2	19/6, 19/6
[R1, R3]	3, 3	7/2, 7/2	3/2, 3/2	13/4, 13/4	9/4, 9/4	5/2, 5/2	8/3, 8/3
[R2, R3]	5/2, 5/2	5/2, 5/2	5/2, 5/2	5/2, 5/2	5/2, 5/2	5/2, 5/2	5/2, 5/2
[R1, R2, R3]	10/3, 10/3	3, 3	2, 2	19/6, 19/6	8/3, 8/3	5/2, 5/2	25/9, 25/9

I believe this game would resolve for hyperrational players by eliminations of choices corresponding to nonequilibrium* strategies in this randomization. Row could eliminate by indirect arguments every choice on Column's part that corresponds to a nonequilibrium* strategy in this randomization. To take one example, Row could reason that Column will not make a choice that would correspond (as far as Row's expectations are concerned) to the strategy ($\frac{1}{2}$C1, $\frac{1}{2}$C2), since if Column did, then C1 would tie with C2. And Column would realize that Row considered C1 and C2 equally probable and C3 not at all probable, and so would certainly employ R1. Hence C1 would uniquely maximize, and not tie with C2.

Taking another example, Row could reason that Column will not make a choice that would correspond (as far as Row's expectations are concerned) to the strategy ($\frac{1}{2}$C2, $\frac{1}{2}$C3), which is an equilibrium strategy but not an equilibrium* strategy. If Column did make such a choice then C2 and C3, but not C1, would maximize. Column would realize that Row considered C2 and C3 equally probable and C1 not at all probable, resulting in each of Row's strategies maximizing and her making an indifferent choice from {R1, R2, R3}. But then Column would consider these strategies of Row's equally probable, with the result that C1 would maximize; in fact, it would uniquely maximize.

The following principle, one that would have game 11 resolve in choices corresponding to (R1, C1) in its randomization, is I think valid.

Theorem 9. *If the randomization of a potentially resolvable game G (i.e., a game whose randomization contains at least one equilibrium*) reduces by discards of nonequilibrium* strategies to the randomization of a game G' that resolves in some way for hyperrational players into indifferent* *choices from (possibly singleton) sets of strategies, then G resolves into those choices.*

Regarding the intent of Theorem 9, I note that discards of nonequilibrium* strategies can be several (indeed all) at one time. Also, included in ways in which game G' may resolve is the way of Theorem 9 itself. It is thus not difficult to see that Theorem 9 subsumes Theorem 8, for if a strategy of a game is ineligible then it is not an equilibrium* strategy in that game's randomization.

Here is another game (with randomization) that would, I think, resolve for hyperrational players in the way of Theorem 9:

Game 12

	C1	C2	[C1, C2]
R1	1, 1	0, 0	1/2, 1/2
R2	0, 0	0, 0	0, 0
[R1, R2]	1/2, 1/2	0, 0	1/4, 1/4

While both (R1, C1) and (R2, C2) are equilibria, only (R1, C1) is an equilibrium*. Row could therefore eliminate, by indirect arguments, every choice on Column's part other than a definite choice of C1. To illustrate, a definite choice of C2 could be eliminated, since if Column made this choice he would know that, expecting it, Row would be indifferent between R1 and R2. Judging these strategies of Row's to be equally probable, Column would choose C1 for sure, not C2 as supposed – C1 would uniquely maximize. An indifferent choice from {C1, C2}, as well as a choice of any available mixed strategy, could be similarly eliminated. (A game of this structure is discussed in Sobel 1972, pp. 173–4.)

6.6. There remains much unfinished business. This includes not only consideration of other ways in which games might resolve for hyperrational players (e.g., further eliminations may be possible for some symmetrical games), but also ministration to the unease to which I have confessed, under game 8, with all the ways of resolution defended in Section 6.

I maintain that hyperrational players, in order to settle their own expectations and expected utilities, use their knowledge that their fellows will, in the end, have somehow settled *their* expectations and thus *their* expected utilities. Hyperrational players in two-person games are in effect allowed to settle their expectations for their fellows by elimination of what would be (for their fellows) unratifiable choices: Players are allowed to eliminate possible choices of their fellows that they know that their fellows will not make, because they know their fellows could not know they were making them. But I do not propose this as a way for hyperrational players to suppose that their fellows make up their minds what to do. Similarly, speaking now of agents in general and not just hyperrational ones, I hold that to be rational an action must not only maximize relative to all relevant alternatives, but must also be the subject of a ratifiable choice. But I do not confine relevant alternatives to subjects of possible ratifiable choices, or propose as a method of deliberation that one first eliminate nonratifiable options and then find what maximizes among options that remain (see note 4). Nor do I think that hyperrational players might reasonably do something similar, and find out what they ought to do by eliminating options of their own that they could not be known by them to maximize.

There is a cloud over the ways of resolution I endorse, wherein hyperrational players in two-person games in effect eliminate unratifiable options of their fellows. There is an air, if not of circularity and question-begging, at least of incompleteness or lack of candor about these ways of resolution. It is clear, as maintained in Section 5, that hyperrational games can resolve only in indifferent choices corresponding to equilibria* in their randomizations, but it is not as clear (and one is not as easy about) *how* at least some hyperrational games can resolve in such indifferent choices, or by what processes of reasoning hyperrational players can at least sometimes settle their expectations for one another's actions, as well as their expected utilities for their own actions.

What is to be made of this problem? – perhaps a virtue. Perhaps one should not expect it to be entirely clear how hyperrational players would manage their affairs. Ordinary interactions are facilitated by all manner of natural and cultural props that would be irrelevant to hyperrational interactions; habits, kinds of suggestibility, conventions, shared senses of "salience," precedence, and so forth either have no place, or lack force, in hyperrational communities. So much that is of importance to ordinary interactions would be irrelevant in the hyperrational case that the wonder should be that hyperrational players can interact at all.

Under the sway of these thoughts, I welcome the idea that hyperrational players would manage to interact by narrowing down possibilities for their fellows' actions. Their play, which might be cast as "Labyrinths Declined," in fact strikes me as just right for them, and if not the whole of the story for them, then at least its first and main part. I believe that the kind of eliminative thinking attributed here to hyperrational players is evident in our own procedures, though so covered over by many more direct and effective deductions as to be hardly noticeable. My view is that this kind of thinking is the main part (if not all) of what, at the

88

hyperrational limit, would be left of ordinary interactive thinking. What is remarkable is not that what would be left at this rarefied limit would be, at least in its main part, so roundabout and restrained. What is marvelous, I think, is that anything at all would be left, and that hyperrational players would have at least some ways to interact.

1. Problems – situations in which hyperrational agents would not do well, or with which they simply could not cope – are discussed in the third part of a 1988–9 manuscript entitled "Hyperrational Games." The present chapter comes from the first two parts and does not include discussion of these problems.
2. The main lemma for this result is that, given the basic assumption for mixed-strategy games, the causal expected value of a mixed strategy M is a weighted average of the expected values of degenerate mixed strategies that correspond to the pure strategies mixed in M. For substantiation, consider Row's mixed strategy $[x\mathrm{R}1, (1-x)\mathrm{R}2]$ in the game

	C1	C2
R1	a, e	b, f
R2	c, g	d, h

By the basic assumption, the normal form for this game includes, for that mixed strategy,

	C1	C2
$[x\mathrm{R}1, (1-x)\mathrm{R}2]$	$xa + (1-x)c$	$xb + (1-x)d$

Suppose that Row's probabilities include $P_r(\mathrm{C}1) = p$ and $P_r(\mathrm{C}2) = (1-p)$. Then, by applications of a general partition principle for causal expected values and by elementary algebra,

$$\mathrm{EV}_r[x\mathrm{R}1, (1-x)\mathrm{R}2] = p[xa+(1-x)c]+(1-p)[xb+(1-x)d]$$
$$= x[pa+(1-p)b]+(1-x)[pc+(1-p)d]$$
$$= x\mathrm{EV}(\mathrm{R}1)+(1-x)\mathrm{EV}(\mathrm{R}2).$$

Given this lemma, if (say) $[x\mathrm{R}1, (1-x)\mathrm{R}2]$ is in Row's MAX then (i)

$$\mathrm{EV}_r[x\mathrm{R}1, (1-x)\mathrm{R}2] = \mathrm{EV}_r(\mathrm{R}1) = \mathrm{EV}_r(\mathrm{R}2)$$

(for otherwise the expected value of one or the other of R1 and R2 would exceed that of $[x\mathrm{R}1, (1-x)\mathrm{R}2]$), and so (ii) for every chance y, $[y\mathrm{R}1, (1-y)\mathrm{R}2]$ is in MAX.
3. Take "at" to mean "just before" and the whole as meaning "at any moment prior to the action during a period bounded above by the time of the action,

during which period the agent's credences and preferences do not change." It is a precondition for hyperrational games that for each player there is such a period.

4. Harper explores the consequences for game theory, not of a simple causal maximizing theory, but of a somewhat qualified theory that makes *non*maximizing actions rational in some situations. He observes that patterns of conditional probabilities can sometimes be such that an action that would maximize causal expected value would not still do so after revisions of probabilities for circumstances by conditionalizing on this action. In that case the action, though maximizing, would not be *ratifiable* (Harper 1988, p. 28). The version of causal decision theory that Harper favors prescribes actions that, among ratifiable ones, are maximizing. In Harper's theory, ratifiability is a lexically prior condition to maximizing, and sometimes the rational act is not one that maximizes among all options relative to time-of-choice preferences and probabilities.

It is relevant to my interest in hypotheses concerning early-on conditional probability, and relevant in particular to the fact that I am less interested in them than is Harper, that the somewhat adulterated causal decision theory that I favor makes ratifiability not a prior condition to but rather a coordinate condition with maximizing, and rules that an action is rational if and only if (i) it is ratifiable and (ii) it would maximize (see Sobel 1983, 1990).

5. As stated in Section 3.2, hyperrational players are causal maximizers, nothing more and nothing less. Thus they do not conform completely to the adulterated causal decision theory that I favor. Superrational players – players who were ideally rational according to my adulterated theory – could not cope with a version of game 5 in which even indifferent choices of particular strategies were good predictors early on of like particular strategies on the other player's part (so that early on for Row, e.g., conditional probabilities $P(C1/R1)$ and $P(C2/R2)$ were both nearly 1). Only a restricted form of Theorem 9 would be valid for superrational games.

REFERENCES

Bernheim, B. Douglas (1984), "Rationalizable Strategic Behavior." *Econometrica* 52: 1007–28.

Harper, William L. (1988), "Causal Decision Theory and Game Theory: A Classic Argument for Equilibrium Solutions, A Defense of Weak Equilibria, and a New Problem for the Normal Form Representation." In W. L. Harper and B. Skyrms (eds.), *Causation in Decision, Belief Change, and Statistics*. Dordrecht: Kluwer.

Lewis, David (1969), *Conventions: A Philosophical Study*. Cambridge, MA: Harvard University Press.

Luce, R. Duncan, and Raiffa, Howard (1957), *Games and Decisions: Introduction and Critical Survey*. New York: Wiley.

Moore, Omar K., and Anderson, Alan R. (1962), "Some Puzzling Aspects of Social Interaction." *Review of Metaphysics* 15: 409–33.

Pearce, David P. (1984), "Rationalizable Strategic Behavior and the Problem of Perfection." *Econometrica* 52: 1029–50.

Rabinowicz, Wlodzimierz (1986), "Non-cooperative Games for Expected Utility Maximizers." In P. Needham and J. Odelstad (eds.), *Changing Positions: Essays Dedicated to Lars Lindahl on the Occasion of his Fiftieth Birthday*. University of Uppsala Press.

Rabinowicz, Wlodzimierz (1989), "Tortuous Labyrinth: Non-Cooperative Normal Form Games between Hyperrational Players." Department of Philosophy, University of Uppsala. Reprinted as Chapter 7 in this volume.

Sobel, Jordan Howard (1972), "The Need for Coercion." In J. R. Pennock and J. W. Chapman (eds.), *Coercion: Nomos XIV.* Chicago: Aldine & Atherton.

Sobel, Jordan Howard (1983), "Expected Utilities and Rational Actions and Choices." *Theoria* 49: 159–83.

Sobel, Jordan Howard (1987), "Self-Doubts and Dutch Strategies." *Australasian Journal of Philosophy* 65: 56–81.

Sobel, Jordan Howard (1989a), "Partition-Theorems for Causal Decision Theories." *Philosophy of Science* 56: 70–93.

Sobel, Jordan Howard (1989b), "Utility Theory and the Bayesian Paradigm." *Theory and Decision* 26: 263–93.

Sobel, Jordan Howard (1990), "Maximization, Stability of Decision, and Actions in Accordance with Reason." *Philosophy of Science* 57: 60–77.

Von Neumann, John, and Morgenstern, Oskar (1953), *Theory of Games and Economic Behavior,* 3rd ed. New York: Wiley.

6

Equilibria and the dynamics of rational deliberation

BRIAN SKYRMS

A first fundamental merit of von Neumann and Morgenstern's work lies in the decisive push it has given for the establishment of the criterion of maximization of expected utility as a universal criterion of decision. . . . Actually, it seems that the best advice in order to make good (and never bad) use of game theory lies in going back to such decision theory.

B. de Finetti, "John von Neumann e Oskar Morgenstern,"
Il Maestri Dell'Economica Moderna

1. INTRODUCTION

De Finetti (1970) and Ramsey (1931) showed how rational decision is founded on *coherence*. This should be true in game-theoretic contexts just as much as in other ones. If followed to their logical conclusion, considerations of coherence can lead to dynamic deliberation and in this context introduce a natural equilibrium concept. Such deliberational equilibria are intimately connected with game-theoretic equilibria in games played by dynamic deliberators.

In the simplest cases deliberation is trivial; one calculates expected utility and maximizes. But in more interesting cases, the very process of deliberation may generate information that is relevant to the evaluation of the expected utilities.[1] Then, processing costs permitting, a Bayesian

This paper is a somewhat modified version of "Correlated Equilibria and the Dynamics of Rational Deliberation," *Erkenntnis* 31 (1989), 347–64. I would like to thank Michael Teit Nielsen for pointing out a flaw in the formulation of "seeks the good" in an earlier draft; Maria-Carla Galavotti for calling my attention to de Finetti's essay on von Neumann and Morgenstern and Maria Dimaio for translating it; Kenneth Arrow and David Lewis for calling my attention to Aumann (1987) prior to its publication; Greg Kavka for discussion of my perverse comments on Hobbes; Richard Jeffrey, Bas van Fraassen, and Bill Harper for discussions about games and rational deliberation, and the John Simon Guggenheim foundation and the Humanities Council and Department of Philosophy of Princeton University for generous support. Issues raised in this paper are discussed at greater length in Skyrms (1990a).

deliberator will feed back that information and recalculate the expected utilities in light of the new knowledge.[2]

In this type of decision problem, deliberation can be modeled as a dynamic system. The decision maker starts in a state of indecision; calculates expected utility; moves in the direction of maximum expected utility; feeds back the information generated and recalculates; and so forth. In this process, his probabilities of doing the various acts evolve until, at the time of decision, his probabilities of doing the selected act become virtually 1.

Dynamic deliberation carries with it an equilibrium principle for individual decision. The decision maker cannot decide to do an act that is not an equilibrium of the deliberational process. If he is about to decide to do it, deliberation carries him away from that decision. This sort of equilibrium requirement for individual decision can be seen as a consequence of the expected-utility principle. It is usually neglected only because the process of informational feedback in deliberation is usually neglected. In cases in which there is no informational feedback, simply choosing the act having initial maximum expected utility automatically fulfills the equilibrium requirement.

This becomes relevant to game theory if we consider games played by deliberators with shared initial subjective probabilities, whose deliberation rules and whose initial subjective probabilities are common knowledge, and who use that common knowledge in the deliberational process to update subjective probabilities by emulation of the other players' deliberation. Under these assumptions, we can show that joint deliberational equilibrium on the part of all the players corresponds to a Nash equilibrium point of the game. In this model, choice of a Nash equilibrium is a consequence of the expected-utility principle together with these assumptions of common knowledge.

This analysis provides a foundation for basic solution concepts in classical noncooperative game theory by embedding game theory in the setting of deliberational dynamics. The dynamics give a natural interpretation for Nash equilibrium, as well as for certain kinds of correlated equilibrium.

2. DELIBERATIONAL EQUILIBRIUM

Let us model the deliberational situation in an abstract and fairly general way. A decision maker must choose between a finite number of acts: $A_1, \ldots, A_n$. Calculation takes time, although its cost is negligible. We assume that the decision maker is certain that deliberation will end and she will choose some act (perhaps a mixed one) at that time. Her *state of indecision* will be a probability vector assigning probabilities to each of the n acts, which sum to 1. These are to be interpreted as her probabilities, now that she will do the act in question at the end of deliberation. A state of indecision P carries with it an expected utility; the expectation according to the probability vector $\mathbf{P} = \langle p_1, \ldots, p_n \rangle$ of the expected utilities of the acts $A_1, \ldots, A_n$. The expected utility of a state of indecision is thus computed just as that of the corresponding mixed act. Indeed, the adoption of a mixed strategy can be thought of as a way of turning to stone the state of indecision for its constituent pure acts. We will call a mixed act corresponding to a state of indecision its *default mixed act*.

The decision maker's state of indecision will evolve during deliberation. In the first place, on completing the calculation of expected utility, she will believe more strongly that she will ultimately do that act (or one of those acts) that are ranked more highly than her current state of indecision. If her calculation yields one act with maximum expected utility, will she not simply become sure that she will do that act? She will not *on pain of incoherence*[3] if she assigns any positive probability at all to the possibility that informational feedback may lead her ultimately to a different decision. So, she will typically in one step of the process move in the direction of the currently perceived good, but not all the way to decision.

We assume that the decision maker moves according to some simple dynamical rule rather than performing an elaborate calculation at each step.[4] One concrete example of such a rule is what I have called the Nash dynamics (after Nash 1951). Define the *covetability* of an act in a state of indecision p as the difference in expected utility between the act and the state of indecision if the act is preferable to the state of indecision, and as zero if the state of indecision is preferable to the act: $\text{cov}(A) = \max[U(A) - U(p), 0]$. Then the Nash dynamics take the decision maker from state of indecision p to state of indecision p', where each component p_i of p is changed to

$$p_i' = \frac{k p_i + \text{cov}(A_i)}{k + \Sigma_j \text{cov}(A_j)}.$$

Here a bold revision is hedged by averaging with the status quo. The constant k ($k > 0$) is an index of caution. The larger the value of k, the more slowly the decision maker moves in the direction of acts that look more attractive than the status quo.

I am not suggesting here that some dynamical rule in this Nash family is optimal. Rather, the Nash dynamics provide simple representatives of a class of dynamical rules, which share a property of special interest. They "seek the good," in the following modest sense:

95

(1) they raise the probability of an act only if that act has utility greater than that of the status quo; and

(2) they raise the sum of the probabilities of all acts with utility greater than the status quo (if any).

All dynamical rules that seek the good have the same fixed points, that is, states in which the expected utility of the status quo is maximal. Here we assume only that the decision maker's rule seeks the good.

The decision maker's calculation of expected utility and subsequent application of the dynamical rule constitutes new information that may affect the expected utilities of the pure acts by affecting the probabilities of the states of nature which, together with the act, determine the payoff. In typical game-theoretical contexts, they consist of the possible actions of the opposing players; for simplicity, we assume here a finite number of states of nature.

The decision maker's *personal state* is then, for our purposes, determined by two things: her state of indecision, and the probabilities that she assigns to states of nature. Her personal state space is the product space of her space of indecision and her space of states of nature. Deliberation defines a dynamics on this space. We could model the dynamics as either discrete or continuous, but we will discuss only discrete models here. We assume a dynamical function f that maps a personal state $\langle x, y \rangle$ into a new personal state $\langle x', y' \rangle$ in one unit of time. The dynamical function f has two associated rules:

(i) the adaptive dynamical rule D, which maps $\langle x, y \rangle$ onto x'; and

(ii) the informational feedback process I, which maps $\langle x, y \rangle$ onto y' (where $\langle x', y' \rangle = f(\langle x, y \rangle)$).

A personal state $\langle x, y \rangle$ is a *deliberational equilibrium* of the dynamics f if and only if $f(\langle x, y \rangle) = \langle x, y \rangle$. If D and I are continuous, then 0 is continuous and it follows from the Brouwer fixed point theorem that a deliberational equilibrium exists. Let N be the Nash dynamics for some $k > 0$. Then, if the informational feedback process I is continuous, the dynamical function $\langle N, I \rangle$ is continuous and has a deliberational equilibrium. Then, since N seeks the good, for any continuous informational feedback process I, $\langle N, I \rangle$ has a deliberational equilibrium $\langle x, y \rangle$ whose corresponding mixed act maximizes expected utility in state $\langle x, y \rangle$. This is a point from which process I does not move y and process N does not move x. But if process N does not move x, then no other process that seeks the good will move x either (whether or not that process is continuous). Thus we have the general result:

If D seeks the good and I is continuous, then there is a deliberational equilibrium $\langle x, y \rangle$ for $\langle D, I \rangle$. If D' also seeks the good then $\langle x, y \rangle$ is also a deliberational equilibrium for $\langle D', I \rangle$. The default mixed act corresponding to x maximizes expected utility at $\langle x, y \rangle$.

3. NASH EQUILIBRIUM

Suppose that two (or more) deliberators are deliberating about what action to take in a noncooperative, non–zero-sum matrix game. We assume here that each player has only one choice to make, and that choices are causally independent in that there is no way for one player's decision to influence the decisions of the other players. Then, for each player, the decisions of the other players constitute the relevant state that, together with her decision, determines the consequence in accordance with the payoff matrix.

Suppose, in addition, that each player has an adaptive rule D that seeks the good (they need not have the same rule), and that the deliberational rules of each player are common knowledge.[5] Suppose also that each player's initial state of indecision is common knowledge, and that other players take a given player's state of indecision as their own best estimate of what that player will ultimately do. Then there is an initial probability assignment to all the acts for all the players that is shared by all the players and is common knowledge. (We assume here and in the subsequent updating that joint probabilities are product probabilities; that is, each player views all other players' actions as probabilistically independent.)

In this idealized situation, an interesting informational feedback process becomes available. Starting from the initial position, player 1 calculates expected utility and moves by her adaptive rule to a new state of indecision. She knows that the other players are Bayesian deliberators who have just carried out a similar process. And she knows their initial states of indecision and their updating rules. So she can simply go through their calculations to see their new states of indecision and update her probabilities of their acts accordingly. We will call this sort of informational feedback process *updating by emulation*. Suppose that all the players update by emulation. Then, in this ideal case, the new state is common knowledge as well, and the process can be repeated.

The joint deliberation of the players can be modeled as a dynamical process in discrete time. Under the foregoing assumptions of common knowledge and updating by emulation, the deliberational state of the system of deliberators can be characterized by an assignment of probabilities to each of the acts of each of the players. Each player interprets the probabilities assigned to her own acts as representing her state of indecision, and the probabilities assigned to the other players' acts as representing her uncertainty about the state of the world. For each player j, her adaptive dynamical rule maps her old probabilities over her acts to new ones. Taken together, the dynamical rules of the players map the old deliberational state of the whole system into a new one. This dynamical process is continuous if the adaptive dynamical rules of the individual players are (because my adaptive dynamical rule is your informational feedback rule, and conversely). Suppose a finite, non–zero-sum game is played by n Bayesian deliberators who use the Nash dynamics. Then the dynamics of the system is continuous and a deliberational equilibrium for the whole system exists, which is by definition a deliberational equilibrium for each player.

We say that an assignment of an act (pure or mixed) to each player is a *Nash equilibrium* if each player, taking the other players' acts as fixed states of nature, is at an act of maximal expected utility. The relation between the Nash equilibrium of a game and deliberational equilibria of the individual players who deliberate in the manner just explained is now evident:

> In a game played by deliberators with an adaptive rule that seeks the good and with updating by emulation, each player is at a deliberational equilibrium at a state of the system if and only if the assignment of the default mixed acts to each player constitutes a Nash equilibrium of the game.

In a finite, N-person, non–zero-sum game, the availability of mixed strategies guarantees the existence of a Nash equilibrium, because if the game were played by Bayesian players with the Nash dynamics and updating by emulation, deliberation would map the space of states of the system into itself continuously, with the Brouwer fixed point theorem guaranteeing a system state that represents a deliberational equilibrium for each of the players. This is how Nash himself proved the existence of Nash equilibria, although he did not interpret the Nash map as deliberational dynamics.

4. CORRELATED EQUILIBRIUM

There are two different points of view adopted in discussions of "solution concepts" for games. One is the point of view of the players themselves as rational actors. The other is that of a disinterested rational observer or theorist. We have so far discussed deliberational dynamics from the viewpoint of the deliberators themselves, but there are interesting consequences for the external point of view. We illustrate this in two cases where the external point of view is taken by a philosopher or social theorist: the question of the possibility of convention, and the question of the nature of the "state of nature." In each case, rational deliberation generates correlation. This phenomenon can be described generally using the notion of a *correlated equilibrium*.

How is convention possible? Quine (1936, 1960) challenged conventionalist accounts of language to provide a satisfactory account of how the relevant conventions are set up and maintained, an account that does not presuppose linguistic communication or competency. Lewis (1969) replied that convention is possible without communication; the mutual expectations of rational agents can explain the maintenance of a convention at a game-theoretic equilibrium. Consider the pure coordination game:

	L	R
L	1, 1	0, 0
R	0, 0	1, 1

This can be thought of as "the winding road": Two cars approach a blind curve from opposite directions. Each would prefer that they both drive on the left or both on the right. There are two pure equilibria, equally attractive, but if Row chooses one and Column chooses the other, both will end up in trouble. If, however, Row believes Column expects him to drive on the left and believes that Column believes him to believe this, etc., and if Column believes likewise about Row, and if each believes that the other is rational and believes that the other believes that he is, then they each have good reason to drive on the left.

The question of how convention without communication is possible between rational agents has two parts: (i) How can convention without communication be sustained? and (ii) How can convention without communication be generated? Lewis has answered the first question in terms of equilibrium (or stable equilibrium) and common knowledge of rationality. His discussion of the second question – following Schelling (1960) – is framed in terms of *salience,* where a salient coordination equilibrium is "one which stands out from the others in some conspicuous respect." Salience could derive from preplay communication among the players, but it could also arise in other ways; it could arise by precedent. In fact, since salience is a psychological rather than a logical notion, the ways in which salience may arise are as various as the possible psychologies of the players. The informal discussions of salience in Lewis and Schelling are

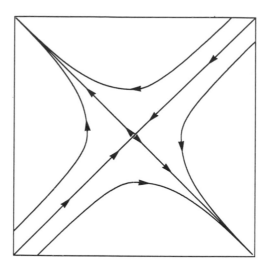

Figure 1. The winding road.

convincing with regard to the plausibility of real-world coordination by salience, but I believe they give only a partial answer to the second question. Here, deliberational dynamics has something to contribute.

Let us model the winding road as a game played by Nash deliberators. Row and Column each have predeliberational probabilities of driving on the left or right; these probabilities can be anything at all. At the onset of deliberation, the players' initial probabilities of driving left or right are announced and become common knowledge. (This idealization will be weakened later.) You – the philosopher – have some probability distribution over the space of Row's and Column's initial probabilities. You needn't think it likely that they are anywhere near an equilibrium. In fact, we will suppose only that your probability distribution is reasonably smooth (i.e., absolutely continuous with respect to Lebesgue measure on the unit square). Then you should believe with probability 1 that the deliberators will converge to one of the pure Nash equilibria, as is evident from Figure 1.

It is not surprising that the players should be led to the state of mutually reinforcing expectations that attend a Nash equilibrium. Coordination is effected by rational deliberation. Precedent and other forms of initial salience may influence the deliberators' initial probabilities, and thus may play a role in determining which equilibrium is selected. The answer to the question of how convention can be generated for Bayesian deliberators has both methodological and psychological aspects.

100

Of course, Bayesian players are not always so lucky as to be involved in pure coordination games. People have conflicting desires and limited altruism. They are roughly equal in their mental and physical powers, and elements of competition intrude. Consequently, Thomas Hobbes (1651) argued that rational, self-interested decision makers in a state of nature, unrestrained by the power of a sovereign, will be engaged in a "war of all against all."

In a fine critical study, Kavka (1983) finds Hobbes's argument inconclusive although, as he points out, many other commentators (e.g., Gautier 1969) appear to regard it as obviously correct. Both Kavka and Gautier model conflict in the state of nature in terms of prisoner's dilemma, but I think the game of "chicken" models Hobbes's premises at least as well:

	Don't	Swerve
Don't	$-10, -10$	$5, -5$
Swerve	$-5, 5$	$0, 0$

Each player would like to profit from his opponent's loss. Each would like to initially appear more aggressive than his opponent, but aggression on the part of both creates an intolerable situation. There are two Nash equilibria in pure strategies: Row swerves and Column doesn't, and Column swerves and Row doesn't. There is also a mixed equilibrium where each player has equal chances of swerving and not swerving. If the players are Bayesian deliberators then coordination can again be achieved by deliberation, just as in the coordination game. For Nash deliberators, every initial point leads to a Nash equilibrium, and almost every initial point leads to a pure Nash equilibrium (see Figure 2).

In this example, it is easy to say which initial points go to which equilibrium. For almost every initial point, one player is initially more likely to swerve; if so, that player ends up swerving while the other player does not. In the case where both players initially are equally likely to swerve, they are carried to a mixed equilibrium where each adopts a random strategy of swerving with chance of $\frac{1}{2}$. Here there is a genuine Hobbesian incentive for initial bellicosity. (There is no such incentive in prisoner's dilemma.) Nevertheless, crashes are almost always avoided as a result of rational deliberation.

This conclusion is a rather surprising result of the kind of analysis possible within the framework of deliberational dynamics. More naive approaches to rational decisions would lead one to expect different results. If each player prepared for the worst and played his security strategy, both would swerve and no equilibrium would be achieved. If both players analyzed the game, identified the equilibria, and chose the act associated

101

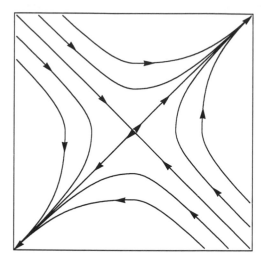

Figure 2. Chicken.

with the equilibrium that had maximum expected utility for him, then neither would swerve and we would indeed have the war of all against all. But if the players deliberate dynamically, that neither would swerve is an unlikely consequence of the mixed equilibrium which itself has negligible prior probability.

Did Hobbes attempt to derive a prisoner's dilemma conclusion from chicken premises? It would be premature to draw this conclusion from such an oversimplified model of the state of nature. A number of complications must be introduced before we could begin to do Hobbes justice. One can find materials for more realistic models in ethological descriptions of varieties of animal conflict. Of particular interest is the prevalence of ritualized aggression in which little real damage is done (Lorenz 1966, Eibl-Eibesfeld 1970). This *is* part of the state of nature, and it does not agree with Hobbes's description.

One might argue that the rationality of humans invalidates the analogy, but analyses of evolutionary game theory (Maynard-Smith and Price 1973, Maynard-Smith 1982, Parker 1974) do not support this objection. For these sorts of game-theoretic models, Bayesian deliberators of the kind considered here will decide by deliberational dynamics in a way analogous to the way that nature decides by evolution. In the games considered by Maynard-Smith and Price, self-interested rational deliberators will play in a decidedly un-Hobbesian way. In general, we must agree with the carefully considered conclusion of Kavka (1986, p. 122): "which strategy is better overall probably cannot be determined a priori for all state of nature situations. Instead, it will depend on the value of a number of important variables and parameters, which will vary according to the version of the state of nature in question."

There is a general conception under which the foregoing examples all fall: Aumann's (1974) notion of a correlated equilibrium. He suggests that mixed strategies, where the chance devices used by different players are assumed independent, be treated as a special case of correlated strategies, where the chance devices may have any joint probability distribution at all. You might think of a referee observing the outcome of some random process – say, the toss of a many-sided die – and communicating to each player which aspect of the process is that player's random variable. For example, one player might get to know the color of the face, another the number of spots showing, and so on. Any correlation of numbers and spots in the random device is allowed.

A random strategy for player i can be thought of as a probability assignment to that player's space of possible actions A_i; that is, a random variable mapping some probability space into A_i. A *joint correlated strategy* can be thought of as any probability assignment on the product space of the action spaces of all players, $A_1 \times A_2 \times \cdots \times A_n$; that is, a mapping of a probability space onto sequences of actions. A correlated strategy can obviously be specified by giving the underlying joint probability space together with a sequence of random variables on it: $f_1, f_2, ..., f_n$, where f_i maps the probability space into the action space of player i. In the special case of ordinary mixed strategies, the probability on the product space is the product measure, and the random strategies are independent. An *i-deviation* from a correlated strategy C consists of the same probability space and the same random variables f_1 etc., except that f_i is replaced by some random variable $g(f_i)$ taking values in A_i. An i-deviation represents player i unilaterally deviating from the original joint correlated strategy, so that when the original strategy tells i to do one thing he does something else, while all other players stick to the original correlated strategy. A *correlated equilibrium* is a joint correlated strategy such that each player i's expected utility on the joint correlated equilibrium strategy is greater than or equal to the expected utility on any i-deviation from it (the expectation being taken according to the underlying probability space). Thus, a correlated equilibrium is a joint correlated strategy from which no player has anything to gain by unilateral deviation.

In certain situations, it might be to all the players' mutual advantage to agree on a joint correlated equilibrium strategy, and either hire a referee or construct a machine to carry out the random experiment and communicate to each player the action selected for him. It might appear that "[f]or strategies to be correlated there must be some mechanism for communicating and contracting between the players" (Shubik 1985, p. 247). But, as we have seen in several examples, rational deliberation can play a powerful role in establishing correlation. Let us consider in a general way the sort of situation sketched at the beginning of this section.

An observer, Theo, knows that n players will be induced to play a certain n-person, noncooperative game. Theo knows that the players are all dynamic deliberators whose dynamical rules seek the good and who update by emulation, and that this fact will be common knowledge to the players at the onset of deliberation, as will their dynamical rules and prior probabilities. Theo has analyzed the game and knows that in it the dynamics always converge to a Nash equilibrium. Theo may or may not know who the players are. He does not know what their initial probabilities for their possible actions will be, but rather has his own probability measure over the possible initial states of indecision of the system. With respect to Theo's probability measure, the players are at a correlated equilibrium.

When the true initial state of indecision is selected, a recommendation for action is delivered to each player by deliberational dynamics. Since the dynamics lead from each initial state to a Nash equilibrium, no player has anything to gain by deviating from that recommendation. Thus no i-deviation from the joint correlated strategy defined by Theo's probability is preferable to it for player i, so that joint correlated strategy is (by definition) a correlated equilibrium. This is true regardless of Theo's probability measure over the space of initial states of indecision. This correlated equilibrium is a general result of the players' common knowledge and dynamic deliberation.

Returning to the first-person point of view, can we have a situation where, prior to deliberation, the players themselves are each in a position analogous to the role of Theo? This raises some delicate questions. Theo is uncertain about the players' prior probabilities, but for the players themselves these probabilities are assumed to be common knowledge at the onset of deliberation. The players have a different source of prior uncertainty: They are uncertain where deliberation will lead because they cannot immediately "see through" the computations. How can they then be sure that deliberation will converge? In general they cannot. But in some cases they may know that they are in a class of cases for which their deliberational rules will converge, without knowing the ultimate outcome for the particular case at hand.

This scenario brings us close to the point of view of Aumann (1987), where he argues that correlated equilibrium is a consequence of a common prior probability together with common knowledge of Bayesian rationality. Starting with a common prior, each of the players receives some private information; then each chooses an action. The assumption of Bayesian rationality assures that each player chooses an action that maximizes (posterior) expected utility, and this is taken to be common knowledge. The common prior includes prior probabilities over what each player will ultimately choose. The common prior then is itself interpreted as the probability setting up the correlated strategy, with the joint maximization of expected utility assuring that it is an equilibrium. In our current scenario, the information that the players receive is generated by deliberation. It is not private, however, since each player can update by emulation. Thus players' ultimate choices constitute a Nash equilibrium. If players at the onset of deliberation know this and know the Nash equilibria of the game, then their initial uncertainty will concern which Nash equilibrium the deliberation will lead to, and the correlated equilibrium according to their prior probabilities will be of a special kind – a mixture of Nash equilibria.

Aumann's viewpoint is somewhat different from ours in that he does not explicitly consider the process of deliberation, but only its result. There is no analysis of how the players jointly arrive at decisions where each maximizes his expected utility, and the difficult question of convergence is, in a way, hidden behind the assumption of Bayesian rationality. Considerations of deliberational dynamics thus add a further dimension to the theory of correlated equilibria, and provide an account of one way in which correlated equilibria may be generated.

NOTES

1. We are in the realm of what Good (1977) calls "dynamic probability." On the treatment of information generated by computation, see also Hintikka (1975) and Lipman (1989).
2. See the discussion in Skyrms (1987c, 1990b).
3. This is dynamic coherence, as treated in Skyrms (1987a,b).
4. Hence we are looking here at a special sort of bounded Bayesian deliberation. Bayesian modeling of more sophisticated deliberators is also of interest, but is outside the scope of this chapter.
5. See Lewis (1969) and Aumann (1974).

REFERENCES

Aumann, R. J. (1974), "Subjectivity and Correlation in Randomized Strategies." *Journal of Mathematical Economics* 1: 67–96.
Aumann, R. J. (1975), "Agreeing to Disagree." *The Annals of Statistics* 4: 1236–9.
Aumann, R. J. (1987), "Correlated Equilibrium as an Expression of Bayesian Rationality." *Econometrica* 55: 1–18.

de Finetti, B. (1970), "John von Neumann e Oskar Morgenstern." In *I Maestri Dell'Economica Moderna*. Milano: Franco Agnelli.

Eibl-Eibesfeld, I. (1970), *Ethnology: The Biology of Behavior*. New York: Holt, Rinehart & Winston.

Gautier, D. (1969), *The Logic of the Leviathan*. Oxford: Oxford University Press.

Good, I. J. (1977), "Dynamic Probability, Computer Chess, and the Measurement of Knowledge." In E. W. Elcock and D. Michie (eds.), *Machine Intelligence*. New York: Wiley, pp. 139-50.

Hintikka, J. (1975), "Impossible Possible Worlds Vindicated." *Journal of Philosophical Logic* 4: 475-84.

Hobbes, T. (1651), *Leviathan*. Reprinted (1958) Oxford: Clarendon Press.

Kavka, G. (1983), "Hobbes War of All Against All." *Ethics* 93: 291-310.

Kavka, G. (1986), *Hobbesian Political and Moral Theory*. Princeton, NJ: Princeton University Press.

Lewis, D. (1969), *Convention*. Cambridge, MA: Harvard University Press.

Lipman, B. L. (1989), "How to Decide How to Decide How to . . . : Limited Rationality in Decisions and Games." Working Paper, Carnegie-Mellon University.

Lorenz, K. (1966), *On Aggression*. London: Methuen.

Maynard-Smith, J. (1982), *Evolution and the Theory of Games*. Cambridge: Cambridge University Press.

Maynard-Smith, J., and Price, G. R. (1973), "The Logic of Animal Conflict." *Nature* 246: 15-18.

Parker, G. A. (1974), "Assessment Strategy and the Evolution of Fighting Behavior." *Journal of Theoretical Biology* 47: 223-43.

Nash, J. F. (1951), "Noncooperative Games." *Annals of Mathematics* 54: 289-95.

Quine, W. V. (1936), "Truth by Convention." in O. H. Lee (ed.), *Philosophical Essays for A. N. Whitehead*. New York: Longmans.

Quine, W. V. (1960), *Word and Object*. New York: Wiley.

Ramsey, F. P. (1931), "Truth and Probability." In R. B. Braithwaite (ed.), *The Foundations of Mathematics and Other Logical Essays*. London: Routledge & Kegan Paul.

Schelling, T. (1960), *The Strategy of Conflict*. Oxford: Oxford University Press.

Shubik, M. (1985), *Game Theory for the Social Sciences*. Cambridge, MA: MIT Press.

Skyrms, B. (1987a), "Dynamic Coherence." In I. B. MacNeill and G. Umphrey (eds.), *Advances in the Statistical Sciences, vol. II: Foundations of Statistical Inference*. Dordrecht: Reidel.

Skyrms, B. (1987b), "Dynamic Coherence and Probability Kinematics." *Philosophy of Science* 54: 1-20.

Skyrms, B. (1987c), "On the Principle of Total Evidence with and without Observation Sentences." In *Logic, Philosophy of Science and Epistemology: Proceedings of the 11th International Wittgenstein Symposium*. Vienna: Holder-Pichler-Tempsky, pp. 187-95.

Skyrms, B. (1990a), *The Dynamics of Rational Deliberation*. Cambridge, MA: Harvard University Press.

Skyrms, B. (1990b), "The Value of Knowledge." In C. W. Savage (ed.), *Scientific Theories*. Minneapolis: University of Minnesota Press, pp. 245-66.

7

Tortuous labyrinth:
Noncooperative normal-form games
between hyperrational players

WLODZIMIERZ RABINOWICZ

> It is characteristic of social interaction that I cannot maximize my own utility without "taking into account" (in some vague sense which we understand only darkly) what *you* are up to. I must have some theories or intuitions about how you are likely to behave, how you will respond to my actions, and the like. Worse than that, I must be aware that you are likely doing the same thing in regard to me, and I have to take *that* possibility into account as well. But of course you may know that I am aware of this possibility, and . . . adjust your behaviour accordingly. And so it goes – we are both involved in a tortuous labyrinth of relations . . . .
>
> Moore and Anderson (1962, p. 413)[1]

In this chapter I shall discuss noncooperative normal-form games between extremely rational ("hyperrational") players. What is characteristic for such players is that they are supposed to maximize their expected utility on the basis of probability assignments that they somehow extract from the available nonprobabilistic information – from the common knowledge of the matrix of the game plus the assumptions of independence and rationality. In order to arrive at these probability assignments, the players are supposed to make use of the principle of insufficient reason.

I shall define the notion of a potential solution of such a hyperrational game: a so-called complete probabilistic equilibrium. It will turn out that some of the games of this kind lack a solution whereas others have more than one. Then I shall describe an iterative deliberation method that, in some cases, would allow the players to settle on a definite solution among

I am indebted to the participants in the Castiglioncello workshop on "Knowledge, Belief, and Strategic Interaction" for their stimulating suggestions and criticism. Needless to say, I have a special debt to Howard Sobel but for whom this chapter would not have been written.

The preparation of this paper has been supported by a generous grant from the Anders Karitz Foundation.

the several possible ones. In the end, however, it will turn out that this method is open to a number of serious objections.

1. HYPERRATIONAL GAMES

We consider noncooperative normal-form games between hyperrational players. Normally, we shall suppose that there are only two of them, Row and Column. Following Sobel (1988b), we shall assume the following to be common knowledge among the hyperrational players (something that they all know, know all the players to know, know all the players to know all the players to know, etc.):

(a) what the strategies available to each player are and what the value matrix of the game looks like;
(b) that the strategy choices by different players are causally independent of each other;
(c) that each of them is "rational" in the sense of being "a self-conscious and deliberate [expected-utility] maximizer"; and
(d) that each of them is wholly indifferent "when confronted with several 'tied' maximizing strategies" and therefore equally likely to choose any one of them.[2]

The last part of assumption (d) depends heavily upon the "principle of insufficient reason," a principle that transforms the equally possible into the equally probable. This move, which generates probability assignments from nonprobabilistic information and thereby transforms decision making under uncertainty into decision making under risk, leads to well-known difficulties: Different partitionings of possibilities result in different probability distributions. Here, we avoid this problem by supposing that all the players structure possibilities (strategies available to each player) in the same way and that they all know this.[3]

2. COMPLETE PROBABILISTIC EQUILIBRIA

Our problem is: What are the solvability conditions for games of this kind? The solution of the game is an assignment of strategy sets to players such that each player's strategy set contains all of the strategies that maximize expected utility.[4] Clearly, in order to identify such a solution, we must identify each player's probability assignments to the strategies of the other players. If it is possible to determine these probability assignments from the common knowledge of the game, then we shall say that the game has a *definite* (unique) solution.

108

In view of the assumption (b) of causal independence, a player can never raise maximal expected utility by playing a mixed strategy rather than a pure one: The expected utility of a probability mixture of a set of strategies is never higher than the expected utility of each strategy in the set. Therefore, when we consider games between hyperrational players, it is perfectly safe to concentrate on pure-strategy solutions. In what follows, we shall use the terms "strategy," "action," and "pure strategy" interchangeably.

As we shall argue, a game G between hyperrational players has a *potential* solution if and only if it contains a "complete probabilistic equilibrium."[5] In the two-person case, a complete probabilistic equilibrium (c.p.e.) is an assignment (X_R, X_C) of nonempty strategy sets, one to each player (Row and Column), such that X_R (X_C) is the set of all strategies of R (C) that maximize R's (C's) expected utility on the assumption that C (R) will perform one of the actions in X_C (X_R), with all of these actions being equally likely.[6]

Assuming, as we shall do, that each player i has only a finite set S_i of (pure) strategies to choose from, it is easy to identify all the c.p.e.'s in a given game. Thus, in the two-person case, it is enough to check all the pairs (X_R, X_C) such that X_R and X_C are arbitrary nonempty subsets of S_R and S_C, respectively. Clearly, the number of such pairs must be finite.

It is easy to show that some games (including some zero-sum games) lack a c.p.e.,[7] whereas other games (of the non–zero-sum variety) contain more than one. Thus, there is no guarantee that a game will have a potential solution, nor that such a solution (if there is one) will be unique.

In what way does a complete probabilistic equilibrium (X_R, X_C), if it exists, constitute a potential solution for a given game? In such a game, probability assignments of each player to the other player's strategies arise exclusively from the common knowledge of the game. Thus, every player must come to know each player's probability assignments. In addition, since each player is known to be an expected-utility maximizer who is equally likely to choose any of his maximizing strategies, it is common knowledge in the game that each player assigns positive (and equal) probabilities only to those strategies of the other player that maximize the other player's expected utility. This presupposes that the game *allows* appropriate probability assignments. That is, there must exist probability assignments P_r and P_c, one for each player, such that P_r is a uniform distribution over some subset X_C of C's actions, P_c is a uniform distribution over some subset X_R of R's actions, and the sets X_R and X_C comprise all the strategies of R and C (respectively) that maximize their expected utility with respect to their probability assignments P_r and P_c. But this is just to say that (X_R, X_C) is a c.p.e.

109

A complete probabilistic equilibrium – although a potential solution – need not be the definite solution of the game, if only because a given game may contain several equilibria of this kind. In such a situation, there exist several alternative probability assignments available to each of the players, and it is unclear how to settle on any particular one.

As an illustration, let us consider one of the games described by Sobel (1988b):

Game 1

	C_1	C_2	C_3
R_1	5, 5	4, 4	0, 0
R_2	4, 4	2, 2	3, 3
R_3	0, 0	3, 3	2, 2

In this symmetrical game, R and C have exactly the same interests. Also, it is easy to see that the game contains a unique Nash equilibrium. This equilibrium is a strong one and consists of pure strategies: (R_1, C_1).

At the same time, however, there are two complete probabilistic equilibria in the game: $(\{R_1\}, \{C_1\})$ and $(\{R_2, R_3\}, \{C_2, C_3\})$. Note that the first equilibrium offers each player higher expected utility than the second; the relevant values are 5 and 2.5, respectively. Unlike Sobel, I earlier believed that this consideration should be sufficient for the definite solution of the game.[8] I am no longer sure of that. Row would have a reason to treat this Pareto consideration as relevant to her probability assignments only if she had reasons to believe that Column would treat this consideration as relevant. But Row knows that Column would do this only insofar as Column had reasons to believe that Row would have a reason to do likewise. Thus, we are back to where we started. As so often occurs in such games, we seem to be drawn into a "tortuous labyrinth."

Actually, Sobel seems to believe that even games with a unique c.p.e. may create insuperable difficulties for hyperrational players. In the following variant of the previous game,

Game 2

	C_1	C_2	C_3
R_1	5, 5	4, 4	1, 1
R_2	4, 4	2, 2	3, 3
R_3	1, 1	3, 3	2, 2

there exists only one c.p.e.: $(\{R_1\}, \{C_1\})$. The second probabilistic equilibrium, $(\{R_2, R_3\}, \{C_2, C_3\})$, is no longer *complete:* On the assumption that Column is equally likely to play C_2 or C_3, Row's maximizing strategies are not only R_2 and R_3 but also R_1. In connection with such games, Sobel (1988b) writes:

... the problem is to see how hyperrational players who knew about each other only what could be gathered from their common knowledge of their situation could know that these strategies [R_1 and C_1] were uniquely maximizing. It is true, as Row must know, that R_1 is maximizing for Row *if* he is confident that C_1 is maximizing for Column. And Row could see that C_1 is maximizing for Column if Column is confident that R_1 is maximizing for Row. And so on. But *if* this is all they can figure out about one another, then they are drawn into that "tortuous labyrinth" and we have no way out. It *seems* that neither can settle that *his* "first strategy" is maximizing without settling *already* and previously that the other player's "first strategy" is maximizing. The *problem* to which I cannot even begin to imagine a solution is to see *how – by what non-question-begging reasoning –* hyperrational maximizers in this game . . . could settle their expectations for one another's strategies.

In the next section I shall sketch a method – an iteration process – that could be used by hyperrational players to "settle their expectations" and thereby sometimes reach definite solutions in games that contain one or more c.p.e.'s. However, I am not really sure whether this method also is not question-begging in some way. What is more, as will be shown in the final section, the iteration method is open to serious objections.

4. THE ITERATION METHOD

The inspiration for this method comes from the following suggestion in Sobel (1988b):

... there is at least a strong pull to the idea that "early on" [in the process of deliberation], when by hypothesis the players have no clear ideas what they will do, their *probabilities* should make all strategies in each given player's strategy set *equally probable.*

I make use of this idea in the following way. I assume that the players start their deliberation process with tentative equiprobability assignments to each player's strategies – the assignments they construct on behalf of all the players – and that they subsequently revise these assignments either until they reach a point where further revisions are neither possible nor necessary, or until they discover that this revision process would never end (that they are caught in a cycle). In the former case, and only then, the iteration method yields a particular c.p.e. as the definite solution of the game.

111

Let me now describe this iteration method in more detail. For simplicity, let us again assume that there are only two players, Row and Column. Let S_R and S_C stand for the sets of all the (pure) strategies that are available to Row and Column, respectively. At each step of the process, the players construct probability assignments to the strategies available to each player. Since each player is believed to be an expected-utility maximizer and to be equally likely to choose any of his maximizing strategies, the assignments in question must distribute the probability equally among some subset of the strategies open to the player.

Consequently, we can represent these assignments simply by strategy subsets: If P is such a uniform probability assignment to player i's strategies, we can represent P by the subset X consisting of all i's strategies to which P assigns positive probability. Clearly, just as X is definable from P, P could be recovered from X: P is the assignment that uniformly distributes all the probability among the members of X. Thus, for any strategy open to i, if this strategy does not belong to X then its P-probability is 0; if it does belong to X then its P-probability is $1/\text{card}(X)$. (Remember that X must be finite, since we have assumed that S_i is finite for every player i.) This means that at each stage n of the iteration process, the relevant probability assignments to the strategies of Row and Column (respectively) could be represented by the pair (X_{Rn}, X_{Cn}).

As we have assumed, we start the process with probability assignments that are perfectly uniform. Thus, if (X_{R0}, X_{C0}) constitutes the starting point then $X_{R0} = S_R$ and $X_{C0} = S_C$. At the next stage, these initial probability assignments are used in order to determine, for each player, the set of maximizing strategies with respect to the probability assignments in question. Let these sets be X_{R1} and X_{C1}, respectively. The assumption that these are the maximizing sets, when coupled with the common knowledge that each player is equally likely to play each of his maximizing strategies, gives rise to a new pair of probability assignments: (X_{R1}, X_{C1}). From these probability assignments we determine new sets of maximizing strategies for each player, X_{R2} and X_{C2}, and consequently a new pair of probability assignments: (X_{R2}, X_{C2}).

The iteration process terminates if, at some stage n, we reach a *fix-point;* that is, if (X_{Rn}, X_{Cn}) turns out to be identical to (X_{Rn+1}, X_{Cn+1}). This means that no further revisions are possible. Nor are they necessary; it is easy to see that, if (X_{Rn}, X_{Cn}) is a fixpoint, then it must be a c.p.e. It should also be obvious that, until we reach such a fixpoint, we encounter no c.p.e. in this iteration process.

Clearly, the iteration process in some games need not terminate; it may continue without end. But since we have assumed that the players have only finitely many strategies at their disposal, the nonterminating process must sooner or later reach a stage n such that (X_{Rn}, X_{Cn}) is identical with a pair that has been reached at some earlier stage. In other words, we get caught in a cycle. This means that, for finite games, it is an easy matter to determine whether the iteration process terminates or not. For example, in the two-person case, the maximal number of iteration steps needed to discover a fixpoint or a cycle equals $[\text{card}(S_R)^2 - 1]$ multiplied by $[\text{card}(S_C)^2 - 1]$.[9]

Clearly, if a game does not contain any c.p.e. then the iteration process does not terminate. As may be seen from the following example, the same could well happen in some games that contain several c.p.e.'s:

Game 3

	C_1	C_2	C_3
R_1	0, 0	2, 2	1, 0
R_2	2, 2	0, 0	0, 0
R_3	0, 1	0, 0	0, 0

In this game, there are three c.p.e.'s:

$$E1 = (\{R_1\}, \{C_2\}), \quad E2 = (\{R_2\}, \{C_1\}), \quad E3 = (\{R_1, R_2\}, \{C_1, C_2\}).$$

But the iteration process leads to a cycle:

Stage 0	$(\{R_1, R_2, R_3\}, \{C_1, C_2, C_3\})$
Stage 1	$(\{R_1\}, \{C_1\})$
Stage 2	$(\{R_2\}, \{C_2\})$
Stage 3	$(\{R_1\}, \{C_1\})$
	and so on

It may be of some interest to note that, in this game, c.p.e.'s $E1$ and $E2$ are *symmetric images* of each other. Thus, it is intuitively quite impossible for hyperrational players to reach one of them as the definite solution of the game. No consideration that would tell for $E1$ could be conclusive, because there would exist a competing equally strong and perfectly analogous consideration that tells for $E2$. Thus, $E3$ is the only serious candidate for the definite solution. It is not simply the only c.p.e. in this game that lacks a symmetric image;[10] what is more, $E3$ is all-inclusive: Its constituent sets include all the c.p.e. strategies at the players' disposal. (A *c.p.e. strategy* is any strategy that belongs to some c.p.e. in the game. Clearly, any all-inclusive c.p.e. must lack a symmetric image, but the converse does not hold. There are games that contain nonsymmetric c.p.e.'s none of which is all-inclusive.) One might expect that the all-inclusiveness

113

of a c.p.e. should be a relevant consideration when players try to reach the solution. But, as we have seen, the iteration method does not allow us to reach such an all-inclusive c.p.e. in game 3.

But what if a certain game contains just one c.p.e.? Will it always be possible to reach this c.p.e. by means of the iteration method? I shall return to this question in the next section. At present, let us just note that it will often be possible to reach such a unique c.p.e. For example, in game 2, the unique c.p.e. will be reached at stage 1. Thus, using the iteration method would allow the players to solve some of the games that Sobel seems to treat as unsolvable.

For game 1, which contains two c.p.e.'s (one that is all-inclusive but Pareto inferior, and one that is Pareto optimal), the iteration method picks out the Pareto optimal one. This observation leads naturally to the following question: Suppose that a game G contains a c.p.e. that is Pareto superior to all the other c.p.e.'s in G; must this c.p.e. be chosen by the iteration method? The answer is: not necessarily. In the following game,

Game 4

	C_1	C_2
R_1	1, 1	2, 0
R_2	0, 2	3, 3

there are three c.p.e.'s:

$$E1 = (\{R_1\}, \{C_1\}), \quad E2 = (\{R_2\}, \{C_2\}), \quad E3 = (\{R_1, R_2\}, \{C_1, C_2\}).$$

Although $E2$ is Pareto superior to both $E1$ and $E3$, the iteration method picks out the all-inclusive $E3$ instead (at stage 0).

Some games contain a c.p.e. that, instead of being all-inclusive, exhibits the opposite property: It is *included in* every other c.p.e. in the game. Would such an "all-included" c.p.e. be favored by the iteration method? Once again, the answer is: not necessarily. Thus, in the following game,

Game 5

	C_1	C_2	C_3
R_1	3, 5	3, 4	0, 0
R_2	0, 3	4, 3	1, 0
R_3	2, 0	4, 1	0, 2

114

there are two c.p.e.'s: $E1 = (\{R_1\}, \{C_1\})$ and $E2 = (\{R_1, R_3\}, \{C_1, C_2\})$. $E1$ is all-included (and also Pareto superior to $E2$, insofar as it makes Column better off without making Row worse off), but the iteration method picks out the all-inclusive $E2$ (at stage 1).

5. TROUBLES

A. *Principle of insufficient reason and repartitioning*
of possibilities

As mentioned in Section 1, our approach to hyperrational games makes an essential use of the principle of insufficient reason – both in the argument for c.p.e.'s as solutions of such games and in the argument for the iteration method. This approach is thereby unsettlingly sensitive to the way in which possibilities are partitioned by the players. For example, in the following game,

Game 6

	C_1	C_2
R_1	2, 2	0, 0
R_2	0, 0	2, 3

the iteration method (at stage 2) picks out the Pareto optimal c.p.e. $(\{R_2\}, \{C_2\})$. But if we were to change the matrix by splitting R_1 into two exactly similar versions,

Game 6'

	C_1	C_2
R_{11}	2, 2	0, 0
R_{12}	2, 2	0, 0
R_2	0, 0	2, 3

the iteration method would give us a very different solution: $(\{R_{11}, R_{12}\}, \{C_1\})$. Note that the difference between R_{11} and R_{12} is irrelevant not only from Row's point of view but also from Column's.

B. *The unique complete probabilistic equilibrium*

Suppose that a given game contains just one c.p.e. Will the players reach this c.p.e. using the iteration method? Unfortunately, the answer must again be: not necessarily. Consider the following game:

115

Game 7

	C_1	C_2
R_1	2, 5	2, 4
R_2	0, 3	3, 1
R_3	1, 0	3, 2

As is easily seen, $(\{R_1\}, \{C_1\})$ is the only c.p.e. in this game. But players using the iteration method are caught in a cycle:

Stage 0	$(\{R_1, R_2, R_3\}, \{C_1, C_2\})$	
Stage 1	$(\{R_1, R_3\}, \{C_1\})$	
Stage 2	$(\{R_1\}, \{C_2\})$	
Stage 3	$(\{R_2, R_3\}, \{C_1\})$	
Stage 4	$(\{R_1\}, \{C_1, C_2\})$	
Stage 5	$(\{R_1, R_3\}, \{C_1\})$	(repeats stage 1)
	and so on	

One might want to say that at last we have here an example of a game with a unique c.p.e. that (to use Sobel's words) the players cannot solve by any "non–question-begging reasoning." Alternatively, such cases might lead one to suspect that games that are unsolvable by the iteration method perhaps could be solved by other means.

C. Alternative starting points

In the iteration method, the players start from zero, so to speak. To begin with, they treat all the strategies of the other players as equally possible. Now, this perfectly neutral starting point may well be challenged. Suppose, for example, that some of the strategies of my opponent are (strictly) dominated. Because I know my opponent is rational, I can be sure in advance (when I begin my deliberation) that she is not going to play such dominated strategies.[11] But then, perhaps, the proper starting point for the iteration method need not be perfectly neutral. We can start with tentative assignments that concentrate all probability on the undominated strategies only, and treat only such undominated strategies as equiprobable. In what follows, I shall refer to this approach as the iteration method with the undominated starting point.[12]

As a matter of fact, the starting point may be even more restrictive. Using Levi's terminology, we may call a strategy *E-admissible* if and only if it maximizes the player's expected utility with respect to some probability assignment to the other players' strategies (Levi 1980, §4.8; Levi 1986, §7.4). Clearly, every E-admissible strategy is undominated, but the converse does not hold. Now, since each player knows that the other players are expected-utility maximizers, he can be sure in advance that they will not choose E-inadmissible strategies. Thus each player can start by distributing probabilities equally among the E-admissible strategies only. This gives rise to a new variant of the iteration method: the iteration method with the *E-admissible starting point.*[13]

A strategy will be called *strictly E-admissible* if it maximizes the player's expected utility with respect to some probability assignment (to the other player's strategies) that gives equal probabilities to all the strategies that are taken to be positively probable. A probability assignment of this sort arises when applying the principle of insufficient reason. Since each player knows that the other players make use of this principle, each can be sure in advance that the strategies the others will use will be strictly E-admissible; each player can therefore adjust his starting point accordingly. In this way, we obtain the iteration method with the *strictly E-admissible starting point.*

In fact, as we have seen in Section 2, each player can be sure that none of the other players will use a strategy that does not belong to some c.p.e. in the game. Each such c.p.e. strategy must be strictly E-admissible, but the converse need not hold. The starting point of the iteration method might therefore be a probability assignment that concentrates all probability to c.p.e. strategies and divides it equally among them.[14] This gives rise to the iteration method with the *c.p.e.-restricted starting point.*[15]

Could the starting point be even more restricted? – well, of course. For example, as shown in Section 4, a player can be sure in advance that the solution of the game cannot be a c.p.e. that has another c.p.e. as its symmetric image. Therefore, the player may from the start want to assign zero probabilities to those c.p.e. strategies that do *not* belong to some c.p.e. which *lacks* a symmetric image. In this way, the starting point may become more and more "loaded."

It is easy to see that some games, which are unsolvable by our original (perfectly neutral) iteration method, become solvable by iteration methods with more restricted starting points.[16] Thus, the originally unsolvable game 7 can be easily solved if we choose the c.p.e.-restricted starting point (the less restricted starting points won't do): At stage 0, we concentrate all probability to the c.p.e. strategies R_1 and C_1, respectively, and we are done.

To take another example, consider the originally unsolvable game 3:

Game 3

	C_1	C_2	C_3
R_1	0, 0	2, 2	1, 0
R_2	2, 2	0, 0	0, 0
R_3	0, 1	0, 0	0, 0

117

In this game, strategies R_3 and C_3, while undominated, are still E-inadmissible. If we choose the E-admissible starting point (or for that matter the c.p.e.-restricted starting point), thereby assigning zero probability to these strategies from the beginning, then the iteration method will yield $(\{R_1, R_2\}, \{C_1, C_2\})$ as the solution (at stage 0). Recall that this solution is the all-inclusive c.p.e. in game 3. It is easy to see that the iteration method that starts from the c.p.e.-restricted starting point must always pick out the all-inclusive c.p.e. (at stage 0), insofar as such c.p.e. exists in the game.

As a matter of fact, for any starting point that we have considered, one could give an example of a game that cannot be solved by the iteration method taking its departure from this starting point, but that can be solved if we make further restrictions at the start. This dependence on the starting point might not be especially troublesome if the successive restrictions of the starting point simply expanded the set of solvable hyperrational games; that is, if they would transform some antecedently unsolvable games into solvable ones. However, the situation is not that simple. In some cases, restrictions of the starting points change the solutions.

Thus, consider the following three examples:

Game 8

	C_1	C_2	C_3
R_1	2, 2	0, 0	0, 4
R_2	0, 0	4, 4	0, 1
R_3	4, 0	1, 0	2, 2

where the c.p.e.'s are $(\{R_2\}, \{C_2\})$ and $(\{R_3\}, \{C_3\})$, and R_1 and C_1 are dominated;

Game 9

	C_1	C_2	C_3
R_1	0, 0	0, 0	0, 2
R_2	0, 0	2, 2	0, 0
R_3	2, 0	0, 0	1, 1

where the c.p.e.'s are $(\{R_2\}, \{C_2\})$ and $(\{R_3\}, \{C_3\})$, and R_1 and C_1 are E-inadmissible;[17]

Game 10

	C_1	C_2	C_3
R_1	2, 2	3, 0	1, 4
R_2	0, 3	4, 4	0, 0
R_3	4, 1	0, 0	3, 3

118

where the c.p.e.'s are $(\{R_2\}, \{C_2\})$ and $(\{R_3\}, \{C_3\})$, and all strategies are strictly E-admissible.

In game 8, our original (neutral) iteration method leads to $(\{R_3\}, \{C_3\})$. However, R_1 and C_1 are dominated by R_3 and C_3, respectively. As a result, the iteration method with the undominated starting point (or with the even more restricted starting points) would take its point of departure from the assignments that divide all probability equally between R_2 and R_3 for Row, and between C_2 and C_3 for Column. Given this starting point, the iteration process leads to a different fixpoint: $(\{R_2\}, \{C_2\})$.

In game 9, the neutral and the undominated starting points yield $(\{R_3\}, \{C_3\})$ as the solution. But since strategies R_1 and C_1 are E-inadmissible, the more restricted starting points assign to them zero probability and consequently they all yield $(\{R_2\}, \{C_2\})$. Finally, in game 10, all starting points that are weaker than the one restricted to c.p.e. strategies yield $(\{R_3\}, \{C_3\})$. But the c.p.e.-restricted starting point assigns zero probability to R_1 and C_1, thereby yielding $(\{R_2\}, \{C_2\})$ instead.

We seem to have reached an impasse. There are too many possible starting points. The iteration method yields different solutions depending on the chosen point of departure, and it is not at all clear which point of departure is the proper one. Nor is it clear if there is any "non–question-begging reasoning" that allows players to settle on a particular starting point in their deliberation process.

D. Succession of dogmatic expectations

Even if we could solve the problem of the starting point, we would still have to face another difficulty: There seems to be something very unreasonable in the way players are supposed to frame their changing probability assignments in the course of the iteration process.

At stage $n+1$, Row identifies the set X_{Cn+1} of Column's maximizing strategies with respect to Column's probability assignment (to Row's strategies) at stage n, and then Row takes it to be certain that Column will use one of the strategies in X_{Cn+1}. All other strategies of Column receive zero probability in Row's probability assignment at stage $n+1$. This is rather astonishing, as Row should be aware of the possibility that the iterative deliberation process may well continue, beyond the present stage, to another stage at which Column's maximizing strategies will be different. Therefore, Row's present expectations about the other player's choice of strategy should not be so dogmatic. There should be room for some uncertainty about whether Column really is going to use some of the strategies in X_{Cn+1}. In other words, it seems unreasonable to go through the deliberation process by jumping from one set of dogmatic expectations to another.

119

This objection is formulated by Sobel (1988b), who suggests that the problem could perhaps be dealt with if, at each stage $n+1$, we let some of the probability go to the strategies outside the maximizing set, doing so in such a way that the probability assigned to the nonmaximizing strategies progressively decreases to zero as we continue through the stages of deliberation. There is a similarity between this proposal and Skyrms's (1990) conception of "Bayesian deliberation dynamics."

Obviously, it is difficult to evaluate this idea before it has been worked out in some detail. But one can already see that avoidance of dogmatism creates additional potential for arbitrariness in applying the iteration method. The recommendation to reserve a low (but positive) probability for nonmaximizing strategies immediately invites a series of questions: How low? How should it be distributed? How fast is it supposed to decrease? And so on. It may well happen that different answers to some of these questions would lead to different game solutions.

1. I found this quotation in Sobel (1988a).
2. In order to be equally likely to choose any one of them, an agent, besides being indifferent, must not be a creature of habit. Ullmann-Margalit and Morgenbesser (1977) have pointed out that a habit may allow one to extricate oneself from a pure "picking situation" (a situation where you are wholly indifferent between several options) without having to transform oneself into a random device:
   > [W]hen, due to some habit, a resolution of a picking situation *is* achieved . . ., the habit cannot be said to have supplied the agent with a *reason* for his selection. . . . At the same time the habit can in a case of this sort be considered to have played a causal role in extricating the agent from the picking situation, and as such it may contribute to an *explanation* of his picking act. (p. 772)
3. Alternatively, one could try to defend use of the principle of insufficient reason by assuming, as Sobel does, that all hyperrational players in a given game will structure the possibilities in the same way *because* they are all able to identify the "finest" possibility partition: They are able to identify the most specific strategies open to each player, or – to use Sobel's (1988b) terminology – the range of each player's "precise choices."
4. This notion of the solution as an assignment of strategy *sets* to different players might be compared with the standard game-theoretical solution concept: an assignment of a unique strategy to each player. Given that the players are expected-utility maximizers, the standard solution concept is clearly inadequate. Because of possible ties, the set of maximizing strategies that are open to a given player need not be a unit set.
5. For this notion, see Rabinowicz (1986). There, however, the notion of a complete probabilistic equilibrium is defined in a more general way, without excluding the mixed strategies.
6. It is easy to see how to generalize this definition to n players, $1, \ldots, n$. A *complete probabilistic equilibrium* $(X_1, \ldots, X_n)$ assigns to each i a set X_i of strategies such that X_i is the complete set of the (pure) strategies of i that maximize

expected utility, on the assumption that each other player j will perform one of the actions in X_j, with all of these actions of j being equally likely and with all of the actions of different players being probabilistically independent of each other. To put it somewhat differently, X_i is the complete set of the strategies of i that maximize expected utility with respect to the uniform probability distribution on the Cartesian product $X_1 \times \cdots \times X_{i-1} \times X_{i+1} \times \cdots \times X_n$. If we remove the demand of completeness but keep the remaining part of the definition unchanged, $(X_1, \ldots, X_n)$ may be said to be a *probabilistic equilibrium*.

Sobel (1988b) introduced the notion of an "equilibrium in the random extension of a pure strategy game," a concept closely related to the notion of a probabilistic equilibrium. A *random strategy* based on a set of strategies is the mixed strategy in which all the strategies in the set are assigned equal chances. Clearly, every strategy may be seen as a random strategy based on the unit set consisting of this strategy alone. We obtain the *random extension* of a pure strategy game if we let each player's options be all the random strategies that are based on the different subsets of the set of that player's pure strategies. Consider a sequence $(s_1, \ldots, s_n)$ of random strategies, one for each player. As is well known, $(s_1, \ldots, s_n)$ is a (Nash) *equilibrium* if each player i maximizes expected utility by playing s_i when other players play their strategies in $(s_1, \ldots, s_n)$. Now, for every random strategy s_i, let X_i be the set of pure strategies that s_i is based on. We shall say that $(s_1, \ldots, s_n)$ is a *strong equilibrium* if, for each i, expected utility would diminish if instead of playing s_i he played some *pure* strategy not belonging to X_i when other players play their strategies in $(s_1, \ldots, s_n)$. It is easy to see that $(s_1, \ldots, s_n)$ is a (strong) equilibrium in the random extension of a pure strategy game G if and only if $(X_1, \ldots, X_n)$ is a (complete) probabilistic equilibrium in G.

Sobel (1988b) demands from any solution of a hyperrational pure-strategy game that it be an equilibrium in its random extension. He does not explicitly demand that it be a strong equilibrium in such an extension, but his discussion (regarding "Game 6") suggests that he might be prepared to impose this stronger demand as well. He argues that a certain equilibrium solution is impossible for hyperrational players because the solution in question does not constitute a strong equilibrium.

(After reading an earlier version of this chapter, Sobel decided to impose the stronger demand. However, instead of the somewhat misleading term "strong equilibrium," Sobel used a different name – equilibrium* – for the same concept.)

7. Perhaps the simplest example is a 2-by-2 zero-sum game, where Row's payoffs for playing strategy R_1 (R_2) are 1 (4) and 3 (2) depending on whether Column plays C_1 (C_2). Although this game lacks a c.p.e., it contains a unique Nash equilibrium in mixed strategies: $((\frac{1}{2}R_1, \frac{1}{2}R_2), (\frac{1}{4}C_1, \frac{3}{4}C_2))$.

8. See Rabinowicz (1986). There, I suggested that a game is (definitely) solvable if it contains a unique "optimal" complete probabilisitc equilibrium $(X_1, \ldots, X_n)$; that is, if $(X_1, \ldots, X_n)$ is the only c.p.e. in the game that satisfies the following condition: for every other c.p.e. $(Y_1, \ldots, Y_n)$ in this game and for every player i, $(X_1, \ldots, X_n)$ offers i at least as high expected utility as $(Y_1, \ldots, Y_n)$ does. Note that this is equivalent to the game containing a unique Pareto optimal c.p.e., where a c.p.e. is *Pareto optimal* in the set of c.p.e.'s if no other c.p.e. offers some players higher expected utility without making any player worse off in expected-utility terms.

9. It should be clear how the iteration method can be extended to n-person cases. We start with the probability assignment $(X_{10}, \ldots, X_{n0})$, where $X_{i0} = S_i$ for each i. Then we determine, for each player i, the corresponding set X_{i1} of i's maximizing strategies. (When calculating the expected values of i's different strategies, we assume that the actions of the other players are probabilistically independent of each other. Thus, the probabilities of their combinations are simply the products of the probabilities of the constituents. See note 6.) Then we repeat the procedure with respect to $(X_{11}, \ldots, X_{n1})$, and continue in this way until we reach a fixpoint or get caught in a cycle.

10. What does this mean, more precisely? In what sense are $E1$ and $E2$ symmetric images of each other, while $E3$ lacks such an image?

For a given game G, a simultaneous permutation p on the set of players in G and on the set of all their (pure) strategies in G shall be said to be an *automorphism* on G if and only if (i) for every player i and for every strategy α of i, $p(\alpha)$ is a strategy of player $p(i)$; and (ii) for every player i and for every combination $(\alpha_1, \ldots, \alpha_n)$ of strategies containing one strategy for each player, G assigns to i in $(\alpha_1, \ldots, \alpha_n)$ and to $p(i)$ in $(p(\alpha_1), \ldots, p(\alpha_n))$ the same utility values. Let u be the value function of G, or (equivalently) the value matrix of G. Clause (ii) could then be stated as follows: G's value matrix u remains invariant under p.

For example, in game 3, consider the permutation p_1 that permutes Row into Column, C_1 into R_1, C_2 into R_2, R_3 into C_3, and vice versa. It is easy to see that p_1 is an automorphism on this game.

Actually, if we abstain from interpersonal comparisons of the players' utility values, clause (ii) in the definition of an automorphism should be somewhat weakened: In order to be an automorphism, a permutation p on players and on their strategies need not keep the matrix u of the game unchanged. It is sufficient that p keeps unchanged some matrix u' that is related to u by a set of independent positive linear transformations, one transformation for each player. That is, u' is to satisfy the following condition: Let u_i be the restriction of u to a player i. That is, $u_i(\alpha_1, \ldots, \alpha_n) = u(i, (\alpha_1, \ldots, \alpha_n))$ for every i and $(\alpha_1, \ldots, \alpha_n)$. Define u'_i accordingly. Then, for every i, we demand that u'_i be a positive linear transformation of u_i.

Now we are in a position to define the notion of a symmetric image. A sequence $E = (X_i, \ldots, X_n)$, containing one set of strategies for each player, will be said to have a *symmetric image* $E' = (X'_i, \ldots, X'_n)$ in G if and only if E and E' are distinct from each other and there exists an automorphism p on G such that, for every player i, $p(X_i)$ – the set of p-images of the strategies in X_i – equals $X'_{p(i)}$. We express this by saying that p transforms E into E': $p(E) = E'$. It is easy to see that, in game 3, every automorphism transforms $E3$ into $E3$ itself; thus $E3$ lacks a (distinct) symmetric image. On the other hand, p_1 transforms $E1$ into $E2$ and so $E2$ is a symmetric image of $E1$.

11. Note that I can only be sure that she will not play a strictly dominated strategy; weakly dominated strategies may sometimes be rational. Thus, suppose that the game is as follows:

	C_1	C_2
R_1	1, 1	0, 1
R_2	1, 1	1, 0

This game contains a unique c.p.e.: $(\{R_1, R_2\}, \{C_1\})$; the iteration method would pick out this c.p.e. at stage 3. In such a situation, Column cannot be sure that Row will not play R_1, even though this strategy of Row is weakly dominated by R_2.

12. It should be noted that assigning zero probability to the dominated strategies at the starting point is not the same as *reducing* the game by removing the strategies in question. Being certain that a player will not use a strategy is not the same as treating this strategy as if it were unavailable to the player. Furthermore, if we remove the dominated strategies then (in the reduced game) some new strategies may become dominated, so that we need to continue the reduction process until we reach a reduct in which all strategies are undominated.

Suppose that we reduce a game G by such a "sequence of strictly dominated discards" (cf. Sobel 1988b for details) and then apply our original iteration method to the reduced game G'. Would we then reach the same solution as when we apply the iteration method with the undominated starting point to the unreduced game G? Not necessarily. In order to see this, consider the following example:

	C_1	C_2	C_3
R_1	2, 5	0, 0	0, 5
R_2	0, 0	4, 4	2, 1
R_3	5, 0	1, 2	1, 2

The unique c.p.e. is $(\{R_2\}, \{C_2\})$. The undominated strategies are R_2, R_3, C_1, C_2, and C_3. In this game, the original (perfectly neutral) iteration method would yield the unique c.p.e. as the solution. But the iteration method with the undominated starting point leads to a cycle:

Stage 0	$(\{R_2, R_3\}, \{C_1, C_2, C_3\})$
Stage 1	$(\{R_3\}, \{C_2\})$
Stage 2	$(\{R_2\}, \{C_3\})$
Stage 3	$(\{R_3\}, \{C_2\})$
	and so on

But if we first reduce this game by a sequence of dominated discards (first R_1, then C_1, and finally R_3 and C_3), we reach the degenerate game in which each player has only strategy 2 at his disposal; the solution to this game is therefore immediate.

This suggests, by the way, that we might consider an iteration method with a more restricted starting point. Instead of assigning zero probabilities to the dominated strategies only, we might assign zero probability to each "indirectly dominated" strategy – where a strategy is indirectly dominated if it does not appear in the final reduct G' of the original game G.

13. Just as we have defined (in note 12) the notion of an indirectly dominated strategy, we could define the notion of a strategy that is "indirectly E-inadmissible": one that does not appear in the final reduct of the original game obtained by a sequence of E-inadmissible discards. Assigning zero probabilities to all such indirectly E-inadmissible strategies would give us an even more restricted starting point.

123

14. There may be some difficulties here in the application of the principle of insufficient reason. Suppose that a game contains two c.p.e.'s, one of which contains two strategies for my opponent to choose from and the other only one. Should I assign to the latter strategy the probability of one third (in view of the fact that my opponent may choose one of three c.p.e. strategies) or rather one half (because, in advance, it is equally probable that the iteration process will pick out the latter c.p.e. as the former)?

15. This iteration method is open, however, to the following objection: It is a reasonable demand on an iteration method for the strategies that are assigned zero probability at the starting point, and for this reason are excluded from stage 0, not to reappear at subsequent stages. Such reappearance would seem unacceptable, since the strategies were excluded at the start – not just tentatively but for good, so to speak. They were given zero probability because it was certain they would not be used by the players.

Although this nonreappearance demand is satisfied by the iteration methods with less restricted starting points, it is violated if we start from the c.p.e. strategies only. Here is an example:

	C_1	C_2
R_1	7, 6	0, 0
R_2	4, 4	4, 5
R_3	0, 0	6, 7

The two c.p.e.'s are $(\{R_1\}, \{C_1\})$ and $(\{R_3\}, \{C_2\})$. If we apply the iteration method with the c.p.e. restricted starting point (or any other starting point, for that matter), we end up with $(\{R_3\}, \{C_2\})$:

Stage 0	$(\{R_1, R_3\}, \{C_1, C_2\})$
Stage 1	$(\{R_2\}, \{C_2\})$
Stage 2	$(\{R_3\}, \{C_2\})$
Stage 3	$(\{R_3\}, \{C_2\})$

As may be seen, the non-c.p.e. strategy R_2 reappears at stage 1. This is a rather unsettling characteristic of the iteration method with the c.p.e.-restricted starting point.

The reappearance problem suggests that we should at the start *remove* the non-c.p.e. strategies from the game, instead of merely assigning to them zero probabilities. Then these strategies would never be able to resurface in the course of the iteration process. In other words, perhaps we should first reduce the game in an appropriate way and only then apply the iteration method. Our starting point in the reduct could then be perfectly neutral; all the restrictions that we have a reason to impose would be taken care of by the reduction. Sobel (1988b) suggests that such a procedure might be more plausible.

I am not sure whether such a modification is defensible. As remarked in note 12, there is a difference between being certain that your opponent will not use a particular strategy and considering this strategy to be unavailable to him; there is a difference between "will not" and "cannot." This potentially important difference is ignored if we try to solve the original game by looking at the solutions for its reduct.

124

There is also an additional difficulty with the suggested modification: By moving to the reduct we may sometimes lose solutions. Thus, in the game just described, removing R_2 gives us a reduct in which the two c.p.e.'s become symmetrical images of each other. Unlike the original game, the reduced game is completely unsolvable.

16. Sometimes, however, a game that is solvable when we start from a less restricted starting point could become *un*solvable from a more restricted point of departure. For an example, see note 12, where we consider a game that is solvable when we make no restrictions yet unsolvable when we use the undominated starting point.

17. This example was constructed by Ryszard Sliwinski.

REFERENCES

Levi, Isaac (1980), *The Enterprise of Knowledge.* Cambridge, MA: MIT Press.

Levi, Isaac (1986), *Hard Choices.* Cambridge: Cambridge University Press.

Moore, Omar K., and Anderson, Alan R. (1962), "Some Puzzling Aspects of Social Interaction." *Review of Metaphysics,* pp. 409–33.

Rabinowicz, Wlodzimierz (1986), "Non-cooperative Games for Expected Utility Maximizers." In Paul Needham and Jan Odelstad (eds.), *Changing Positions: Essays Dedicated to Lars Lindahl.* Uppsala: Department of Philosophy, pp. 215–33.

Skyrms, Brian (1990), *Deliberation Dynamics.* Cambridge, MA: Harvard University Press.

Sobel, Howard (1988a), "Game Theory: An Introduction to Some of Its Ideas." Scarborough College, University of Toronto.

Sobel, Howard (1988b), "Hyperrational Games." Scarborough College, University of Toronto. [See also Chapter 5 of this volume.]

Ullmann-Margalit, Edna, and Morgenbesser, Sydney (1977), "Picking and Choosing." *Social Research* 44: 757–85.

8

On consistency properties of some strongly implementable social choice rules with endogenous agenda formation

STEFANO VANNUCCI

1. INTRODUCTION AND SUMMARY OF RESULTS

A social choice rule (SCR) is best seen as a conflict resolution device. The most common interpretation goes as follows. The players agree on some general principles: say, some criteria of efficiency and symmetry. Then they identify an SCR that meets such criteria. Finally, they agree to use it as a collective decision device in all relevant circumstances. Two major game-theoretic problems arise in this connection:

(1) Are the socially best outcomes likely to be selected when the players, and their coalitions as well, behave strategically?

This is the implementation problem. An SCR is *implementable* with respect to (w.r.t) a suitable equilibrium concept if and only if a game form exists such that, for any relevant profile of players' characteristics, the SCR-optimal outcomes at that profile arise as the equilibrium outcomes of the induced game.

(2) To what extent are the socially best outcomes robust w.r.t changes in the agenda – that is, in the set of actually feasible alternatives?

This is a subtle question, because in a sense socially best outcomes do obviously depend on the specification of the set of feasible alternatives. However, one would like to rule out any change in the optimal set due to mere addition or deletion of suboptimal alternatives. The main reason for this requirement is, again, a strategic one. If it is the case that such "suboptimal" changes in the agenda do modify the optimal set, then the choice of the agenda may be a source of conflict (perhaps of a conflict stronger than the one over the final choice). Thus, an SCR vulnerable to that kind of agenda manipulation would lose much of its significance.

There is no easy solution to these problems, especially when tackled at the same time. In his classic work, Arrow (1951) deliberately disregarded

127

the implementation problem. On the other hand, the combination of the Arrowian "independence" with the "social ordering" requirement is best understood as an attempt to overcome the agenda manipulation problem. The same is true for later mixtures of independence and consistency (or collective-rationality) requirements in a choice-functional format (see, e.g., Fishburn 1973 or Sen 1983). The impressive sequence of impossibility results discovered along this line of research shows that escaping the agenda manipulation problem is a very difficult task, if the SCR is to be fairly symmetric and efficient. On the positive side, Nakamura's theorem (see Nakamura 1979, Moulin 1983, or Peleg 1984) and its extensions to smooth manifolds of alternatives (see Schofield 1986) entail that restrictions on the admissible pairs of societies and outcome sets must be introduced, provided that exclusive reliance on unanimity rules is avoided.

Another strand of the social-theoretic literature deals instead with the problem of implementation. Fixed-agenda models are typically used in this research area. Thus, the agenda manipulation problem is tacitly disregarded (but see Moulin 1983 for explicit observations on this point). A major positive result in the implementation literature, due to Moulin and Peleg (1982), relies heavily on effectivity functions (EFs). An EF is a formal representation of the allocation of power among coalitions under a given game form. Symmetry properties of an EF can be defined in a natural way. At any profile of individual preferences, the core of an EF can also be easily defined. The Moulin–Peleg theorem entails that there are symmetric EFs such that: (i) at any profile of linear orderings, their core is nonempty; and (ii) the resulting core correspondence – an SCR, by definition – is implementable w.r.t. strong Nash equilibria (or, simply, "strongly implementable"; see Moulin and Peleg 1982, Moulin 1983, and Peleg 1984). The main limitation of this important "possibility" theorem is that the just-mentioned symmetric EFs embody a strong restriction on the cardinalities of the players' set and of the agenda. Namely (staying with the finite case), these cardinalities are bound to be a pair of relatively prime numbers. This is the point of departure of the present chapter. In fact, the aim of the following analysis is to scrutinize the credentials of core correspondences as a solution to the agenda manipulation problem.

First, we need to assess the relevance of the foregoing restrictions on the cardinalities of both the outcome set and the players' set. Actually, it seems to us that this restriction is only disturbing if the agenda formation process is assumed to be totally independent of the parameters of the ensuing voting game.

If some modeling of the agenda formation process is allowed, however, the dependence of the outcome set's cardinality on the number of players seems to be a sensible assumption. One may introduce an explicit agenda formation rule (AFR) that specifies, for any set of players, a set of admissible agendas.

Then a compelling criterion of *symmetry-w.r.t-players* (for an AFR) dictates that their number – and only their number – counts in order to establish the admissible agendas. A similar extension to an AFR of a natural criterion of *symmetry-w.r.t.-outcomes* requires that two agendas of the same cardinality are bound to be either both admissible or both inadmissible.

An AFR that satisfies the first criterion of symmetry, and a weakened version of the second, is the following. For any society of n players, the admissible agendas are the sets of outcomes containing exactly $kn + 1$ outcomes, to be interpreted as k proposals for each player (k being a positive integer) plus one "status quo" outcome. To keep things as simple as possible, we shall in fact stick to the $k = 1$ case, that is, to the AFR that entitles each player to one proposal.

Under the present modeling of the collective choice process – that is, via an SCR plus an AFR – the most prominent issue is ascertaining whether the agenda formation process can be expected to be efficient. Therefore, a (mild) criterion of efficiency for the agenda formation process is formulated in Section 2. It requires that, whenever a player discovers a feasible outcome that is both (a) preferred by that player to any alternative in the previous choice set as well as (b) Pareto improving w.r.t. the previous proposal, then substituting the new outcome for the old proposal can only be to the player's own advantage. Hence, the criterion is labeled "dramatic improvements pay" (DIP).

The main result of this chapter is that the core correspondences of stable EFs meet the DIP criterion. As a consequence, the core correspondences of (maximal, monotonic) stable and symmetric EFs do qualify as a solution to the agenda manipulation problem.

The chapter is organized as follows: Section 2 contains the formal description of the model and the results. Section 3 is devoted to a brief discussion of the results of Section 2 and to comments on some related literature.

2. DEFINITIONS AND RESULTS

We denote by Ω (the set of natural numbers) the universal set of players, and by X the universal set of outcomes. A finite subset of Ω is also called a society. For any society N we denote by N^* the set $N \cup \{\Omega\}$, and by $X^*(N)$ the set $X \times N^*$ of extended outcomes. We shall also write X^* for $X^*(N)$ whenever the underlying society is clearly identified.

129

An agenda for N is a nonempty subset of $X^*(N)$. An agenda forma-
tion rule (AFR) is a function $f: PF(\Omega) \to \bigcup_{N \in PF(\Omega)} 2^{X^*(N)}$ such that, for
any society N, $\emptyset \neq f(N) \subset 2^{X^*(N)}$. Throughout this chapter we denote by
$P(Y)$, $PF(Y)$, and 2^Y (respectively) the power set of Y, the set of all finite
subsets of Y, and the set of all nonempty subsets of Y, for any set Y.

Thus, an AFR specifies – for any society – a set of admissible agendas.
Each agenda consists of a (possibly empty) set of proposals for any player
of the given society, plus a (possibly empty) set of exogenously fixed or
status quo outcomes, indexed by $\{\Omega\}$.

An AFR f is *intersociety-anonymous* if and only if $f(N) = f(M)$ whenever $\#N = \#M$, for any pair of societies N, M. An AFR f is *intrasociety-
anonymous* iff for any society N, for any $i, j \in N$, and for any agenda
$A \in f(N)$, $A_i = A_j$ (where $A_h = \{x \in A \mid x \in X \times \{h\}\}$ for any $h \in N$). Moreover, an AFR is said to be *strictly anonymous* whenever it is both intersociety-anonymous and intrasociety-anonymous.

An AFR f is *neutral* (resp., *nearly neutral*) iff a $B \in 2^{X^*(N)}$ exists such
that, for any society N, $f(N) = \{A \subset X^*(N) \mid \#A = \#B\}$ (resp., $f(N) =
\{A \subset X^*(N) \mid \#A = \#B$ and $x \in A\}$ for some $x \in X \times \{\Omega\}$).

For any positive integer k and for any status quo outcome $s \in X \times \{\Omega\}$,
a nearly neutral and strictly anonymous AFR $f(k, s)$ can be defined as
follows: $f(k, s)(N) = \{A \subset X^*(N) \mid A_\Omega = \{s\}$ and $\#A_i = k$ for all $i \in N\}$.

Hence, under $f(k, s)$ a status quo s is fixed, and each player is entitled
to k distinct proposals to be chosen in X. (In the sequel s will also be used
as a shorthand for $s \times \{\Omega\}$.)

Let N be a society. We assume without loss of generality (w.l.o.g.) that
$N = \{1, \ldots, n\}$. We denote by $L = L(X, N)$ the set of all total preorderings
on $X^*(N)$ (a *total preordering* is a total and transitive binary relation).
Moreover, the following notation will be used: For any total preordering
$r \in L$, $I(r)$ denotes the symmetric part of r and $P(r)$ denotes the asymmetric part of r. We shall restrict our analysis to the subset $R \subset L$ defined
as follows: $R = \{r \in L \mid (x, i), (y, j) \in I(r)$ iff $x = y$ for any $(x, i), (y, j) \in
X^*(N)\}$.

All this amounts to denying any welfare implication to the identity of
proponents, in spite of the fact that the latter has been previously used in
order to distinguish the relevant outcomes. This admittedly roundabout
procedure is introduced as a matter of technical convenience, as will be
apparent in what follows.

A *profile* for N is a function $r^N: N \to R$. We denote by R^N the set of all
profiles for N. Let $D \subset 2^{X^*(N)}$ be a nonempty set of agendas. Then a social
choice rule (SCR) for N on $R^N \times D$ is a function $F: R^N \times D \to 2^{X^*(N)}$ such
that $F(r^N, A) \subset A$ for any $(r^N, A) \in R^N \times D$. In words, an SCR selects a
nonempty subset of optimal extended outcomes for any profile of indi-

vidual preferences and for any admissible agenda. If f is an AFR then the SCR $F: R^N \times D \to 2^{X^*(N)}$ is consistent with f iff $D = f(N)$. SCRs with a fixed agenda correspond to the case $D = \{Y \times N^*\}$, for any $Y \subset X$ (i.e., they are consistent with any AFR f such that $f(N) = \{Y \times N^*\}$). Unconstrained SCRs, such as those most commonly used in the literature focusing on issues of consistency and agenda manipulation, correspond to the case $D = 2^{X^*(N)}$. Hence unconstrained SCRs are consistent with any AFR f such that $f(N) = 2^{X^*(N)}$. In the sequel, we shall be mainly concerned with an intermediate case – namely, with SCRs that are consistent with the AFR $f(k, s)$ (as defined previously).

We now turn to the implementation of our model. To begin with, it is tacitly assumed that the players are aware of each other's preferences (perhaps owing to a small-dimension effect or to a repetition effect). Also, binding agreements are feasible (perhaps owing to a repetition effect). Thus, a cooperative environment is assumed to prevail. As a consequence, we shall be concerned with coalitional – or strong (Nash) – implementation.

A *game form* for N and $A \in 2^{X^*(N)}$ is a $(n+1)$-tuple $G = ((S_i)_{i \in N}, g)$, where S_i denotes the set of strategies of player i $(i = 1, \ldots, n)$ and g: $S_1 \times \cdots \times S_n \to A$ is the outcome function.

A *strong equilibrium* of G at profile $r^N \in R^N$ is a multistrategy $s \in S_1 \times \cdots \times S_n$ such that, for any nonempty coalition $T \in 2^N$, and for any $s'_T = (s_i)_{i \in T}$ (where $s_i \in S_i$ for any $i \in T$), $(g(s), g(s'_T, s_{N-T})) \in r_j$ for some player $j \in T$. We denote by $SE(G, r^N)$ the set of all strong equilibria of G at r^N.

An SCR (for N) $F: R^N \times D \to 2^{X^*(N)}$ is *strongly implementable* iff for any $A \in D$ a game form G (for N and A) exists such that, for all $r^N \in R^N$, $g(SE(G, r^N)) = F(r^N, A)$. Obviously, a strongly implementable SCR is Pareto efficient as well.

An *effectivity function* (EF) on N and $A \in 2^{X^*(N)}$ is a function E: $P(N) \to P(P(A))$ such that:

  (i) $E(\emptyset) = \emptyset$;
 (ii) $\emptyset \notin E(S)$ for any $S \in 2^N$;
(iii) $A \in E(S)$ for any $S \in 2^N$; and
 (iv) $B \in E(N)$ for any $B \in 2^A$.

We denote by $EF(N, A)$ the set of all EFs on N and A for any $A \in 2^{X^*(N)}$.

The *core* of $E \in EF(N, A)$ at profile $r^N \in R^N$ – written $\text{core}(E, r^N)$ – is the set of all (extended) outcomes $x \in A$ satisfying the following condition: for no $B \subset A - \{x\}$ and $S \in 2^N$ do both

(1) $B \in E(S)$ and
(2) for all $y \in B$, $(y, x) \in \bigcap_{i \in S} P_i(r^N)$

hold true.

131

An $E \in \mathrm{EF}(N, A)$ is said to be *(core-)stable* on R^N iff $\mathrm{core}(E, r^N) \neq \emptyset$ for any profile $r^N \in R^N$. An $E \in \mathrm{EF}(N, A)$ is said to be *monotonic* iff it is both *N-monotonic* (i.e., for all $S, T \in 2^N$ and for all $B \in 2^A$, $S \subset T$ and $B \in E(S)$ entail $B \in E(T)$) and *A-monotonic* (i.e., for all $S \in 2^N$ and for all $B, C \in 2^A$, $B \subset C$ and $B \in E(S)$ entail $C \in E(S)$).

An $E \in \mathrm{EF}(N, A)$ is *maximal* iff, for all $S \in 2^N$ and $B \in 2^A$, $B \notin E(S)$ entails $A \setminus B \in E(N \setminus S)$. The typical symmetry properties of an EF are defined as follows: $E \in \mathrm{EF}(N, A)$ is *anonymous* iff, for all $S, T \in 2^N$ and all $B \in 2^A$, $\#S = \#T$ and $B \in E(S)$ entail $B \in E(T)$; and $E \in \mathrm{EF}(N, A)$ is *neutral* iff, for all $T \in 2^N$ and for all $B, C \in 2^A$, $\#B = \#C$ and $B \in E(S)$ entail $C \in E(S)$. Clearly enough, an anonymous (neutral) EF treats the players (the outcomes) in a fairly equal way. An anonymous and neutral EF will also be called *symmetric*.

Two EFs, $E \in \mathrm{EF}(N, A)$ and $E' \in \mathrm{EF}(N, A')$, are said to be *equivalent* (w.r.t a bijective function $\pi: A \to A'$) iff, for all $S \in 2^N$ and all $B \in 2^A$, $B \in E(S)$ iff $\pi B \in E'(S)$. If $E \in \mathrm{EF}(N, A)$ and $E' \in \mathrm{EF}(N, A')$ are two equivalent EFs then we also write $E' = E(A' \setminus A, \pi)$, $E = E'(A \setminus A', \pi^{-1})$.

The relevance of stable EFs to the strong implementation problem is apparent from the following fundamental result.

Theorem 1 (Moulin and Peleg 1982). *Let A be a finite agenda (i.e., $A \in PF(X^*(N)) \setminus \{\emptyset\}$), and let $E \in \mathrm{EF}(N, A)$ be monotonic and stable. Then the SCR (with fixed agenda) $\mathrm{core}(E, \cdot): R^N \times \{A\} \to 2^A$ is strongly implementable iff E is maximal as well.*

On the other hand, EFs that are stable, monotonic, maximal, and symmetric do actually exist (the typical example being the EFs attached to some suitable voting-by-veto method; see Moulin 1983 and Peleg 1984). Hence, the foregoing theorem may be regarded as an outstanding positive result with respect to the implementation problem. Apart from the finiteness of A, the main limitation of this positive result is that the symmetry requirement entails the restriction to pairs of societies and agendas whose cardinalities are relatively prime numbers. However restrictive, this requirement is consistent with a variable-agenda setting. Actually, it is easily shown that the following proposition holds true.

Proposition 1. *Let $f = f(k, s)$ be the strictly anonymous and nearly neutral AFR defined previously (for some arbitrarily fixed $s \in X \times \{\Omega\}$ and integer $k > 0$). Let $A \in f(N)$ and let $\pi = \{\pi_B\}_{B \in f(N)}$, a family of bijective functions $\pi_B: A \to B$ such that $\#\pi = \#f(N)$ (i.e., π contains exactly one function for each $B \in f(N)$). Moreover, let $E \in \mathrm{EF}(N, A)$ be stable, monotonic, maximal, anonymous, and neutral, and let $H = R^N \times f(N) \to 2^{X^*(N)}$*

be the function defined by the following rule: For any $r^N \in R^N$ and any $B \in f(N)$,

$$H(r^N, B) = \text{core}(E(B \setminus A, \pi_B), r^N).$$

Then H is a well-defined and strongly implementable SCR.

Proof. It is easily shown that if E, E' are equivalent EFs and E is monotonic (resp., maximal, anonymous, neutral) then E' is monotonic (resp., maximal, anonymous, neutral) as well; the proof is straightforward and is left to the reader. Also, any E' equivalent to E is stable. In fact, we know from the previous statement that E' is monotonic, maximal, anonymous, and neutral. On the other hand, we know from Moulin and Peleg (1982) that: (i) a monotonic, maximal, anonymous, and neutral $E \in$ EF(N, B) exists iff $(\#N, \#B)$ is a pair of relatively prime numbers; (ii) furthermore, if this is the case then there is only one stable, monotonic, maximal, anonymous, and neutral $E \in$ EF(N, B), defined as follows: For all $S \in 2^N$ and all $C \in 2^B$, $C \in E(S)$ iff $(\#S/\#N) > 1 - (\#C/\#B)$. Such an EF E is also called the *proportional* EF of EF(N, B).

Thus, if E is the proportional EF of EF(N, A) then it follows, by definition of equivalence, that E' is the proportional EF of EF(N, B). Therefore E' is stable as well. As a consequence, for any $B \in f(k, s)$, $\text{core}(E(B \setminus A, \pi_B), \cdot)$ is a well-defined and strongly implementable SCR. Hence, the proposition follows. $\square$

It should be noticed that, for any agenda $B \in f(N)$, $\text{core}(E(B \setminus A, \pi_B), \cdot)$ retains the nice symmetry properties of $\text{core}(E, \cdot)$. Therefore, the foregoing proposition makes precise the obvious suggestion that the Moulin–Peleg solution to the strong implementation problem is by no means confined to fixed-agenda SCRs.

In other words, we are now confronted with an SCR (namely, H of Proposition 1) that may be envisaged as a positive solution to the strong implementation problem in a variable-agenda setting. It is now worth asking about the performance of H with respect to the agenda-manipulation problem.

The most relevant issue is the efficiency of the agenda formation problem under $H = \bigcup_{B \in f(k, s)} \text{core}(E(B \setminus A, \pi_B), \cdot)$. We assume that each player has partial information about feasible alternatives. Borrowing Simon's (1972) favorite terminology, we may well say that the players have a state-description of any alternative in X, whereas they have only a process-description of alternatives in a proper (possibly small) subset of X.

As a minimal guarantee against socially inefficient agenda manipulation, we require that – whenever a player happens to discover a new

133

feasible alternative that is both (1) Pareto improving with respect to an old proposal and (2) regarded as better than any current optimal outcome – then substituting the new outcome for the old proposal is a profitable move.

All this motivates the following definitions.

Definition (d.i. pairs). Let $F: R^N \times D \to 2^{X^*(N)}$ be an SCR for N. A pair $(x, y) \in (X \times \{i\})^2$ is a *dramatic improvement* (d.i.) at $(r^N, A) \in R^N \times D$ and w.r.t the SCR F iff (i) $x \notin A$ and $y \in A$; (ii) $(x, y) \in \bigcap_{i \in N} r_i$; and (iii) $(x, \dot{a}) \in P_i(r^N)$ for any $a \in F(r^N, A)$.

Definition (DIP criterion). Let $F: R^N \times D \to 2^{X^*(N)}$ be an SCR for N. F is said to satisfy the *dramatic improvements pay* (DIP) criterion iff for all $(r^N, A) \in R^N \times D$, for all $i \in N$, and for all $(x, y) \in (X \times \{i\})^2$: that (x, y) is a d.i. pair w.r.t. F at (r^N, A) implies that $(z, w) \in r_i$ for any $z \in F(R^N, A \cup \{x\} \setminus \{y\})$ and any $w \in F(r^N, A)$.

According to the classification of social choice axioms proposed by Fishburn (1973), DIP is an active intraprofile condition. Broadly speaking, DIP amounts to a consistency condition, since it is a restriction on choice sets when the agenda changes while the profile remains fixed. However weak, the DIP criterion ensures that the agenda-formation process enjoys a minimal degree of dynamic efficiency.

A few remarks are in order here. First, it should be noticed that the consequent of the DIP criterion amounts to the most typical condition imposed on extensions of orderings to powersets (see, e.g., Gardenfors 1979). Furthermore, the consequent of the DIP criterion closely resembles the "ideal" criterion of nonmanipulability invoked and regretfully dismissed by Demange (1987).

We are now ready to prove the main result of this chapter.

Theorem 2. *Let $E \in EF(N, A)$ be stable (on R^N), let $f = f(k, s)$ be an AFR as in Proposition 1, and let $A \in f(N)$ be an f-admissible agenda. Let $H = H(E)$ be an SCR for N as defined in Proposition 1. Then H is DIP.*

Proof. Let $(x, y) \in (X \times \{i\})^2$ be a d.i. pair at $(r^N, A) \in (R^N, f(N))$ and w.r.t H (see the preceding definition), and let $A' = A \cup \{x\} \setminus \{y\}$. We denote by $\pi: A \to A'$ the bijective function defined by the following rules: $\pi(y) = x$; and $\pi(a) = a$ for any $a \in A \setminus \{y\}$. The following notation will also be used: $H_1 = \text{core}(E, r^N)$; $H_2 = \text{core}(E(A' \setminus A, \pi), r^N)$. Thus, it must be shown that $(z, w) \in r_i$ for all $z \in H_2 \setminus H_1$ and for all $w \in H_1 \setminus H_2$.

134

We must distinguish the following cases:

(i) $z = x$, $w = y$ – In this case, there is nothing to prove because $(x, y) \in \bigcap_{i \in N} r_i$ by definition of a d.i. pair.

(ii) $z = x$, $w \in A \setminus \{y\}$ – Since $w \in H_1$, it follows that $(z, w) \in P_i(r^N)$ by definition of a d.i. pair.

(iii) $z \neq x$ – Since $z \in H_2 \setminus H_1$, it follows that a pair (S, B) exists such that $S \in 2^N$, $B \in 2^A$, $B \in E(S)$, and $(b, z) \in \bigcap_{i \in S} P_i(r^N)$ for all $b \in B$.

Case (iii) has two subscases: either $w = y \notin B$ or $w = y \in B$. If $y \notin B$ then $B \in E'(S)$ (where $E' = E(A' \setminus A, \pi)$), by definition of E'. If $y \in B$ then $B' = B \cup \{x\} \setminus \{y\} \in E'(S)$. Moreover, by hypothesis, by definition of a d.i. pair, and by transitivity of individual preferences, $(x, z) \in \bigcap_{i \in S} P_i(r^N)$. Hence, in either subcase, a $C \in 2^A$ exists (namely, $C = B$ or $C = B'$) such that $C \in E'(S)$ and $(c, z) \in \bigcap_{i \in S} P_i(r^N)$ for any $c \in C$. As a consequence, $z \notin \text{core}(E', r^N) = H_2$, contrary to the hypothesis. Therefore, case (iii) is to be excluded. It follows that the theorem holds true. $\square$

As mentioned previously, Theorem 2 entails that the SCR consisting of the core correspondences of nicely stable EFs enjoys an encouraging degree of robustness against inefficient agenda manipulation.

3. Discussion

The suggested interpretation of Theorem 2 is as follows:

The agenda formation process associated to the SCR consisting of the core correspondences of stable EFs (henceforth, EF-cores-SCR) may be expected to be efficient. Thus, such an SCR does qualify as a solution to the agenda manipulation problem, provided that the involved EFs enjoy a suitable set of nice properties.

A proper assessment of the previous claim entails an evaluation of the performance of such an SCR with respect to the typical criteria invoked as a guarantee against "bad" agenda manipulation. The relevant criteria are the following:

(1) Independence of Irrelevant Alternatives (IIA): A SCR $F: R^N \times D \to P(X^*(N)) \setminus \{\emptyset\}$ satisfies IIA iff, for any $A \in D$ and for any pair (r^N, q^N) of profiles in R^N such that $r_i \cap A^2 = q_i \cap A^2$ for all $i \in N$ and $(R^N, A) = F(Q^N, A)$.

(2) Weak Axiom of Revealed Preference (WARP): A SCR $F: R^N \times D \to P(X^*(N)) \setminus \{\emptyset\}$ satisfies WARP iff, for all profiles in R^N and for all

135

$A, B \in D$, $A \subset B$ and $F(r^N, B) \cap A \neq \emptyset$ imply $F(r^N, A) = F(r^N, B) \cap A$. (It should be recalled that WARP amounts to a contraction-consistency condition, $F(r^N, B) \cap A \subset F(r^N, A)$, plus an expansion-consistency condition, $F(r^N, A) \subset F(r^N, B) \cap A$.)

IIA rules out the relevance of any information about (possibly feasible) outcomes that are not in the actual agenda. Hence, there is no way to manipulate the choice set through a change in the description of the feasible set, whenever the actual agenda remains fixed.

Moreover, under the usual unconstrained-domain assumption, WARP entails the "rationalizability" of social choices through an ordering; that is, for any agenda the choice set may be represented as the set of elements of the agenda that are maximal w.r.t. that ordering (see Hansson 1968 for a more general result). As a consequence, addition or deletion of outcomes that are suboptimal with respect to the "social" ordering cannot affect the choice set.

It is easily checked that the following proposition holds true.

Proposition 2. *The* EF-cores-SCR *H of Theorem 2 satisfies both* IIA *and* WARP.

The fact that H satisfies both IIA and WARP is a straightforward consequence of the strong restriction imposed by the agenda formation rule $f(k, s)$ on the set of admissible agendas: namely, only trivial (i.e. non-proper) subsets of admissible agendas are admissible as well.

Hence IIA and WARP are trivially satisfied. Therefore, in this highly particular variable-agenda setting, both IIA and WARP lose any force as a safeguard against bad agenda manipulation. Hence the role and the significance of other criteria such as DIP follow.

Another implication of the present analysis is the utmost importance of "structural" domain restrictions. Of course, the general relevance of structural restrictions is well known; it should be noted, however, that the proposed model focuses on a new class of domain restrictions that are explicitly rationalized as the outcome of a suitable agenda formation rule. This is at variance with the standard treatment of structural domain restrictions as a completely exogenous matter.

The very definition of our EF-cores-SCR embodies the foregoing restrictions, as dictated by the underlying AFR. In fact, the proposed AFR, $f(k, s)$, imposes a strong regularity condition on the admissible agendas and thus on the relevant EFs. This is why any disturbing volatility of choice sets can be ruled out. The possibility that, in principle, EF-cores-SCRs may exhibit such a volatility of choice sets was recently emphasized by Strnad (1987), who maintains that the alleged volatility of the choice set under EF-cores-SCRs shows that they are only viable when the agenda is exogenously fixed.

Obviously, this amounts to a strong criticism of EF-cores-SCRs. However, volatility may easily be ruled out by a sensible choice of an explicit AFR. Therefore, Strnad's conclusion should be reversed. Namely, EF-cores-SCRs are exactly the kind of SCRs that strongly need an endogenous agenda formation process guided by a suitable AFR (whereas other relevant SCRs may perhaps admit exogenously fixed agendas as well). Furthermore, the main result of this chapter shows that the relevant AFR can be expected to be fairly efficient.

We conclude with some comments about related literature. First, there is a rich literature on the so-called nonbinary approach to social choice theory, started by Hansson and Fishburn (see Hansson 1969, Fishburn 1973, and Fishburn 1974) and subsequently pursued by Grether and Plott (1982), among others. The common focus of this literature is on the role of structural assumptions on the domain of admissible agendas.

However, two main lines of research are currently pursued under the same "nonbinary social choice" heading. The original topic of nonbinary social choice was the attempt to generalize impossibility theorems of the Arrowian variety through a relaxation of domain conditions. The aim of this research is to single out the axiom (if any) that is the ultimate source of Arrow's theorem. (Incidentally, the most widely accepted conclusion is that such a role should be ascribed to consistency or collective-rationality requirements; see Fishburn 1974 and Grether and Plott 1982.)

As a consequence of this concern for suitable generalizations of Arrow's theorem, no agenda is really excluded (either binary or not). The emphasis is on restrictive sufficient conditions for the admissibility of agendas; typically, these conditions establish the admissibility of any agenda that contains at least k distinct alternatives (for some integer $k > 2$).

More recently, another strand of the nonbinary social choice–theoretic literature has addressed the problem of finding positive results by introducing explicit restrictions on the domain of admissible agendas (see Grether and Plott 1982). The underlying rationale of these restrictions consists in the observation that, in a standard economic environment, feasible sets of outcomes are typically infinite.

Moreover, in this literature one can hardly find attempts to provide an explicit model of the agenda formation process. Hence the admissible agendas are straightforwardly identified with the conceivable feasible sets. As a result, all finite agendas (not only the binary ones) are excluded from the relevant domain. This is the most striking difference with respect to the model proposed in the present chapter, in spite of the common focus on structural restrictions on the domain of agendas.

Concern for the outstanding role of the status quo in social choice problems is the major motivation of some other recent work on agenda domain restrictions; see Richelson (1984) and Gibbard, Hylland, and Weymark (1987). These works point out that some scope for positive results arises when the admissible agendas are bound to contain the status quo outcome. Namely, using such a domain restriction, Richelson identifies a pair of SCRs that approximately satisfy Arrow's conditions; however, the first rule fails on efficiency grounds, whereas the second fails on consistency grounds.

Furthermore, Gibbard et al. show that there is no SCR satisfying the requirements of symmetry, independence, efficiency, and consistency whenever all the agendas containing the status quo are admissible. Once again, a comparison with our own results suggests the central role of agenda domain assumptions.

The literature on agenda control is also relevant to the issue of agenda manipulation. However, the focus of this literature is on the effects of alternative orderings of a given agenda on the final choice, under a sequential elimination procedure. A recent notable exception is Banks and Gasmi (1987), who provide an explicit analysis of the agenda formation process in a highly specialized setting.

An earlier attempt to provide an explicit analysis of the agenda formation process is due to Dutta and Pattanaik (see Dutta and Pattanaik 1975, Pattanaik 1978, and Dutta 1981). These authors envisage a two-stage game: The first stage consists of a sponsoring game that determines the agenda in view of the second stage. The second stage consists of a voting game on the previously chosen agenda. Building on an ingenious example due to Majumdar (1956), Dutta and Pattanaik are able to prove that the agenda formation process is likely to involve manipulative (i.e., "insincere") sponsoring of outcomes. Namely, they prove that the foregoing result holds true under a wide class of decision procedures, including all the majoritarian ones, and under various assumptions on the relation between individual preferences over alternatives and individual preferences over agendas. To put it in other terms, under such assumptions the players can be expected to sponsor alternatives that are not their preferred ones (in the feasible set). However, Dutta and Pattanaik are very careful to emphasize that this kind of manipulation may well turn out to be an advantageous occurrence.

In a similar vein, Mueller (1978) quotes, as a major virtue of his own version of the voting-by-veto method, the fact that this decision procedure induces players to take into account each other's preferences while making their proposals.

The present chapter reinforces Mueller's contention, but from a different perspective. In fact, the EF-cores-SCRs can be seen as a representation of the equilibrium behavior of suitable voting-by-veto schemes. On the other hand, the results presented here show that the EF-cores-SCRs enjoy another nice strategic property besides their well-known property of strong implementability. Namely, these rules provide players with the incentive to propose any feasible dramatic improvement they envisage. To the extent that such incentives are well captured by the DIP criterion, Theorem 2 of this chapter may well be interpreted as an encouraging positive result toward a simultaneous solution of the problems of implementation and agenda manipulation.

REFERENCES

Arrow, K. J. (1951), *Social Choice and Individual Values*. New York: Wiley.

Banks, J. S., and Gasmi, F. (1987), "Endogenous Agenda Formation in Three-Person Committees." *Social Choice and Welfare* 4: 133–52.

Demange, G. (1987), "Nonmanipulable Cores." *Econometrica* 55: 1057–74.

Dutta, B. (1981), "Individual Strategy and Manipulation of Issues." *Mathematical Social Sciences* 1: 169–76.

Dutta, B., and Pattanaik, P. K. (1975), "On Strategic Manipulation of Issues in Group Decision Making." In P. K. Pattanaik (ed.), *Strategy and Group Choice*. Amsterdam: North Holland.

Fishburn, P. C. (1973), *The Theory of Social Choice*. Princeton, NJ: Princeton University Press.

Fishburn, P. C. (1974), "On Collective Rationality and a Generalized Impossibility Theorem." *Review of Economic Studies* 41: 445–58.

Gardenfors, P. (1979), "On Definitions of Manipulation of Social Choice Functions." In J. J. Laffont (ed.), *Aggregation and Revelation of Preferences*. Amsterdam: North Holland.

Gibbard, A., Hylland, A., and Weymark, J. A. (1987), "Arrow's Theorem with a Fixed Feasible Alternative." *Social Choice and Welfare* 4: 105–15.

Grether, D. M., and Plott, C. R. (1982), "Nonbinary Social Choice: An Impossibility Theorem." *Review of Economic Studies* 49: 143–9.

Hansson, B. (1968), "Choice Structures and Preference Relations." *Synthese* 18: 423–42.

Hansson, B. (1969), "Voting and Group Decision Functions." *Synthese* 20: 526–37.

Majumdar, T. (1956), "Choice and Revealed Preference." *Econometrica* 24: 71–3.

Moulin, H. (1983), *The Strategy of Social Choice*. Amsterdam: North Holland.

Moulin, H., and Peleg, B. (1982), "Cores of Effectivity Functions and Implementation Theory." *Journal of Mathematical Economics* 10: 115–45.

Mueller, D. C. (1978), "Voting by Veto." *Journal of Public Economics* 10: 57–75.

Nakamura, K. (1979), "The Vetoers in a Simple Game with Ordinal Preferences." *International Journal of Game Theory* 8: 55–61.

Pattanaik, P. K. (1978), *Strategy and Group Choice*. Amsterdam: North Holland.

Peleg, B. (1984), *A Game Theoretic Analysis of Voting in Committees*. Cambridge: Cambridge University Press.

Richelson, J. T. (1984), "Social Choice and the Status Quo." *Public Choice* 42: 225–34.

Schofield, N. (1986), "Existence of a 'Structurally Stable' Equilibrium for a Non-Collegial Voting Rule." *Public Choice* 51: 267–84.

Sen, A. K. (1983), "Social Choice Theory." In K. J. Arrow and M. D. Intriligator (eds.), *Handbook of Mathematical Economics,* vol. 3. Amsterdam: North Holland.

Simon, H. A. (1972), "The Theory of Problem Solving." *Information Processing* 71: 261–77.

Strnad, J. (1987), "Full Nash Implementation of Neutral Social Functions." *Journal of Mathematical Economics* 16: 17–37.

9

Algorithmic knowledge and game theory

KEN BINMORE & HYUN SONG SHIN

What ye know, the same do I know also . . . .
                                        Job 13:2

1. INTRODUCTION

In game theory, the question "Who knows what?" is crucial at many levels. In poker, for example, it matters that you do not know what hands your opponents are holding, and they do not know yours. Knowledge and ignorance at this level are studied in the theory of games of *imperfect* information. In such games, the rules and the preferences of the players are taken to be commonly known. In real life, however, this is seldom a realistic assumption. To deal with such situations, one requires the richer theory of games of *incomplete* information.

This chapter is about neither of these informational levels. Its attention is directed at what players know about the results of the reasoning processes of other players. Traditional game theory takes it for granted that rational players can replicate the reasoning of their opponents, so that arguments of the type "If I think that he thinks that I think . . ." can be employed. Before authors became cautious about foundational issues, the underlying assumption would sometimes be expressed by saying, "It is taken to be common knowledge that the players are perfectly rational." This assumption is sometimes referred to as the "implicit axiom of game theory."

What does this implict axiom mean? There are at least three difficulties:

(1) Formal definitions of common knowledge, beginning with that of Aumann [3], require the prior existence of a state space, so that we have events or propositions about which we can claim to have knowledge. In the implicit axiom, an assertion is made about players. What is a player?
(2) The implicit axiom distinguishes perfectly rational players from some lesser breed of player. But what is perfect rationality?

(3) What is one to do if the implicit axiom is refuted during the course of the game?

Of these difficulties, perhaps the most pressing from the practical point of view is the third. The vast "refinements of equilibrium" literature are based on various expedients for evading this question. The natural response is that, if events reveal that the opponent is not perfectly rational, then he or she must be imperfectly rational. But how can we study deviations from perfect rationality if we have no clear idea of what perfect rationality is? No answers to such questions are offered here. Our method of evading the question is simply to confine attention to normal-form games, so that it is impossible for the implicit axiom to be refuted during the game because the game has no duration.

This leaves us with the first two questions. Meggiddo [16] observes that the traditional approach requires that "a fully rational player . . . can even decide undecidable problems." Implicit in such a view is the belief that a player should be modeled as a suitably programmed computing machine. Those who think there is no point even in exploring the implications of such an approach need read no further.

It has been argued elsewhere (Binmore [5; 6] and Anderlini [2]) that computing machines cannot be "perfectly rational" in the sense taken for granted in traditional game theory. Albin [1], Lewis [15], and others have argued similarly in related contexts. The essential point is quite simple. A computing machine that only gives correct answers to relevant problems will sometimes necessarily compute forever. That is to say, certain relevant questions cannot be settled by an effectively computable procedure that gives a definitive answer in finite time.

Anderlini [2] is among those who have emphasized the close connection between such results and the question of what a computer can "know." For example, with a not very demanding definition of "rationality," he observes that a computing machine can "know" an opponent's program in its entirety without simultaneously "knowing" whether the opponent is or is not "rational."

The current chapter is also about the question of what computers can or cannot "know." In particular, what can they know about the knowledge of other computers? In contrast to most discussions, we confine our attention to *algorithmic* knowledge: information that a computer can deduce from its given data using effective procedures. In view of the preceding discussion, we take this as excluding the possibility that a computer can know a whole range of "facts" unless these are somehow built in as a priori postulates or supplied by an "oracle" whose verity need not be checked. However, even after such exclusions, much remains to be studied.

Some preliminary disclaimers are necessary. This chapter is not an attempt to construct a theory of rational play in games where the players are construed as computing machines. Any such theory must specify the relationship between what players know and what they do; only the former is considered here.

The second disclaimer concerns originality. No results that are not well known in mathematical logic are required. Any originality lies in the context in which they are applied. For example, the work referred to previously can be seen as a consequence of Gödel's first incompleteness theorem: that Peano arithmetic cannot be consistent unless it is incomplete. Similarly, Löb's theorem, which we use in the text, follows very quickly from Gödel's second imcompleteness theory. (See the Appendix.)

Perhaps the third disclaimer will serve to encourage those dismayed by the preceding sentences. We have followed Gödel, albeit at a great distance, in trying to be no more formal than seems necessary. Those interested in the formal theory are referred to Boolos [8], Boolos and Jeffrey [9], and Mendelson [17].

2. Knowledge operators

For a more extensive discussion of this subject, see Binmore and Brandenburger [7] or Geanakoplos [14]. Following Aumann [3], discussion of the knowledge operator K usually begins in a game-theoretic context, with a set Ω of possible states of the world. A subset E of Ω is identified with a possible event. The set KE is interpreted as the event that the decision maker knows that E has occurred. The usual assumptions about K are:

$(K0)$ $K\Omega = \Omega$,
$(K1)$ $K(E \cap F) = KE \cap KF$,
$(K2)$ $KE \subseteq E$ (axiom of knowledge),
$(K3)$ $KE \subseteq K^2 E$ (axiom of transparency),
$(K4)$ $(\sim K)^2 E \subseteq KE$ (axiom of wisdom).

The names given to $(K2)$–$(K4)$ are Bacharach's [4].

One can also express the same properties by means of a possibility function P. With this formulation, $P(\omega)$ is interpreted as the set of states that the decision maker regards as possible if the actual state is $\omega \in \Omega$. The assumptions corresponding to $(K2)$–$(K4)$ are:

$(P2)$ $\omega \in P(\omega)$,
$(P3)$ $\zeta \in P(\omega) \Rightarrow P(\zeta) \subseteq P(\omega)$,
$(P4)$ $\zeta \in P(\omega) \Rightarrow P(\zeta) = P(\omega)$.

With these assumptions, $P(\omega)$ will be familiar to game theorists as the "information set" containing ω. Given P, one can define K by $KE = \{\omega : P(\omega) \subseteq E\}$. Given K, one can define P by $P(\omega) = \bigcap_{\omega \in KE} E$.

When doubts are expressed about the appropriateness of the properties attributed to K, attention usually centers on the axiom of wisdom ($K4$). It has the interpretation that: "If I am unaware that I am unaware of something, then I know it." Such doubts are unlikely to disturb game theorists very much if the state space Ω is adequately restricted. For example, if the states ω in Ω correspond to the decision nodes in a game tree, then ($P4$) simply asserts the standard game-theoretic requirement that the information sets partition the set of all decision nodes. One can, in any case, retain some game-theoretic results even when ($P4$) does not hold true of information sets. (See Samet [18], Geanakoplos [14], Shin [19], and Brandenburger and Dekel [10; 11].)

Although game theorists can afford to be complacent about ($K4$) when Ω is the set of decision nodes in a game, they are not on such safe ground when working with more complicated state spaces. In particular, ($K4$) is not a comfortable assumption when Ω is sufficiently rich as to allow statements about what a player can know about another player's knowledge. The literature usually signals that knowledge at this or higher levels is to be admitted by observing that a state ω is to be interpreted as *all-inclusive*. The difficulties to which this chapter are addressed only arise in contexts where this informal assumption has some bite.

However, our concern is not centered on ($K4$). We follow Binmore and Brandenburger [7] and Shin [19] in raising graver doubts about the standard knowledge assumptions. When one insists that what is known is known by virtue of its being the result of applying a properly defined procedure or algorithm, we argue that the status of the axiom of knowledge ($K2$) becomes suspect. That is, concern is necessary when claiming that everything one knows is actually true. Since this is a bitter pill to swallow, we devote the next section to an example from Binmore and Brandenburger [7] that may serve to indicate why such problems arise.

3. Baffling a Turing machine

A Turing machine may be thought of as a program for a regular computer written on the assumption that no limitations on storage space need be observed. Suppose that, for each (all-inclusive) state of the world ω, possibility questions are resolved by a Turing machine $S = S(\omega)$ that sometimes answers NO to:

Is it possible that . . . ?

Unless the answer is NO, possibility is conceded. (Timing issues are neglected.) Consider a specific possibility question concerning the Turing machine N. Let the computer code for this question be $\lceil N \rceil$.

Next, let $\lfloor M \rfloor$ be the computer code for the question:

Is it possible that M will answer NO to $\lceil M \rceil$?

Let T be a Turing machine that outputs $\lfloor x \rfloor$ when its input is $\lceil x \rceil$. Then the program $R = ST$, which consists of first operating T and then operating S, responds to $\lceil M \rceil$ as S responds to $\lfloor M \rfloor$.

Suppose that R responds to $\lceil R \rceil$ with NO. Then S reports that it is impossible that R responds to $\lceil R \rceil$ with NO. Translating this conclusion into the notation of the previous section, we obtain that

$$\omega \notin P(\omega).$$

If $(P2)$ – and therefore the equivalent $(K2)$ – are to be rescued, it is necessary to argue that R will never respond to $\lceil R \rceil$ with NO in any state ω. But we, as observers, will then know that it is *false* that it is possible that R responds to $\lceil R \rceil$ with NO. So why didn't we write a better program than S in the first place, one that accurately reflects what we know? Either our algorithm for determining what is possible is "incomplete," in that it allows events to be possible that we know to be false, or else it is "inconsistent," in that it rejects as impossible events that we know to be true.

This is not the place for a discussion of how the human mind differs from a computer program. However, it should be observed that there is no "easy out" in this direction. If we are able coherently to justify our knowing something, then the justification process must be capable of formalization. Hence, if the Church–Turing thesis is not to be rejected, the justification process can be carried through by a computer. If, on the other hand, we have knowledge for which we cannot provide a coherent justification, why do we call it knowledge?

4. MODAL LOGIC

The remainder of this chapter is an attempt to provide a more satisfactory grounding for the conclusion of the previous section. The current section sets the scene by describing the modal logic G. This differs sharply from the modal logic $S5$, which embodies the properties of the knowledge operator that economists usually take for granted. The modal logic G (for Gödel) is specified in Boolos [8] by the following axioms and rules of inference:

145

(*G*1) $\Box(A \to B) \to (\Box A \to \Box B)$

(*G*2) $A \to B$

$\underline{A \qquad\qquad}$

$\qquad B$ (*modus ponens*)

(*G*3) $\underline{A \quad}$

$\qquad \Box A$ (necessitation)

(*G*4) $\Box(\Box A \to A) \to \Box A$ (Löb)

Items under a horizontal line represent deductions that may be made from the list of items above the line. The modal operator $\Box$ is usually interpreted as "necessitation." It comes with a companion $\Diamond = \sim\Box\sim$, interpreted as "possibility." The interest of G is that, when $\Box$ is particularized as "provability in (Peano) arithmetic," then G characterizes the "provable principles of provability" (Boolos [8], Solovay [20]).

In later sections, we seek to justify a view of algorithmic knowledge within which the knowledge operator can be seen as an instance of the modal operator $\Box$, provided that the formulas to which it is applied are suitably coded. In this section, some of the consequences are explored.

It is reassuring that

$$\Box A \to \Box\Box A$$

is a theorem of G (Boolos [8, p. 30]). One therefore as an analog of the axiom of transparency (*K*3). Thus, from $\Box A$ (A is known or knowable), one can deduce

$$\Box\Box A,\ \Box\Box\Box A,\ \Box\Box\Box\Box A, \ldots,$$

and so on. This not very deep conclusion can be thought of as justifying statements about their being common knowledge of the reasoning processes of algorithmic players. (I know what you would deduce if you had certain information. You know that I would know. And so on.)

Much less reassuring is the fact that

$$\Box A \to A$$

is *not* a theorem of G. But this is the analog of the axiom of knowledge (*K*2). Not only is $\Box A \to A$ not a theorem of G, it cannot be consistently appended to the rules of G. (If $\Box A \to A$ is made (*G*5), then using (*G*3), (*G*4), (*G*2), and (*G*3) in turn yields that A is a theorem.) One can, of course, append *instances* of $\Box A \to A$ to G. But this is not the same as appending the *principle* $\Box A \to A$.

The axiom in the specification of G that creates this difficulty is (*G*4). With a knowledge interpretation, it can be translated as: An algorithmic

146

player cannot know that knowing A makes A true unless he knows A. This does not imply that he may know false things. However, just because something is true (for example: that what I know is true), it does not follow that I know it.

What we have here is therefore a reprise of the conclusion of the previous section, but in a context that is more clearly pertinent to game theory. Of course, the relevance of G to such a context, and (in particular) the relevance of $(G4)$, have yet to be defended. The next section begins this task.

5. COMPUTABILITY AND KNOWLEDGE

If players are to be modeled as computer programs, a prerequisite is that attention be restricted to a domain of inquiry for which each object of interest can be coded in a manner that allows it to be recognized by a computer. This amounts to the requirement that it be coded as a unique natural number $n \in I\!N$. The manner in which each finite game with rational payoffs can be so coded will be familiar. (We do not insist on this point, but it is not essential that consideration be restricted to games with rational payoffs or to mixed strategies with rational probabilities. Neither computers nor humans can operate with real numbers unless they are specified by a finite list of symbols. Cohen's [13] discussion of the Löwenheim–Skolem theorem may serve to clarify this point.)

Only relations between objects in the field of inquiry for which membership can be checked by an effectively computable procedure are to be admitted. A k-nary relation R defined over $I\!N$ is formally a subset of $I\!N^k$. To say that the relation R holds among the natural numbers $m_1, m_2, ..., m_k$ is interpreted as $(m_1, ..., m_k) \in R$. For membership in R to be effectively computable, we require that the characteristic function C_R defined by

$$C_R(m_1, ..., m_k) = \begin{cases} 1 & \text{if } (m_1, ..., m_k) \in R, \\ 0 & \text{if } (m_1, ..., m_k) \notin R, \end{cases}$$

be recursive (i.e., effectively computable as mathematicians normally understand this term). Informally, an algorithmic player will be understood to *know* that the relation R holds among $x_1, x_2, ..., x_k$ if and only if C_R is a recursive function and $(x_1, ..., x_k) \in R$. (It would be more accurate to speak of what is knowable to an algorithmic player, rather than of what is known, but we use the latter terminology since it is less clumsy.)

It is important that not only objects like games or strategies can be coded. Algorithms can themselves be coded as natural numbers, since they can be viewed simply as listings of computer instructions. This allows *players* to be admitted as part of the domain of inquiry. This avenue

147

has been explored by a number of authors – for example, Binmore [5; 6], Anderlini [2], and Canning [12].

6. Arithmetic

Recursive functions have a close affinity with a first-order logical system known as elementary Peano Arithmetic, which we shall denote by PA. Apart from the usual symbols and axioms of first-order logic, the language of PA contains arithmetical symbols. It has a designated element $\bar{0}$, a one-place function symbol $s(\cdot)$, and the two-place function symbols $+$ and $\bullet$. In the standard interpretation of this logic, the symbol $\bar{0}$ denotes the number 0, s denotes the successor function (i.e., the function that adds 1), and the symbols $+$ and $\bullet$ denote (respectively) addition and multiplication.

For those unfamiliar with PA, it is necessary to rehearse the assumptions it embodies so that it is plain that nothing is included that would not normally be regarded as entirely innocuous. In addition to the axioms of first-order logic, PA has the following arithmetical axioms:

(PA1) $\bar{0} \neq s(x)$,
(PA2) $s(x) = s(y) \rightarrow x = y$,
(PA3) $x + \bar{0} = x$,
(PA4) $x + s(y) = s(x + y)$,
(PA5) $x \bullet \bar{0} = \bar{0}$,
(PA6) $x \bullet s(y) = (x \bullet y) + s$,
(PA7) for any formula $F(\cdot)$,

$$[F(\bar{0}) \& \forall x[F(x) \rightarrow F(s(x))]] \rightarrow \forall x F(x).$$

These are the recursion axioms for addition and multiplication, together with the induction axiom (PA7).

We must be scrupulous in distinguishing the claim that a formula is true from the claim that a formula can be proved. This distinction is crucial. We say that a formula of PA is *true* if and only if the interpretation of this formula in the standard interpretation of arithmetic is a correct statement about the natural numbers. For example, the formula

(1) $$s(s(\bar{0})) = s(\bar{0}) + s(\bar{0})$$

is to be interpreted as saying that $2 = 1 + 1$. Since this is a correct statement about the natural numbers, (1) is a true formula of PA.

The notion of proof is concerned purely with the correctness of a symbolic derivation. More precisely, a *proof* of a formula φ is a sequence of formulas in which φ is the final term of the sequence, and each element of the sequence is either an axiom of PA or a consequence of earlier formulas in the sequence via the rules of inference of first-order logic. In other words, a formula φ is provable if it can be derived from the axioms and rules of inference of the logic PA.

Most true formulas of PA are also provable. For example, (1) has a proof that uses axioms (PA3) and (PA4). However, there are formulas which are true but which cannot be proved. The celebrated "Gödel sentence" is an example.

It is the notion of proof in the logic PA that will be our primary concern. This is because proof in PA provides a precise formal counterpart to the notion of recursiveness. Before elaborating on this assertion, we must say something about notation. Let $\bar{m}$ denote the formula $s(s(...s(\bar{0})...))$ in which s appears m times. Here, m is just a natural number, but $\bar{m}$ is the numeral corresponding to the number m. In other words, $\bar{m}$ is the formula in PA that, in the standard interpretation of PA, denotes the number m.

Let f be a function $f: \mathbb{N}^n \to \mathbb{N}$ (which need not be recursive). We say that f is *representable* in PA if there exists a formula $\varphi(x_1, ..., x_n, x_{n+1})$ in PA such that, for any n-tuple of natural numbers $(m_1, ..., m_n)$,

(i) if $f(m_1, ..., m_n) = m_{n+1}$ then $\varphi(\bar{m}_1, ..., \bar{m}_n, \bar{m}_{n+1})$ has a proof in PA;
(ii) $\forall x[\varphi(\bar{m}_1, ..., \bar{m}_n, x) \to x = \bar{m}_{n+1}]$ has a proof in PA.

Less formally, a formula φ in PA represents f if, whenever f is defined for some n-tuple of numbers $(m_1, ..., m_n)$ and takes the value m_{n+1}, then there is a formula φ in PA such that, when the variable terms $x_1, ..., x_n$, x_{n+1} are replaced by the numerals $\bar{m}_1, ..., \bar{m}_n, \bar{m}_{n+1}$, the formula φ has a proof in PA. Part (ii) of the definition states that φ indeed represents a *function*. Not all functions are representable in PA.

Lemma (Mendelson [17, pp. 148, 157]). *Suppose that f is a function f: $\mathbb{N}^n \to \mathbb{N}$. Then f is representable in PA if and only if f is a recursive function.*

Thus, recursiveness has a formal counterpart in the notion of proof in PA. This implies that the abilities of a player in a game with algorithmic powers can be represented by the notion of proof in PA in the following sense. Given a representation of a recursive function f by a formula φ in PA, the statement that a player with computable powers knows that $f(m_1, ..., m_n) = m_{n+1}$ can be translated into the statement that $\varphi(\bar{m}_1, ..., \bar{m}_n, \bar{m}_{n+1})$ has a proof in PA.

By virtue of the lemma, an analysis of games played by algorithmic individuals may proceed as follows. First, some coding of relevant game-

149

theoretic structures is fixed. Second, interesting relations between these coded structures are identified. Only recursive relations are to be permitted if these relations are to form part of what a player can know. Each such relation can be represented in PA by a formula of PA. Moreover, any answer to a question involving the relations can be obtained by finding a proof of the relevant formula in PA.

These considerations lead to our proposing the following definition of knowledge for algorithmic players. If R is a recursive relation and φ is a formula in PA that represents the characteristic function of R, then a player *knows* that R holds among $m_1, m_2, \ldots, m_k$ if and only if $\varphi(\bar{m}_1, \ldots, \bar{m}_k, \bar{1})$ has a proof in PA.

It is important to bear in mind that most of the considerations that normally worry game theorists when studying "who knows what" have been abstracted away into the coding procedure in this definition. Recall that we are not studying the issues for which the theories of games of imperfect or incomplete information were created. Our concern is with the extent to which one player can know how another would reason, *if* he had that player's information.

7. ITERATED KNOWLEDGE

In developing a theory of games played by algorithmic individuals, it is essential to construct a framework that is rich enough to accommodate statements about what one player knows about what other players know about the game and each other. Solution concepts incorporating some iterated deletion of dominated strategies are prime examples of procedures that rely on iterated knowledge about the game. Since we are interpreting the knowledge of a player in the game as the provability of the appropriate formula in PA, to say that some player knows that another player knows some fact about the game is to say that one player can prove in PA that some other player can prove some formula in PA. In other words, we must have some way of expressing statements about proofs in PA *within* the system PA.

This can be accomplished in a similar way to that in which algorithms are coded by the natural numbers. More precisely, the idea is to attach a unique natural number (the Gödel number) to each symbol, expression, and finite sequence of expressions of PA. In effect, the Gödel number serves as the unambiguous "name" for each of these objects. In turn, each Gödel number has a corresponding numeral that can be expressed in the language of PA. In this way, certain expressions in PA can be interpreted as making claims about other expressions in PA.

In particular, a proof of a formula φ in PA is a finite sequence of formulas in PA with φ as the final term, and in which each term is either an axiom of PA or is a consequence of earlier terms in the sequence via the rules of inference. Because each finite sequence of expressions in PA has a Gödel number, each proof of a formula φ has a Gödel number. Crucially, there is an effective procedure for checking whether a number is in fact the Gödel number of a proof of a given formula. That is, the binary relation Π, defined as: $(m, n) \in \Pi$ if and only if m is the Gödel number of a proof of the formula with Gödel number n, has a recursive characteristic function.

The actual construction of the relation Π is a laborious affair, since it must express in number-theoretic functions all relevant features of a proof in the system PA. (Mendelson [17, pp. 121–72] has a full exposition.) However, it is not surprising that Π is a recursive relation, since it is simply a mechanical matter to check each step in the proof to see whether it conforms to the rules for a proof.

From the fact that Π is a recursive relation together with the lemma on the representability of recursive functions, there exists a formula in PA that represents the characteristic function of Π. Let $\Psi(x_1, x_2, x_3)$ be this formula. Then, defining the formula $P(x_1, x_2)$ to be $\Psi(x_1, x_2, \bar{1})$, we have the following:

$$(m, n) \in \Pi \Leftrightarrow P(\bar{m}, \bar{n}) \text{ has a proof in PA.}$$

In other words, the sequence of expressions with code m is a proof of the formula with code n if and only if the formula $P(\bar{m}, \bar{n})$ can be proved in PA. In this way, it is possible to express statements about the proof of a formula of PA *within* the system itself.

It is now possible to discuss iterated knowledge with algorithmic players. By identifying recursiveness with the notion of proof in PA, we can speak coherently about one player proving in PA that some other player is able to prove a formula in PA.

To complete this interpretation, we define $\text{Prov}(x)$ to be the formula in PA given by $\exists y\, P(y, x)$. That is, $\text{Prov}(\bar{n})$ is the formula of PA which states that the formula with Gödel number n has a proof in PA. Also, we use the notation $\lceil \varphi \rceil$ to denote the numeral of the Gödel number associated with the formula φ. Thus, the formula

$$\text{Prov}(\lceil \varphi \rceil)$$

is a formula of PA which states that the formula φ is provable in PA. Observe that, because $\text{Prov}(\lceil \varphi \rceil)$ is itself a formula of PA, such expressions as

$$\text{Prov}(\lceil \text{Prov}(\lceil \varphi \rceil) \rceil) \quad \text{or} \quad \text{Prov}(\lceil \text{Prov}(\lceil \varphi \rceil) \rightarrow \varphi \rceil)$$

are well defined.

151

The relevance of the discussion to Section 4 will now be clear. After a suitable coding, algorithmic knowledge (as conceived of in this chapter) is isomorphic with provability in PA. In particular, the ugly duckling ($G4$) of the modal logic G is justified by an appeal to Löb's theorem, which asserts that $\text{Prov}(\lceil \varphi \rceil) \to \varphi$ is provable in PA if and only if φ is provable in PA. (See the Appendix.)

8. Conclusion

Three points require particular emphasis.

1. There is a precise sense in which algorithmic players with computable powers have common knowledge of those features of the game that are recursive. These include standard properties of the payoff function in the normal form of the game. This implies that algorithmic players are capable of conducting the sort of reasoning that gives rise to constructive solution concepts such as rationalizability.

2. However, this common knowledge is restricted to features of the game that can be coded in the language of arithmetic. Thus, vague assertions such as: "It is common knowledge that all players are perfectly rational" are meaningless in the context of this chapter, unless the sense in which individuals are perfectly rational can be written down (in coded form) as a formula in arithmetic. If the sense in which players are rational boils down to the prosaic claim that players do not play dominated strategies, then we can take this in our stride. However, if the claim is more exotic then we may not be able to accommodate it. Thus, the significance of the discussion here is that it enables us to distinguish those features of the game situation that are amenable to the language of common knowledge from those features that are not. The criterion is: "Can this concept be coded as a formula of arithmetic?"

3. There are definite bounds to what algorithmic individuals can know. One of these limits is that they cannot take for granted the principle that knowledge implies truth. However, the significance of this feature for the analysis of games is unclear. This feature would only become a deficiency if we were to insist that players in a game be able to justify not only what they do, but also what they know. This may be unreasonable, not only for computers but for people also. In any case, there seem ample grounds for caution before adopting a framework within which knowledge is discussed unless the framework recognizes that difficulties on this score exist.

Appendix

In this appendix, we provide a proof of Löb's theorem from Gödel's second incompleteness theorem.

152

A logical system is consistent if, for any sentence S in the language of this system, it is not the case that both S and $\sim S$ are provable in the system. Equivalently, a logical system is consistent if there is at least one sentence in the language of this system that is not provable in the system.

Gödel's second incompleteness theorem states that, for any consistent extension PA^+ of Peano arithmetic, any sentence asserting the consistency of PA^+ cannot be proved in PA^+. In particular, let PA^+ be the extension of Peano arithmetic obtained by augmenting Peano arithmetic with the axiom $\sim S$. If S is not provable in Peano arithmetic, then it remains unprovable in Peano arithmetic with the addition of the premise $\sim S$. Thus, a sentence that asserts the consistency of PA^+ is the following sentence, which asserts that S is not provable in Peano arithmetic:

$$\sim \text{Prov}(\lceil S \rceil).$$

From Gödel's second incompleteness theorem, we can conclude that *either* PA^+ is inconsistent *or* $\sim \text{Prov}(\lceil S \rceil)$ is not provable in PA^+. Using this result, we may prove Löb's theorem as follows. For any sentence S of Peano arithmetic, we have the following sequence of inferences:

	$\text{Prov}(\lceil S \rceil) \rightarrow S$ is provable in PA
if and only if	$\sim S \rightarrow \sim \text{Prov}(\lceil S \rceil)$ is provable in PA
if and only if	$\sim \text{Prov}(\lceil S \rceil)$ is provable in PA^+
only if	PA^+ is inconsistent (by Gödel's second theorem)
if and only if	S is provable in PA.

Thus we have

$$\text{Prov}(\lceil S \rceil) \rightarrow S \text{ is provable in PA only if } S \text{ is provable in PA.}$$

Since the converse implication is trivial, this is the statement of Löb's theorem. □

REFERENCES

[1] Albin, P. S. (1982), "The Metalogic of Economic Predictions, Calculations and Propositions." *Mathematical Social Sciences* 4: 329–58.

[2] Anderlini, L. (1988), "Some Notes on Church's Thesis and the Theory of Games." Cambridge Economic Theory Discussion Paper, Faculty of Economics, Cambridge University.

[3] Aumann, R. (1976), "Agreeing to Disagree." *Annals of Statistics* 4: 1236–9.

[4] Bacharach, M. (1987), "A Theory of Rational Decision in Games." *Erkenntnis* 27: 17–55.

[5] Binmore, K. (1984), "Equilibria in Extensive Games." *Economic Journal* 95: 51–9.

[6] Binmore, K. (1987), "Modeling Rational Players," I and II. *Economics and Philosophy* 3: 179–214 and 4: 9–55.

[7] Binmore, K., and Brandenburger, A. (1990), "Common Knowledge and Game Theory." In K. Binmore (ed.), *Essays on the Foundations of Game Theory.* Oxford: Basil Blackwell.

[8] Boolos, G. (1979), *The Unprobability of Consistency.* Cambridge: Cambridge University Press.

[9] Boolos, G., and Jeffrey, R. (1974), *Computability and Logic.* Cambridge: Cambridge University Press.

[10] Brandenburger, A., and Dekel, E. (1987), "Common Knowledge with Probability 1." *Journal of Mathematical Economics* 16: 237–45.

[11] Brandenburger, A., and Dekel, E. (1985), "Hierarchies of Beliefs and Common Knowledge." Research Paper No. 841, Graduate School of Business, Stanford University.

[12] Canning, D. (1988), "Rationality and Game Theory when Players are Turing Machines." ST/ICERD Discussion Paper 88/183, London School of Economics.

[13] Cohen, P. (1966), *Set Theory and the Continuum Hypothesis.* Amsterdam: Benjamin.

[14] Geanakoplos, J. (1988), "Common Knowledge Without Partitions, with an Application to Speculation." Cowles Foundation Discussion Paper No. 914, Yale University.

[15] Lewis, A. (1985), "On Effectively Computable Realizations of Choice Functions." *Mathematical Social Sciences* 10: 43–80.

[16] Meggiddo, N. (1986), "Approximating Common Knowledge by Common Beliefs." Research Paper RJ5270, IBM, Armonk, NY.

[17] Mendelson, E. (1964), *Introduction to Mathematical Logic.* New York: Van Nostrand.

[18] Samet, D. (1987), "Ignoring Ignorance and Agreeing to Disagree." M.E.D.S. Discussion Paper No. 749, KGSM, Northwestern University. Forthcoming in *Journal of Economic Theory.*

[19] Shin, H. (1987), "Logical Structure of Common Knowledge," I and II. Nuffield College, Oxford.

[20] Solovay, R. (1976), "Provability Interpretation of Modal Logic." *Israel Journal of Mathematics* 25: 287–304.

154

10

Possible worlds, counterfactuals, and epistemic operators

Maria Luisa Dalla Chiara

In spite of its metaphysical appearance, the idea of possible worlds has found natural applications in the logical analysis of a variety of concrete experiences. In practical life, all of us are accustomed to comparing the actual world with a number of possible worlds that we consider more or less attractive. Our choices and actions seem to depend on such systematic comparisons. I will consider two examples that represent respectively a kind of successful and unsuccessful application of possible-worlds semantics: the theory of counterfactual conditionals and the semantics of epistemic logics. Both examples seem to play a relevant role in some game-theoretical problems.

1. COUNTERFACTUALS

As is well known, counterfactual arguments are not particularly appreciated in certain domains of knowledge. For instance, historians frequently repeat that "one cannot make history with ifs!" At the same time, in physics and in experimental sciences in general, counterfactual statements can hardly be avoided. Most physical laws have a counterfactual form, in the sense that they refer to boundary conditions that are generally not satisfied in our actual laboratories.

At first sight, counterfactual conditionals seem to behave in a silly way, because they violate some fundamental properties that we are accustomed to associate with our basic idea of implication. One of these properties is represented by transitivity, which notoriously constitutes the deep structure of the syllogistic argument. As a counterexample, let us consider the following odd inference.

Premise 1: If Wojtyla were a communist, then he would not be the pope.
Premise 2: If Wojtyla were not the pope, then Poland would have a less important role on the international stage.

155

Conclusion: If Wojtyla were a communist, then Poland would have a less important role on the international stage.

In the framework of the standard possible-worlds semantics, one can give a satisfactory analysis of counterfactuals that explains why the transitive property generally breaks down. Let me here recall the basic idea of this analysis, due essentially to Lewis [9]. Consider a sentence of the form

If α were the case then β would be the case.

(For instance, if Wojtyla were a communist, then Wojtyla would not be the pope.) When are we inclined to valuate our sentence as true? One must suppose a kind of thought-experiment: We imagine a "sheaf" (or "sphere," in Lewis's terminology) of possible worlds where it is the case that Wojtyla is a communist. Further, we recognize that in such worlds Wojtyla is not the pope. From an intuitive point of view, such possible worlds represent "small variations" of the actual world: Wojtyla might have in these different worlds different roles; for instance, he might be a man like Jaruzelski or alternatively a Trotskyist flown from abroad. What our thought-experiment excludes is the possibility (imagined in some movies) of a Marxist pope.

On this basis, the failure of transitivity can be explained as follows. When we assert the truth of our second premise (if Wojtyla were not the pope, then Poland would have a less important role on the international stage), we refer to a particular sheaf of possible worlds where Wojtyla is not the pope, and this sheaf does not necessarily include the sheaf referred to by the first premise (if Wojtyla were a communist, then Wojtyla would not be the pope). As a consequence, both premises may be true and the conclusion may be false.

As a particular case, one can refer to sheaves that are either singletons or empty sets. In such situations, the truth conditions for a counterfactual sentence "If α were the case, then β would be the case" are analyzed as follows. The counterfactual sentence is true in a world w if and only if (hereafter "iff") either the antecedent α is false in any world or the consequent β is true in the closest world to w that verifies α.

This kind of semantical analysis, developed particularly by Stalnaker [11], gives rise to the following question: Is it reasonable to assume that – in any situation where a sentence α is verified in some worlds – any world has a closest world where α is true? How can we determine the closest world to the real world, where Wojtyla is a communist?

Strangely enough, good models for Stalnaker's analysis of counterfactuals have been found in the domain of quantum mechanics. This application is due mainly to Hardegree [8]. According to the standard formalism of quantum mechanics, pure states of micro-objects are represented as unitary vectors in a Hilbert space, and any subspace of that space represents a proposition that may hold true or false with respect to a given state. One can imagine that each state represents a possible world of the micro-universe. In this framework, the truth conditions for a counterfactual sentence of the form "If α were the case, then β would be the case" (e.g., "if the spin value were positive, then the value of the magnetic moment would be positive") can reasonably be given in Stalnaker's sense: The closest world to a pure state ψ where α is true can simply be identified with the normalized projection of ψ over the proposition corresponding to the sentence α. Significantly enough, an apparently unrealistic and ad hoc hypothesis – the existence of a function that introduces a kind of "metric" to the class of possible worlds – receives a plausible mathematical and physical interpretation.

2. Epistemic semantics

Why do the orthodox possible-worlds semantics, which admit of so many successful applications, find crucial difficulties in the case of epistemic logic? The basic reason is simple. Possible worlds are usually dealt with in such a way that they turn out to be closed under logical consequence: If α is true in a world w and β is a logical consequence of α, then β is true in w. Hence, if we try to explain the semantical behavior of the epistemic operators purely in terms of possible worlds, it is very hard to avoid a kind of "logical omniscience" of the subjective agents. Knowledge should be generally closed under logical consequence; knowing the axioms of a theory should be sufficient for knowing all the theorems. This is of course unrealistic, even if we assume a very strong idealization of the concept of subjective knowledge. Let us think only of the fact that, owing to Church's theorem [4], the relation of logical consequence is undecidable.

I will describe an alternative approach to epistemic semantics, one essentially founded on an abstract theory of *intensions*. As a starting point, we suppose that what the members of the scientific community communicate are intensions. Intensions may be dealt with as autonomous abstract entities that admit a mathematical description, in the same way as other more familiar abstract entities (like sets or real numbers) do. As a consequence, we do not try to reduce intensions to more or less complicated systems of extensional objects; a more general hypothesis seems to be deeply adequate with respect to many situations that arise in contemporary science, from computer science to physics.

What, precisely, is an intension? The intuitive idea is that intensions correspond to some intersubjective invariants with respect to the different systems of subjective knowledge. Generally, an intension may be

157

understood, even if the corresponding extension is not known, or is only partially known, or is simply undetermined. One can suppose that the behavior of these intensional objects is governed by a general intensional set theory; for instance, the theory presented in [5], which has some features in common with other theories proposed by Church [4], Feferman [7], Aczel [1], Parsons [10], Bealer and Moennich [2], and Cantini [3]. The aim of our intensional set theory is to describe at the same time *extensional objects* (e.g., sets, classes, individuals, and truth values) and *intensional objects* (e.g., concepts and propositions). Particular cases of propositions, individual concepts, and predicate concepts are represented (respectively) by truth values, individuals, and classes. Concepts and propositions may have an extension. When defined, the extension of a proposition is a truth value, the extension of an individual concept is either an individual or a class, and the extension of an *n*-ary predicate concept is a class of *n*-tuples. A proposition or a concept is called *extensional* when it is the extension of itself and *intensional* otherwise. Truth values, individuals, and classes are extensional.

In the particular case of epistemic logic, we suppose that the semantical analysis of an epistemic sentence like "*a* knows α" or "*a* believes α" concerns a relation that may hold between individuals and propositions. The basic idea is that the objects of our knowledge and belief are not sentences (which strictly depend on the contingent uses of language) but rather propositions.

On this basis, let us sketch a formal semantics for a weak form of epistemic logic, which is supposed to correspond to the following intuitive situation: The subjective agents accept classical logic; nevertheless, they are generally not able to grasp all the classical logical laws nor to decide all the classical logical consequences.

We will refer to a first-order language $\mathcal{L}$ with predicates P_i^n and individual constants a_i. The non-epistemic logical constants are the connectives *not* ($\neg$), *and* ($\wedge$), and the *universal quantifier* ($\forall$). The epistemic operators are the following: *to understand* (U), *to believe* (B), *to know* (K). We will use $x, y, z, \ldots$ as metavariables for individual variables, $t_1, t_2, \ldots$ as metavariables for individual terms, and $\alpha, \beta, \gamma, \ldots$ as metavariables for formulas. The non-epistemic formulas are defined in the usual way. Further, for any term t and any formula α, $Bt\alpha$ (t believes α) and $Kt\alpha$ (t knows α) are formulas; for all terms t and t', predicates P_i^n, and formulas α, Utt', UtP_i^n, $Ut\alpha$ (t understands t', t understands P^n, t understands α) are formulas.

The logical constants *disjunction* ($\vee$), *material conditional* ($\rightarrow$), and *existential quantifier* ($\exists$) are introduced in the usual classical way, via a metalinguistic definition. Note that the language allows nested occurrences of epistemic operators. For example, the following expression is a formula:

$$\exists x\, Kx \,\forall y\, ByPx$$

(somebody knows that everybody believes that he is a playboy).

Let us now define the notions of classical propositional algebra, of epistemic algebra, and of epistemic realization (or possible model) of the language $\mathcal{L}$.

Definition 1 (Classical propositional algebra). A *classical propositional algebra* is a structure

$$\Pi = \langle \mathcal{P}, {}^{\perp}, \sqcap, \sqcap, \mathcal{P}^0, \mathbf{W}, \mathbf{F}, \underline{1}, \underline{0} \rangle,$$

where:

(1.1) $\mathcal{P}$ represents a set of propositions (or thoughts).

(1.2) $\underline{1}, \underline{0}$ are privileged propositions representing (respectively) the truth values *truth* and *falsity*.

(1.3) $\mathbf{W}$ is the set of the true propositions and $\mathbf{F}$ is the set of the false propositions. Any proposition is either true or false and not at the same time true and false ($\mathbf{W} \cup \mathbf{F} = \mathcal{P}$, $\mathbf{W} \cap \mathbf{F} = \emptyset$). The truth is true ($\underline{1} \in \mathbf{W}$) and the falsity is false ($\underline{0} \in \mathbf{F}$).

(1.4) The negation operation $^{\perp}$ transforms any proposition X into a proposition $X^{\perp}$ satisfying the following conditions:

$$X \in \mathbf{W} \text{ iff } X^{\perp} \in \mathbf{F}$$

(a proposition is true iff its negation is false);

$$\underline{1}^{\perp} = \underline{0}$$

(the negation of the truth is the falsity); and

$$X^{\perp\perp} = X$$

(any proposition coincides with its double negation).

(1.5) The conjunction operation $\sqcap$ transforms any pair of propositions X, Y into a proposition $X \sqcap Y$, satisfying the condition

$$X \sqcap Y \in \mathbf{W} \text{ iff } X, Y \in \mathbf{W}$$

(a conjunction is true iff both members are true).

(1.6) $\mathcal{P}^0$ is a privileged family of sets of propositions. For any set of propositions $\mathcal{S}$ in $\mathcal{P}^0$, the infinitary conjunction $\sqcap \mathcal{S}$ is defined for $\mathcal{S}$ and transforms $\mathcal{S}$ into the proposition $\sqcap \mathcal{S}$, satisfying the condition

$$\sqcap \mathcal{S} \in \mathbf{W} \text{ iff } X \in \mathbf{W} \text{ for any } X \text{ in } \mathcal{S}$$

(an infinitary conjunction is true iff all the members are true). Furthermore,

159

$$\mathcal{S} \in \mathcal{P}^0 \Rightarrow [X^\perp / X \in \mathcal{S}] \in \mathcal{P}^0.$$

In any classical propositional algebra Π, finitary and infinitary disjunctions can be defined in the usual way (via de Morgan). The set of the truth values $\mathcal{P}^E = \{\underline{1}, \underline{0}\}$ will be called the set of the *extensional propositions;* $\mathcal{P}^I = \mathcal{P} - \mathcal{P}^E$ will represent the set of the properly *intensional propositions* (or *proper thoughts*).

Definition 2 (Epistemic structure). An *epistemic structure* is a structure

$$\mathcal{E} = \langle \Pi, \mathfrak{D}, \text{ref}, \mathfrak{A}, \mathfrak{M}, \mathfrak{B}, \mathfrak{K} \rangle,$$

where:

(2.1) Π is a classical propositional algebra.

(2.2) $\mathfrak{D}$ represents the set of the *individual concepts*. A possibly empty subset $\mathfrak{D}^E$ of $\mathfrak{D}$ represents the set of the *extensional individuals* (or simply *individuals*). The set $\mathfrak{D}^I = \mathfrak{D} - \mathfrak{D}^E$ will contain the *properly intensional individual concepts*.

(2.3) The *reference function,* ref, is a partial function whose arguments are in $\mathfrak{D}$ and whose values are in $\mathfrak{D}^E$. The reference of any extensional individual is the individual itself (for any $d \in \mathfrak{D}^E$, $\text{ref}(d) = d$).

(2.4) $\mathfrak{A}$ is the set of the *predicate concepts*. Any predicate concept in $\mathfrak{A}$ has an n-arity ($n \geq 1$). An n-ary predicate concept A^n is a function that transforms any n-tuple of individual concepts $d_1, \ldots, d_n$ into a proposition $A^n(d_1, \ldots, d_n)$. Let $\mathfrak{A}^n$ represent the set of all n-ary predicate concepts. We require that $\mathfrak{A}^2$ contain the identity concept Id, which satisfies the following condition:

$$\text{Id}(d_1, d_2) \in \mathbf{W} \text{ iff } d_1 = d_2 \text{ or } \text{ref}(d_1) = \text{ref}(d_2).$$

A predicate concept A^n is called *extensional* iff it is the characteristic function of a set of n-tuples of extensional individuals. In other words:

(a) For any $d_1, \ldots, d_n \in \mathfrak{D}$, $A^n(d_1, \ldots, d_n) = \underline{1}$ or $A^n(d_1, \ldots, d_n) = \underline{0}$.

(b) $A^n(d_1, \ldots, d_n) = \underline{1} \Rightarrow d_1, \ldots, d_n \in \mathfrak{D}^E$.

Let $\mathfrak{A}^E$ represent the set of all extensional predicate concepts. The set $\mathfrak{A}^I = \mathfrak{A} - \mathfrak{A}^E$ will represent the set of the properly intensional predicate concepts. The set $\mathcal{P} \cup \mathfrak{D} \cup \mathfrak{A}$ will be called the *universe* of $\mathcal{E}$; whereas the sets $\mathcal{P}^I \cup \mathfrak{D}^I \cup \mathfrak{A}^I$ and $\mathcal{P} \cup \mathfrak{D} \cup \mathfrak{A}^I$ will represent (respectively) the set of the *proper intensions* and the set of the *intelligible objects* of $\mathcal{E}$.

(2.5) $\mathfrak{U}$, the *understanding* operator, transforms any pair – consisting of an individual concept d and an intelligible object E – into a proposition $\mathfrak{U}(d, E)$ (d understands E).

(2.6) $\mathfrak{B}$ (the *belief* operator) and $\mathfrak{K}$ (the *knowledge* operator) transform any pair – consisting of an individual concept d and a proposition X – into a proposition $\mathfrak{B}(d, X)$ (d believes X) and $\mathfrak{K}(d, X)$ (d knows X), respectively.

(2.7) The notion of *part* of a proposition is inductively defined (in an obvious way). When an individual concept d is a part of a proposition X, we will write $X(d)$. Similarly to the syntactical case, we introduce a substitution operation: $X(d/d')$ will represent the proposition obtained from $X(d)$ by substituting the part d with the part d'.

(2.8) For any individual concept d, the domain of *comprehension* of d (und(d)), of *belief* of d (bel(d)), and of *knowledge* of d (knowl(d)) are defined as follows:

$$\text{und}(d) = [E \in \mathcal{P} \cup \mathfrak{D} \cup \mathfrak{C}^1 / \mathfrak{U}(d, E) \in \mathbf{W}];$$

$$\text{bel}(d) = [X \in \mathcal{P} / \mathfrak{B}(d, X) \in \mathbf{W}];$$

$$\text{knowl}(d) = [X \in \mathcal{P} / \mathfrak{B}(d, X) \in \mathbf{W}].$$

In other words, an intelligible object E belongs to the domain of comprehension of d iff the proposition "d understands E" is true. The other cases are interpreted similarly.

The following conditions are required for the understanding operator, for any $d \in \mathfrak{D}$:

(2.8.1) $\underline{1}, \underline{0} \in \text{und}(d)$.

(2.8.2) $A^n(d_1, \ldots, d_n) \in \text{und}(d) \Rightarrow A^n \in \text{und}(d)$ and $d_1 \in \text{und}(d)$ and ... and $d_n \in \text{und}(d)$.

(2.8.3) $X^\perp \in \text{und}(d) \Rightarrow X \in \text{und}(d)$.

(2.8.4) $X \sqcap Y \in \text{und}(d) \Rightarrow X, Y \in \text{und}(d)$.

(2.8.5) $\bigsqcap_{e \in \mathfrak{D}} \{X(e)\} \in \text{und}(d) \Rightarrow X(e') \in \text{und}(d)$ for at least one e'.

(2.8.6) $\bigsqcap_{e \in \mathfrak{D}} \{X(e)\} \in \text{und}(d)$ and $e' \in \text{und}(d) \Rightarrow X(e') \in \text{und}(d)$.

(2.8.7) $d_1 = \text{ref}(d_2) \Rightarrow \text{und}(d_1) = \text{und}(d_2)$.

In other words, anybody understands the privileged truth and the privileged falsity. In most cases, understanding a proposition implies understanding its parts. However, it seems realistic to admit the possibility of understanding a universal proposition like "all are mortal" ($\bigsqcap_{d \in \mathfrak{D}} X(d)$) without grasping each particular instance ("Socrates is mortal," "Plato is mortal," etc.). At the same time, understanding a universal proposition implies the understanding of at least one instance. Further, whoever understands a universal proposition like "All are mortal" and understands an individual concept such as "Socrates" understands also the instantiation "Socrates is mortal." Finally, an individual concept and its reference have the same domain of comprehension.

161

For the belief operator the following conditions are required, for any $d \in \mathfrak{D}$:

(2.8.8) $\underline{1} \in \mathrm{bel}(d)$; $\underline{0} \notin \mathrm{bel}(d)$.

(2.8.9) $\mathrm{bel}(d) \subseteq \mathrm{und}(d)$.

(2.8.10) $X \sqcap Y \in \mathrm{bel}(d) \Rightarrow X, Y \in \mathrm{bel}(d)$.

(2.8.11) $[X \in \mathrm{bel}(d)$ or $Y \in \mathrm{bel}(d)]$ and $X \sqcup Y \in \mathrm{und}(d) \Rightarrow X \sqcup Y \in \mathrm{bel}(d)$.

(2.8.12) $\bigcap_{e \in \mathfrak{D}}\{X(e)\} \in \mathrm{bel}(d)$ and $e' \in \mathrm{und}(d) \Rightarrow X(e') \in \mathrm{bel}(d)$.

(2.8.13) $X(e') \in \mathrm{bel}(d)$ and $\bigsqcup_{e \in \mathfrak{D}}\{X(e)\} \in \mathrm{und}(d) \Rightarrow \bigsqcup_{e \in \mathfrak{D}}\{X(e)\} \in \mathrm{bel}(d)$.

(2.8.14) $d_1 = \mathrm{ref}(d_2) \Rightarrow \mathrm{bel}(d_1) = \mathrm{bel}(d_2)$.

In other words, everybody believes the privileged truth, and does not believe the privileged falsity. Who believes a proposition, understands it. Believing a conjunction implies believing its members, but generally not the other way around. Whoever believes a universal proposition and understands a particular instance, believes that particular instantiation. Whoever understands a disjunction and believes at least one member, believes the disjunction, but generally not the other way around. Whoever understands an existential proposition and believes a particular instance, believes also the existential proposition.

Finally, for the knowledge operator we require:

(2.8.15) $\mathrm{knowl}(d) = \mathrm{bel}(d) \cap \mathbf{W}$.

This corresponds to a standard definition according to which a proposition is known iff it is at the same time true and believed.

On this basis, we can now introduce the notion of an epistemic realization (or epistemic possible model) of $\mathfrak{L}$.

Definition 3 (Epistemic realization). An *epistemic realization* of $\mathfrak{L}$ is a system

$$\mathfrak{M} = \langle \mathcal{E}, \mathrm{int} \rangle,$$

where

(3.1) $\mathcal{E}$ is an epistemic structure such that, for any proposition $X(d)$,

$$[X(d)/d \in \mathfrak{D}] \in \mathcal{P}^0.$$

(3.2) The interpretation function, int, associates an intensional meaning in $\mathcal{E}$ to the nonlogical constants of $\mathfrak{L}$, according to the following rules:

$$\mathrm{int}(a_i) \in \mathfrak{D};$$

$$\mathrm{int}(P_i^n) \in \mathcal{Q}^n;$$

$$\mathrm{int}(=) = \mathrm{Id}.$$

162

In other words, the intensional meanings of an individual constant, of an *n*-ary predicate, and of the identity predicate are (respectively) an individual concept, an *n*-ary predicate concept, and the identity concept.

The usual semantical machinery can be now introduced, mutatis mutandis.

Given an intensional realization $\mathfrak{M}$ of $\mathcal{L}$, an *interpretation* σ of the variables of $\mathcal{L}$ is a function whose arguments are variables of $\mathcal{L}$ and whose values are individual concepts in $\mathfrak{D}$. Any interpretation σ determines an intensional valuation int^{σ} of the terms and of the formulas of $\mathcal{L}$. The intensional valuation of a term t under $\sigma(\mathrm{int}^{\sigma}(t))$ will be an individual concept in $\mathfrak{D}$, whereas the intensional valuation of a formula α under $\sigma(\mathrm{int}^{\sigma}(\alpha))$ will be a proposition in $\mathcal{P}$.

Definition 4 (Intensional valuation of a term).
$$\mathrm{int}^{\sigma}(x) = \sigma(x);$$
$$\mathrm{int}^{\sigma}(a_i) = \mathrm{int}(a_i).$$

Definition 5 (Intensional valuation of a formula).
$$\mathrm{int}^{\sigma}(P^n t_1 \ldots t_n) = \mathrm{int}(P^n)(\mathrm{int}^{\sigma}(t_1), \ldots, \mathrm{int}^{\sigma}(t_n));$$
$$\mathrm{int}^{\sigma}(\neg\beta) = \mathrm{int}^{\sigma}(\beta)^{\perp};$$
$$\mathrm{int}^{\sigma}(\beta \wedge \gamma) = \mathrm{int}^{\sigma}(\beta) \sqcap \mathrm{int}^{\sigma}(\gamma);$$
$$\mathrm{int}^{\sigma}(\forall x \beta) = \begin{cases} \mathrm{int}^{\sigma}(\beta) & \text{if } \beta \text{ does not contain } x \text{ free,} \\ \prod_{d \in \mathfrak{D}}\{\mathrm{int}^{\sigma[\frac{x}{d}]}(\beta)\} & \text{otherwise,} \end{cases}$$

where $\sigma[\frac{x}{d}]$ is an interpretation of the variables such that
$$\begin{cases} \sigma[\frac{x}{d}](x) = d \\ \sigma[\frac{x}{d}](y) = \sigma(y) \end{cases}$$

for any variable y different from x. Further:
$$\mathrm{int}^{\sigma}(Utt') = \mathfrak{U}(\mathrm{int}^{\sigma}(t), \mathrm{int}^{\sigma}(t'));$$
$$\mathrm{int}^{\sigma}(UtP^n) = \mathfrak{U}(\mathrm{int}^{\sigma}(t), \mathrm{int}(P^n));$$
$$\mathrm{int}^{\sigma}(Ut\alpha) = \mathfrak{U}(\mathrm{int}^{\sigma}(t), \mathrm{int}^{\sigma}(\alpha));$$
$$\mathrm{int}^{\sigma}(Bt\alpha) = \mathfrak{B}(\mathrm{int}^{\sigma}(t), \mathrm{int}^{\sigma}(\alpha));$$
$$\mathrm{int}^{\sigma}(Kt\alpha) = \mathfrak{K}(\mathrm{int}^{\sigma}(t), \mathrm{int}^{\sigma}(\alpha)).$$

On this basis, one can define the extensional valuation of a formula α under a given interpretation σ, $\mathrm{ext}^{\sigma}(\alpha)$. For any α, $\mathrm{ext}^{\sigma}(\alpha)$ is a truth value;

the truth if the intensional valuation of α under σ is a true proposition, the falsity otherwise.

Definition 6 (Extensional valuation).

$$\text{ext}^\sigma(\alpha) = \begin{cases} \underline{1} & \text{if } \text{int}^\sigma(\alpha) \in \mathbf{W}, \\ \underline{0} & \text{otherwise.} \end{cases}$$

As expected, one can prove a coincidence lemma. This justifies the following definition of intension (extension) of a sentence α:

$$\text{int}(\alpha) = \text{int}^\sigma(\alpha) \text{ and } \text{ext}(\alpha) = \text{ext}^\sigma(\alpha),$$
where σ is any interpretation of the variables.

Definition 7 (Satisfaction and truth of a formula α in $\mathfrak{M}$).
(7.1) An interpretation σ satisfies a formula α in $\mathfrak{M}$ ($\sigma \models_{\mathfrak{M}} \alpha$) iff $\text{ext}^\sigma(\alpha) = \underline{1}$.
(7.2) α is true in $\mathfrak{M}$ ($\models_{\mathfrak{M}} \alpha$) iff, for any interpretation σ, $\sigma \models_{\mathfrak{M}} \alpha$.

One can also define, in a natural way, the notion of extension of an individual constant, of a predicate, and of a formula.

Definition 8.
(8.1) The extension $\text{ext}(a_i)$ of an individual constant a_i is not necessarily defined. If defined, it coincides with the reference of the intension of a_i:

$$\text{ext}(a_i) \approx \text{ref}(\text{int}(a_i)).$$

(8.2) The extension $\text{ext}(\alpha(x_1, \ldots, x_n))$ of a formula $\alpha(x_1, \ldots, x_n)$ (containing exactly n free variables) is the set of all n-tuples of extensional individuals that satisfy the formula

$$\text{ext}(\alpha(x_1, \ldots, x_n)) = [\langle d_1, \ldots, d_n \rangle / d_1 \in \mathfrak{D}^E \text{ and } \ldots \text{ and } d_n \in \mathfrak{D}^E$$
$$\text{and, for any } \sigma, \sigma[{}^{x_1}_{d_1} \cdots {}^{x_n}_{d_n}] \models_{\mathfrak{M}} \alpha(x_1, \ldots, x_n)]$$

(8.3) The extension of a predicate P^n is the extension of the formula $P^n(x_1, \ldots, x_n)$.

Finally, one can define, in the expected way, the notions of epistemic validity and of epistemic logical consequence.

Definition 9.
(9.1) A formula α is *epistemically valid* iff α is true in any epistemic realization.

(9.2) A formula α is an *epistemic logical consequence* of a set of formulas T iff, for any epistemic realization $\mathfrak{M}$ and any interpretation σ,

$$[\text{for any } \beta \in T : \sigma \vDash_{\mathfrak{M}} \beta] \Rightarrow \sigma \vDash_{\mathfrak{M}} \alpha.$$

In conclusion, let us sum up some characteristic features of our epistemic semantics.

(1) Logical omniscience is avoided: Knowledge and belief systems are not generally closed under logical consequence. The following situation is possible:

$$\vDash_{\mathfrak{M}} Ka\alpha; \ \beta \text{ is a logical consequence of } \alpha; \ \nvDash_{\mathfrak{M}} Ka\beta.$$

(2) Truths as well as knowledge and belief systems do not generally depend on extensions. The following situation is possible:

$$\vDash_{\mathfrak{M}} Pa; \ \vDash_{\mathfrak{M}} KbPa; \text{ the extension of } a \text{ is not determined.}$$

(3) Identity is interpreted in an extensional sense. A sentence like "Hesperus = Phosphorus" is true iff the two names have the same reference.

(4) Substitutivity for identity breaks down. The following situation is possible:

$$\vDash_{\mathfrak{M}} a = b; \ \vDash_{\mathfrak{M}} KcPa; \ \nvDash_{\mathfrak{M}} KcPb.$$

(5) *De dicto–de re* epistemic modalities are distinguished. The following situation is possible:

$$\vDash_{\mathfrak{M}} Ka \exists x \, Px; \ \nvDash_{\mathfrak{M}} \exists x \, KaPx.$$

(6) Knowledge does not generally imply knowledge of knowledge. It is possible that

$$\vDash_{\mathfrak{M}} Ka\alpha \text{ and } \nvDash_{\mathfrak{M}} KaKa\alpha.$$

(7) A form of common knowledge can be reasonably defined and turns out to be semantically possible. Let us first introduce the notion of a *universal epistemic chain* – that is, a denumerable sequence $\alpha_1 \ldots \alpha_n \ldots$ of sentences such that:

$$\begin{cases} \alpha_1 \text{ is a non-epistemic sentence;} \\ \alpha_{n+1} = \forall x \, Kx\alpha_n. \end{cases}$$

A sentence α is called *common knowledge* in a realization $\mathfrak{M}$ iff α is the first element of a universal epistemic chain and all elements of the chain are true in $\mathfrak{M}$.

(8) Intensions remain constant through iterated applications of epistemic operators. For instance, α receives the same intensional interpretation

165

in the case of both sentences $Ka\alpha$ and $KbKa\alpha$. As a consequence any multiplication of semantical entities is avoided, in accordance with Dummett's revision of Frege's epistemic modalities.

This epistemic logic can be axiomatized in a natural way. One also can prove a soundness and a completeness theorem with respect to our semantics.

REFERENCES

[1]   Aczel, P. (1988), "Algebraic Semantics for Intensional Logics I." In G. Chierchia, B. Partee, and R. Turner (eds.), *Property Theories, Type Theories, and Semantics.* Dordrecht: Reidel.
[2]   Bealer, G., and Moennich, U. (1989), "Property Theories." In D. Gabbay and F. Guenther (eds.), *Handbook of Philosophical Logic,* vol. IV. Dordrecht: Reidel.
[3]   Cantini, A. (1983), *Proprietà e Operazioni.* Naples: Bibliopolis. (English version to appear, North Holland.)
[4]   Church, A. (1951), "A Formulation of the Logic of Sense and Denotation." In P. Henle, H. H. Kallen, and S. K. Langer (eds.), *Structure, Method, and Meaning.* New York: The Liberal Arts Press.
[5]   Dalla Chiara, M. L. (1987), "An Approach to Intensional Semantics." *Synthese* 73: 479–96.
[6]   Feferman, S. (1984), "Toward Useful Type-Free Theories I." *Journal of Symbolic Logic* 49: 75–111.
[7]   Feferman, S. (1985), "Intensionality in Mathematics." *Journal of Philosophical Logic* 14: 41–55.
[8]   Hardegree, G. M. (1976), "The Conditional in Quantum Logic." In P. A. Suppes (ed.), *Logic and Probability in Quantum Mechanics.* Dordrecht: Reidel.
[9]   Lewis, D. (1973), *Counterfactuals.* London: Basil Blackwell.
[10] Parsons, C. (1982), "Intensional Logic in Extensional Language." *Journal of Symbolic Logic* 47: 289–328.
[11] Stalnaker, R. (1968), "A Theory of Conditionals." In N. Rescher (ed.), *Studies in Logical Theory.* London: Basil Blackwell.

166

11

Semantical aspects of quantified modal logic

GIOVANNA CORSI & SILVIO GHILARDI

This chapter is designed to outline some techniques, results, and new trends in quantified modal logics. Since it is directed to readers not specialized in the field of modal logic, we will start "from the beginning" and so discuss some basic material; at the same time, we intend to offer an idea of some of the recent research in the area and of possible directions for future research.

Quantified modal logics contain – in addition to classical connectives and quantifiers – a unary operator, the "box" operator $\Box$, whose meaning can be variously interpreted depending on the context and on the applications. Here is a list of possible readings of $\Box A$ (taken from [15]):

It is necessarily true that A;
It will always be true that A;
It ought to be that A;
It is known that A;
It is believed that A;
It is provable in Peano arithmetic that A;
After the program terminates, A.

A natural semantic demand is that the truth value of a sentence such as $\Box A$ be determined (according to some of the above readings) once the truth value of A is known in a suitable set of "instants of time" or "states of affairs" that are considered as alternative to the actual one. Consequently, a semantics for quantified modal logic should contain a mathematical formalization of the following entities: (a) the different "worlds"; (b) the relation of "being conceivable as an alternative"; and (c) the objects or individuals "existing" in them and the connections between these. Such a framework gives rise to several philosophical discussions and also to interesting mathematical developments; we will not touch upon the former aspects (there is a large literature – see [3], [4], [6], and [9]), but will concentrate on some of the latter topics. In particular, we here offer

167

a basic exposition of Kripke-style semantics (Section 1), investigate some problems connected with completeness and interpolation (Sections 2–7), and present some recent generalizations of Kripke semantics related to categorical notions (Sections 8–11) that seem particularly worthy of attention.

The modal language $\mathcal{L}$ we consider contains, as usual, the logical constants $\rightarrow$ (if... then), $\forall$ (for all), $\Box$ (necessarily); the symbol of falsehood $\bot$; individual constants $c_1, c_2, \ldots$ and individual variables $x_1, x_2, \ldots$; and n-ary predicate letters P_i^n ($1 \leq i < \omega$, $0 \leq n < \omega$). Individual constants and variables are *terms*. The notion of (well-formed) *formula, wff*, is defined as usual and a *sentence* is a formula containing no free variables. For brevity we may indicate lists of terms by the notation $\vec{t}, \vec{u}, \ldots$ (possibly with indices). The substitution of a term u for a term t in a formula A is indicated by $A(t/u)$. $\neg A \overset{\text{df}}{=} A \rightarrow \bot$, $A \vee B \overset{\text{df}}{=} \neg A \rightarrow B$, $A \wedge B \overset{\text{df}}{=} \neg(A \rightarrow \neg B)$, and $\Diamond A \overset{\text{df}}{=} \neg \Box \neg A$. By $\mathcal{L}^=$ we denote the language $\mathcal{L}$ plus the identity relation. As to the semantics for $\mathcal{L}$ ($\mathcal{L}^=$), we will present several types of semantics, starting with the Kripke semantics (K-semantics).

1. KRIPKE SEMANTICS

Kripke semantics was the first semantics to be introduced for quantified modal logic; its attraction is due to the extreme simplicity with which it fulfills requirements (a), (b), and (c).

A *Kripke model* (*K-model*) $\mathfrak{M}$ for $\mathcal{L}^=$ is a quadruple $\langle W, R, D, \mathfrak{I} \rangle$, where

(1) $W \neq \emptyset$ (W is the set of "worlds");
(2) $R \subseteq W^2$ (R is the relation "being conceivable as an alternative");
(3) D is a function associating with every $w \in W$ a nonempty set D_w such that if wRv then $D_w \subseteq D_v$ (D_w is the set of the individuals "existing" at w); and
(4) $\mathfrak{I}$ is an interpretation function such that for every $w \in W$ and n-ary predicate P_j^n, $\mathfrak{I}_w(P_j^n) \subseteq (D_w)^n$ and $\mathfrak{I}_w(=) = \{\langle d, d \rangle \mid d \in D_w\}$; and such that for every individual constant c, $\mathfrak{I}_w(c) \in D_w$ and if wRv then $\mathfrak{I}_w(c) = \mathfrak{I}_v(c)$.

$\langle W, R \rangle$ ($\langle W, R, D \rangle$) is said to be the *frame* (the *extended frame*) on which $\mathfrak{M}$ is based. Let $w \in W$: μ is said to be a *w-assignment* if μ is a function whose domain is the set of terms of $\mathcal{L}$, term($\mathcal{L}$), whose codomain is D_w and $\mu(c_i) = \mathfrak{I}(c_i)$. If μ is a w-assignment and $d \in D_w$, by $\mu^{(x_i/d)}$ we denote the w-assignment so defined; if $y \neq x_i$ then $\mu^{(x_i/d)}(y) = \mu(y)$, and if $y = x_i$ then $\mu^{(x_i/d)}(y) = d$. Note that if μ is a w-assignment and wRv then μ is a v-assignment.

Given a model $\mathfrak{M} = \langle W, R, D, \mathfrak{I} \rangle$ and a w-assignment μ, the notion of a wff being true in $\mathfrak{M}$ at w under μ, $\mathfrak{M}^\mu \vDash_w A$, is defined as follows (where "iff" and "s.t." stand, respectively, for "if and only if" and "such that"):

$$\mathfrak{M}^\mu \vDash_w P_j^n(t_1, \ldots, t_n) \quad \text{iff} \quad \langle \mu(t_1), \ldots, \mu(t_n) \rangle \in \mathfrak{I}_w(P_j^n);$$

$$\mathfrak{M}^\mu \nvDash_w \bot;$$

$$\mathfrak{M}^\mu \vDash_w A_1 \to A_2 \qquad \text{iff if } \mathfrak{M}^\mu \vDash_w A_1 \text{ then } \mathfrak{M}^\mu \vDash_w A_2;$$

$$\mathfrak{M}^\mu \vDash_w \forall x_i A \qquad \text{iff for all } d \in D_w, \; \mathfrak{M}^{\mu(x_i/d)} \vDash_w A;$$

$$\mathfrak{M}^\mu \vDash_w \Box A \qquad \text{iff for all } v \text{ s.t. } wRv, \; \mathfrak{M}^\mu \vDash_v A.$$

We say that

A is true at w in $\mathfrak{M}$, $\mathfrak{M} \vDash_w A$, iff for all w-assignments μ, $\mathfrak{M}^\mu \vDash_w A$;
A is true in $\mathfrak{M}$, $\mathfrak{M} \vDash A$, iff for all $w \in W$, $\mathfrak{M} \vDash_w A$;
A is valid on the extended frame $\mathfrak{F}$, $\mathfrak{F} \vDash A$, iff for all models $\mathfrak{M}$ based on $\mathfrak{F}$, $\mathfrak{M} \vDash A$;
A is valid on a class $\mathbf{H}$ of extended frames, $\mathbf{H} \vDash A$, iff for all $\mathfrak{F} \in \mathbf{H}$, $\mathfrak{F} \vDash A$; and
A is valid iff for all extended frames $\mathfrak{F}$, $\mathfrak{F} \vDash A$.

Because of the condition on the domains – that is, wRv only if $D_w \subseteq D_v$ – the semantics and models just introduced are called "nested domains semantics" and "nested domains models" respectively; if $D_w = D_v$ for all $w, v \in W$, then we speak of "constant domains semantics" and "constant domains models." Main features of the K-semantics are the following:

(i) the classical laws of quantification are valid;
(ii) $\Box \forall x_i A \to \forall x_i \Box A$ is valid;
(iii) if A is true in a model $\mathfrak{M}$ then $\Box A$ is true in $\mathfrak{M}$;
(iv) $t = s \to \Box(t = s)$ is valid; $t \neq s \to \Box(t \neq s)$ is also valid, because of the definition of $\mathfrak{I}$ (it would not be valid if identity were interpreted, for instance, on a "growing congruence"); and
(v) $\forall x_i \Box A \to \Box \forall x_i A$ is true in every model with constant domains; $\forall x_i \Box A \to \Box \forall x_i A$ is called the *Barcan formula, BF.*

Notice, moreover, that if μ is not a w-assignment then $\mathfrak{M}^\mu \vDash_w A$ is meaningless; that is, we cannot "talk" at w of individuals that do not exist at w.

Several versions of Kripke-style semantics have been proposed in the literature since Kripke [19] presented the first, and each is intended to capture some philosophical insight on what "possible worlds" or "possible individuals" are or ought to be. A different presentation of the semantics adopted here is the truth-value gaps semantics of Hughes and Cresswell [17]. The main differences are that in [17]: (i) the assignments of the terms are not relativized to the elements of W; that is, an assignment is a map μ with codomain $\mathbf{U} = \bigcup_{w \in W} D_w$; and (ii) if $\mu(x) \notin D_v$ then $\mathfrak{M}^\mu \models_w A$ is "undefined." But these differences are not fundamental, insofar as the notion of "model" is essentially the same and, given a K-model $\mathfrak{M}$ and a wff A, A is true in $\mathfrak{M}$ according to our definition iff it is true in the Hughes–Cresswell semantics.

The classical alternative to nested domains semantics is the "varying domains" semantics introduced by Kripke in [19] and reelaborated by Fine in [3]. The differences with respect to the K-semantics are that:

(1) the function D is not subject to any restriction at all, so that D_w may even be empty (D_w is intended, as before, as the set of individuals existing at w and coincides with the domain of variation of the quantifiers);

(2) $\mathfrak{I}_w(P_j^n) \subseteq \mathbf{U}^n$, $\mathfrak{I}_w(c_i) \in \mathbf{U}$; and

(3) an assignment is a map from term($\mathfrak{L}$) into $\mathbf{U}$.

Thus each world has an "inner" domain D_w and an "outer" domain $\mathbf{U}$ (which is, of course, the same for every world). Main features of this semantics are:

(i) the quantificational laws of classical logic are not valid – in particular, $\forall x_i A(x_i) \rightarrow A(x_i)$ is not valid;

(ii) the quantificational laws of free logic are valid;

(iii) $\Box \forall x_i A \rightarrow \forall x_i \Box A$ is not valid; and

(iv) $\forall x_i \Box A \rightarrow \Box \forall x_i A$ is not valid.

This semantics represents a generalization with respect to the nested domain semantics, and has great appeal from a philosophical point of view; still, it has the consequence that in every world it is possible "to talk with truth or falsity" of every object of the universe, even of those objects existing in worlds not related to the one under consideration. We believe that a varying domain semantics and a semantics that admits empty domains are better dealt with in the semantics presented in Section 9.

2. QUANTIFIED MODAL LOGIC

The basic quantified modal logic (or system) we consider is $Q - K$, that is, the classical quantified extension of the minimum normal propositional logic K; it is determined by the following axiom schemata and rules:

1. propositional tautologies;

2. $\Box(A \rightarrow B) \rightarrow (\Box A \rightarrow \Box B)$;

3. $\forall x_i A \rightarrow A(x_i/t)$, where t is a term free for x_i in A; and

4. the rules modus ponens (MP), necessitation (N), and universal generalization (UG):

$$\text{(MP)} \ \frac{A \ \ A \rightarrow B}{B}, \quad \text{(N)} \ \frac{A}{\Box A}, \quad \text{(UG)} \ \frac{B \rightarrow A}{B \rightarrow \forall x_i A},$$

where x_i is not free in B.

$Q - K^=$ is the system $Q - K$ plus the following axiom schemata:

5. $t = t$;
6. $t = s \rightarrow (A(x_i/t) \rightarrow A(x_i/s))$; and
7. $t \neq s \rightarrow \Box(t \neq s)$.

In general, by a quantified modal logic we intend a set of wffs of $\mathcal{L}$ (or $\mathcal{L}^=$) that contains all the theorems of $Q - K$ ($Q - K^=$) and is closed under uniform substitution and under the rules (MP), (N), and (UG). Notice that, given a class $\mathbf{H}$ of extended frames, the set of wffs valid on $\mathbf{H}$ is a quantified modal logic. Given a propositional modal logic L, by $Q - L$ we denote the quantified modal logic obtained by adding to $Q - K$ the axiom schemata corresponding to the propositional axioms of L, and by $X_1. \cdots . X_n Q - L. Y_1. \cdots . Y_m$ we denote the logic obtained by adding to $Q - L$ the quantificational axiom schemata $X_1, ..., X_n$ and the propositional axiom schemata $Y_1, ..., Y_m$.

One of the main problems connected with quantified modal logics is the characterization problem. Given a quantified modal logic L, is there a class of extended frames $\mathbf{H}$ such that the theorems of L are just the formulas valid on $\mathbf{H}$? This problem splits naturally into two parts: Is there a class $\mathbf{H}$ of extended frames such that (i) (validity or consistency part) the theorems of L are valid on every $\mathfrak{F} \in \mathbf{H}$? and (ii) (completeness part) every nontheorem of L is nonvalid on some $\mathfrak{F} \in \mathbf{H}$? If the answer to the first question is "yes" then L is said to be *valid* with respect to $\mathbf{H}$; if "yes" to the second question then L is said to be *complete* w.r.t $\mathbf{H}$. If the answer to both questions is "yes" then L is said to be *characterized by* $\mathbf{H}$. The difficult part is completeness. Given that an extended frame $\mathfrak{F}$ is said to be a *frame for L* iff all the theorems of L are valid on $\mathfrak{F}$, the characterization problem can be reformulated as follows: Given a nontheorem A of L, is there a frame for L that falsifies A?

In the simplest cases, the class of extended frames for a logic is determined by a first-order condition on the relation R, for example, the class of frames $\mathfrak{F} = \langle W, R, D \rangle$ where R is reflexive (or R is transitive, etc.).

All the notions mentioned here are relative to the Kripke semantics just introduced, so to be more precise we should talk of K-frames, K-completeness, K-validity, and so on; this specification will turn out to be important in the sequel. For the moment we use the expressions frame, validity, completeness, and K-frame, K-validity, and K-completeness as interchangeable. If we limit ourselves to propositional modal logic, then there are some beautiful (even if not many) general results on completeness, such as Sahlqvist's theorem. With quantified modal logics, general completeness theorems are very rare, and even completeness or incompleteness theorems for particular systems are often difficult to find. In particular, we will see that there is no regular connection, for a propositional modal logic L, between the completeness property of its quantified extensions $Q-L$ and $BF.Q-L$. In the following, we will concentrate only on very simple systems, being more interested in the methods than in particular results. We will describe three ways of proving completeness theorems: Henkin's method, the subordination method, and the diagram method. For the sake of simplicity, we will outline these methods for logics not containing the identity symbol.

3. Preliminary definitions

We need to expand the original language $\mathcal{L}$ with some countable set of new constants. New languages obtained in this way are indicated with the letters $\mathcal{L}_1, \mathcal{L}_2, \ldots$. As to derivability, it is clear that syntactic proofs are relative to a language and to a logic formulated in it, so we write (for instance) $\vdash_{\langle \mathcal{L}_1, L \rangle} A$ in order to express the fact that the wff A of $\mathcal{L}_1$ is provable in the system $\langle \mathcal{L}_1, L \rangle$ axiomatized by means of a set of axioms for L in $\mathcal{L}_1$. Let Ψ be a set of sentences of $\mathcal{L}_1 = \mathcal{L} + C$. Then:

$\Psi \vdash_{\langle \mathcal{L}_1, L \rangle} A$ iff there are wffs $B_1, \ldots, B_n$ in Ψ s.t. $\vdash_{\langle \mathcal{L}_1, L \rangle} (B_1 \wedge \cdots \wedge B_n) \to A$;
Ψ is $\langle \mathcal{L}_1, L \rangle$-*deductively closed* iff for every sentence A, if $\Psi \vdash_{\langle \mathcal{L}_1, L \rangle} A$ then $A \in \Psi$;
Ψ is $\vdash_{\langle \mathcal{L}_1, L \rangle}$-*consistent* iff $\Psi \nvdash_{\langle \mathcal{L}_1, L \rangle} \perp$;
Ψ is $\mathcal{L}_1$-*maximal* iff for all sentences A of $\mathcal{L}_1$, $A \in \Psi$ or $\neg A \in \Psi$;
Ψ is $\mathcal{L}_1$-*inductive* iff if $\Psi \vdash A(c)$ for all c in $\mathcal{L}_1$, then $\Psi \vdash \forall x A(x)$; and
Ψ is $\langle \mathcal{L}_1, L \rangle$-*saturated* iff Ψ is $\langle \mathcal{L}_1, L \rangle$-deductively closed, $\langle \mathcal{L}_1, L \rangle$-consistent, $\mathcal{L}_1$-maximal, and $\mathcal{L}_1$-inductive.

Here is a list of formulas that we consider in the sequel:

$D: \Box A \to \Diamond A$; $T: \Box A \to A$;

$B: A \to \Box \Diamond A$; $E: \Diamond A \to \Box \Diamond A$;

$1: \Box \Diamond A \to \Diamond \Box A$; $2: \Diamond \Box A \to \Box \Diamond A$;

$3: \Box(A \wedge \Box A \to B) \vee \Box(B \wedge \Box B \to A)$; $4: \Box A \to \Box \Box A$;

$Grz: \Box(\Box(A \to \Box A) \to A) \to \Box A$; $X: \Box \Box A \to \Box A$;

Alt_n: $\Box A_1 \vee \Box(A_1 \to A_2) \vee \cdots \vee \Box(A_1 \wedge \cdots \wedge A_n \to A_{n+1})$;

ML^*: $\Box \forall x \Diamond \Box A(x) \to \Diamond \Box \forall x A(x)$.

4. Henkin's method

This is the classical method for proving the completeness theorem for classical logic. It has been generalized to the modal case by Scott and Lemmon (see [22]).

Lemma 4.1 (Henkin). *Let $L \supseteq Q - K$ and let Ψ be an $\langle \mathcal{L}_1, L \rangle$-consistent set of sentences. Then there is a set of sentences Ω such that $\Psi \subseteq \Omega$ and Ω is $\langle \mathcal{L}_2, L \rangle$-saturated, where $\mathcal{L}_2 = \mathcal{L}_1 + C$ for some countable set C of new constants.*

Definition 4.2. Let C be a countable set of new constants[1] and let $L \supseteq Q - K$. The canonical model $\mathfrak{M}_L = \langle W, R, D, \mathcal{I} \rangle$ for L is defined as follows:
  (i) W contains all the pairs $w = \langle \Psi_w, C_w \rangle$, where
      (i1) $C_w \subset C$ and $C \backslash C_w$ is countable, and
      (i2) Ψ_w is a set of sentences of $\mathcal{L}_w = \mathcal{L} - C_w$ that is $\langle \mathcal{L}_w, L \rangle$-saturated
           (in the following, for brevity, Ψ_w is indicated simply by w);
  (ii) wRv iff $\Box^-(w) \subseteq v$, where $\Box^-(w) = \{A \mid \Box A \in w\}$;
  (iii) $D_w = \text{Const}(\mathcal{L}_w)$, the set of constants of $\mathcal{L}_w$; and
  (iv) $\mathcal{I}_w(c) = c$ and $\mathcal{I}_w(P_j^n) = \{\langle c_1, \ldots, c_n \rangle \mid P_j^n(c_1, \ldots, c_n) \in w\}$.

Lemma 4.3 (Canonical model). *Let $\mathfrak{M}_L = \langle W, R, D, \mathcal{I} \rangle$ be the canonical model for L. Then, for any wff A of $\mathcal{L}$ and for any w-assignment μ,*

$$\mathfrak{M}_L^\mu \models_w A(x_1, \ldots, x_n) \quad \textit{iff} \quad A(x_1/\mu(x_1), \ldots, x_n/\mu(x_n)) \in w,$$

where all the variables occurring free in A are among $x_1, \ldots, x_n$.

The crucial step in the proof of this lemma is to show that, if $\Box A$ is not in w, then there is a v such that A is not in v and $\Box^- w$ is contained in v. Because $w \in W$, $\Box^-(w) \cup \{\neg A\}$ is $\langle \mathcal{L}_w, L \rangle$-consistent and Henkin's lemma guarantees the existence of the desired v (the countable set of new constants required in the Henkin lemma is chosen in such a way that its union with C_w still has a countable complement in C). From the lemma of the canonical model, the following is immediate.

Theorem 4.4. *If $Q - K \subseteq L$ and $L \nvdash A$ then $\mathfrak{M}_L \nvDash A$, where $\mathfrak{M}_L$ is the canonical model for L.*

Notice that Theorem 4.4 holds for every extension of $Q-K$ and provides a model that is *generic,* in the sense that it falsifies "in one go" every non-theorem of L. Now, in order to prove that a logic L is complete, it is sufficient to show that the frame of the canonical model is a frame for L. Since any extended frame is a frame for $Q-K$, the validity and completeness of $Q-K$ with respect to the class consisting of *all* the extended frames follows. As an example of another system, consider $Q-K.4$. Because any transitive frame is a frame for $Q-K.4$, it will suffice to show that the frame of the canonical model for $Q-K.4$ is transitive – that is, that if wRv and vRz then wRz, which means that if $\Box^-(w) \subseteq v$ and $\Box^-(v) \subseteq z$ then $\Box^-(w) \subseteq z$. Now if $A \in \Box^-(w)$ then $\Box A \in w$ and so, thanks to Axiom 4 and the fact that w is $\langle \mathcal{L}_w, Q-K.4 \rangle$-saturated, $\Box\Box A \in w$, whence from the hypotheses follows that $\Box A \in v$ and $A \in z$.

Theorems 4.5 (Completeness).
$Q-K$ *is complete w.r.t the class of all frames with nested domains;*
$Q-K.D$ *is complete w.r.t. the class of all serial (i.e., such that $\forall w \exists v$*
wRv) frames with nested domains;
$Q-K.T$ *is complete w.r.t. the class of all reflexive frames with nested domains;*
$Q-K.4$ *is complete w.r.t. the class of all transitive frames with nested domains; and*
$Q-S4 (=Q-K.T.4)$ *is complete w.r.t. the class of all reflexive and transitive frames with nested domains.*

Henkin's method is easily adaptable for systems containing *BF* as a theorem. As is well-known, *BF* imposes the condition on the extended frames that the domains are constant; that is, $\langle W, R, D \rangle \models BF$ iff D is a constant function. Let $L \vdash BF$. The *canonical model with constant domains* $\mathfrak{M}_L = \langle W, R, D \rangle$ for L is defined as in Definition 4.2, except that (i1) is replaced by
(i1') $C_w = \mathbf{C}$ for all w.
Given $w \in W$, by making use of *BF* it is easy to prove that $\Box^-(w)$ is $\mathcal{L}_w$-inductive; this is all we need, because any $\mathcal{L}_w$-inductive set of sentences can be extended to one that is $\mathcal{L}_w$-saturated.

Theorems 4.6 (More completeness).
$BF.Q-K$ *is complete w.r.t. the class of all frames with constant domains;*
$BF.Q-K.D$ *is complete w.r.t. the class of all serial frames with constant domains;*
$BF.Q-K.T$ *is complete w.r.t. the class of all reflexive frames with constant domains;*

174

BF.Q−K.4 is complete w.r.t. the class of all transitive frames with constant domains;

BF.Q−S4 (= BF.Q−K.T.4) is complete w.r.t. the class of all reflexive and transitive frames with constant domains;

Q−K.B is complete w.r.t. the class of all symmetric frames with constant domains;[2]

Q−S5 (Q−K.T.E = Q−K.T.4.B) is complete w.r.t. the class of all symmetric, reflexive, and transitive frames with constant domains; and

BF.Q−S4.3 is complete w.r.t. the class of all reflexive, symmetric, transitive, and connected frames with constant domains.

The canonical model technique fails for systems like $Q−S4.2$, where – in order to show that the canonical model is based on a frame for the system – we need to show that, given two points w and v such that zRw and zRv for some z, there exists a u such that wRu and vRu. To this aim we need to know that $\Box^-(w) \cup \Box^-(v)$ is $\langle \mathcal{L}_1, Q−S4.2 \rangle$-consistent, where $\mathcal{L}_1 = \mathcal{L} + C_w + C_v$. In order to prove this, it is essential to have precise information about the constants occurring in w and v. The completeness of $Q−S4.2$ can be proved by the subordination method. This method also provides refinements of most of the completeness theorems (like $Q−S4$) quoted in the present section, in the sense that the logic is shown to be complete with respect to a proper subclass of the class of reflexive and transitive frames, and in particular to the class consisting of just the subordination frame.

5. THE SUBORDINATION METHOD

The subordination method [18] gives us a procedure for building models for a logic L by using pairs $\langle w, T_w \rangle$, where w ranges over a suitable set of indices and T_w is a $\langle \mathcal{L}_w, L \rangle$-saturated set of wffs (for an appropriately chosen language $\mathcal{L}_w$). Contrary to what happens in the canonical model, it might well be that the same set of wffs is associated with two or more distinct indices; moreover, the subordination model does not falsify "in one go" all nontheorems of L.

In general, a subordination model is built up in two steps: First we choose a set of indices W ordered by a binary relation R such that $\langle W, R \rangle$ is a frame for the propositional part of the logic under examination; then we define a function T on W that associates to every $w \in W$ an $\langle \mathcal{L}_w, L \rangle$-saturated set of sentences. The obvious conditions to be satisfied are that if wRv then $\Box^-(T_w) \subseteq T_v$, and that if $\Diamond A \in T_w$ then there is a v such that wRv and $A \in T_v$. A *privileged* set of indices is the set N^* of finite sequences of natural numbers ordered by the following relation S:

175

$$vSw \quad \text{iff} \quad v = v * \langle n \rangle \text{ for some } n \in N,$$

where $*$ is the operation of concatenation. The frame $Z = \langle N^*, S \rangle$ is called the *canonical subordination* frame.

Definition 5.1. Let Ψ be an $\langle \mathcal{L}_1, L \rangle$-saturated set of sentences, where $\mathcal{L}_1 = \mathcal{L} + C$ for a set C of new constants. A canonical subordination model with nested domains for Ψ is a triple $\langle N^*, S, T \rangle$ such that $\langle N^*, S \rangle$ is the canonical subordination frame and T is a function assigning to each $w \in N^*$ an $\langle \mathcal{L}_w, L \rangle$-saturated set T_w of sentences satisfying the following conditions:

(i) for all $w \in N^*$, $\mathcal{L}_w = \mathcal{L}_1 + C_w$ for some set C_w of constants;
(ii) $\mathcal{L}_{\langle \rangle} = \mathcal{L}_1$ and $T_{\langle \rangle} = \Psi$;
(iii) for all $v \in N^*$ and $n \in N$, $\mathcal{L}_v \subseteq \mathcal{L}_{v * \langle n \rangle}$ and $\Box^-(T_v) \subseteq T_{v * \langle n \rangle}$; and
(iv) for all $v \in N^*$, if $\Diamond A \in T_v$ then there is an $n \in N$ such that $A \in T_{v * \langle n \rangle}$.

Lemma 5.2. *If $L \supseteq Q - K.D$, then there is a canonical subordination model with nested domains for every $\langle \mathcal{L}_1, L \rangle$-saturated set of sentences Ψ of $\mathcal{L}_1 = \mathcal{L} + C$, where C is a set of new constants.*

Sketch of the proof. Define T by: (1) $T_{\langle \rangle} = \Psi$. (2) Let T_w be already defined and suppose that $\Diamond A_1, \Diamond A_2, \ldots$ is a list of all diamond sentences in T_w. (Notice that, thanks to Axiom D, there are certainly infinitely many such sentences – e.g., all the $\Diamond A$ where A is a tautology.[3]) Let $C_1, C_2, \ldots$ be a denumerable list of countable sets of new constants and let

$$\mathcal{L}_{v * \langle n \rangle} = \mathcal{L}_w + C_n.$$

Define $T_{w * \langle n \rangle}$ as an $\langle \mathcal{L}_{w * \langle n \rangle}, L \rangle$-saturated extension of $\Box^-(T_w) \cup \{A_n\}$.

Lemma 5.3. *Let Ψ be an $\langle \mathcal{L}_1, L \rangle$-saturated set of sentences. For every canonical subordination model $\langle N^*, S, T \rangle$ for Ψ, a Kripke model $\mathfrak{M} = \langle N^*, S, D, \mathfrak{I} \rangle$ can be defined such that, for all sentences A of $\mathcal{L}_1$ and $w \in N^*$,*

$$\mathfrak{M} \models_w A \quad \text{iff} \quad A \in T_w.$$

Sketch of the proof. As usual, let $D_w = \text{Const}(\mathcal{L}_w)$ and

$$\mathfrak{I}_w(P_j^n) = \{\langle c_1, \ldots, c_n \rangle \mid P_j^n(c_1, \ldots, c_n) \in T_w\}.$$

It follows immediately that $Q - K.D$ is characterized by the canonical subordination frame, since any serial extended frame is a frame for $Q - K.D$.

Consider now the canonical subordination frame of width *at most n*, $Z_n = \langle \{1, \ldots, n\}^*, S \rangle$, where $\{1, \ldots, n\}^*$ is the set of finite sequences from

176

$\{1, \ldots, n\}$. Z_n is a frame for $Q-K.D.Alt_n$; moreover, for every $\langle \mathcal{L}_1,$ $Q-K.D.Alt_n \rangle$-saturated set of sentences Ψ there is a model for Ψ based on Z_n. To wit, perform the construction of Lemma 5.2 as follows: Suppose that $T_{w*\langle j \rangle}$, $j = 1, \ldots, n$, has already been defined and that all the sentences up to $\Diamond A_r$ have already been examined. Consider the sentence $\Diamond A_{r+1}$, and define $T_{w*\langle j+1 \rangle}$ to be an $\langle L_{w*\langle j+1 \rangle}, Q-K.D.Alt_n \rangle$-saturated extension of $\Box^-(T_w) \cup \{A_{r+1}\}$ if $\neg A_{r+1}$ belongs to all $T_{w*\langle h \rangle}$, $h = 1, \ldots, j$; otherwise, consider the wff A_{r+2}. It can be seen, by using Axiom Alt_n, that if $j = n$ then there is an h, $h = 1, \ldots, n$, such that A_{r+1} belongs to $T_{w*\langle h \rangle}$. Of course, after exhausting all the diamond sentences in T_w, if not all the $T_{w*\langle h \rangle}$ for $h = 1, \ldots, n$ have been defined then we can arbitrarily define the remaining ones using the fact that, by Axiom D, $\Box^-(T_w)$ is consistent. It follows that $Q-K.D.Alt_n$ is characterized by the canonical subordination frame of width n and with nested domains. It is interesting to note that if we add axioms T and 4 to $Q-K.D.Alt_n$, the resulting system is not Kripke complete.

Let $\leq$ ($<$) be the reflexive and transitive (transitive) closure of S; $\langle N^*, \leq \rangle$ ($\langle N^*, < \rangle$) is said to be the (strict) subordination frame. It can be easily shown that $Q-S4$ ($Q-K.D.4$) is characterized by the (strict) subordination frame with nested domains.

The Ω-subordination frame is the frame $\langle N \times N^*, \leq \rangle$, where $\leq$ is defined by $\langle n_1, v_1 \rangle \leq \langle n_2, v_2 \rangle$ iff either $n_1 < n_2$ or $[n_1 = n_2$ and there is a $w \in N^*$ such that $v_1 * w = v_2]$. Notice that the Ω-subordination frame is a directed frame. The notion of Ω-subordination model is defined as expected. It can be shown that $Q-S4.2$ is characterized by the Ω-subordination frame with nested domains; see [2]. $BF-S4.2$ is not Kripke complete (see [27]), and this is the only example we have found so far of a logic that is complete with respect to a class of frames with nested domains but whose extension (by adding BF) is not complete with respect to that class of frames with constant domains.

It is immediate to realize that, at each level of the construction of a subordination model, we exhaust all the constants available at that level. This fact may become an obstacle when dealing with logics valid with respect to linearly ordered frames with nested domains. To wit, in building a countermodel for a nontheorem of $Q-S4.3$, we might be in the position of having to add a new point z between two points v and w, already constructed, where wRv; but in order to instantiate an existential formula occurring in z we may need to use a constant that has not been used up to this point of the construction, not used even in w. From this evolves the necessity of dealing with finite sets of wffs at each level of the construction. The method of diagrams helps us.

6. The method of diagrams

The method of diagrams was first introduced by Fine in [3] and is applicable to quite a large range of logics. We will show an application of that method to the logic $Q - S4.3$, taken from [1]. The construction of a countermodel for a nontheorem of $Q - S4.3$ is done in such a way that at every level of the construction we deal with only a finite number of points and a finite number of sentences associated to each point.

Consider the frame $\langle Q^+, \leq \rangle$, where Q^+ is the set of positive rational numbers (zero included) and $\leq$ is the numerical relation "less than or equal to." $\langle Q^+, \leq \rangle$ is easily seen to be a frame for $S4.3$. With each $w \in Q^+$ we associate a countable nonempty set C_w of individual constants such that, for all $v, w \in Q^+$, if $v \neq w$ then $C_w \cap C_v = \emptyset$. For any $w \in Q^+$, let $D_w = \bigcup_{v \leq w} C_w$, $\mathcal{L}_w = \mathcal{L} + D_w$, and Fm_w be the set of sentences of $\mathcal{L}_w$. Let $\mathcal{P} = \bigcup_{w \in Q^+} \{\langle w, A \rangle \mid A \in Fm_w\}$ be the set of pairs each of which is determined by a rational number w and a sentence of $\mathcal{L}_w$.

Definition 6.1. A diagram Δ is a subset of $\mathcal{P}$.

A diagram Δ is (roughly speaking) a subset of rational numbers, ordered by $\leq$, with sentences attached to them. If a sentence is attached to rational number w (i.e., if $\langle w, A \rangle \in \Delta$), then the intended meaning is that A is true at w. We build up new diagrams step by step by adding either a new rational number (together with a wff) to the diagram built so far, or a formula to a rational number that is already present in the diagram. The construction must be such that the final diagram is a model with nested domains. The crucial notion is that of coherence.

Definition 6.2. Let Δ be a diagram. The support of Δ, Supp(Δ), is the set $\{w \mid \langle w, A \rangle \in \Delta$ for some wff $A\}$, ordered by the numerical relation $\leq$.

For any $w \in$ Supp(Δ), let $\Delta(w) = \{A \mid \langle w, A \rangle \in \Delta\}$. When Δ is finite, the support of Δ is denoted by $\langle v_1, \ldots, v_n \rangle$, where $v_i < v_{i+1}$ $(1 \leq i < n)$; moreover, we let

$$\Delta_i = \bigwedge_{\langle v_i, A \rangle \in \Delta} A.$$

Definition 6.3. Let Δ be a finite diagram whose support is $\langle v_1, \ldots, v_n \rangle$. Δ is said to be *coherent* iff

$$\nvdash \Box \forall \vec{x}_1 [\Delta_1(\vec{c}_1/\vec{x}_1) \rightarrow \Box \forall \vec{x}_2 [\Delta_2(\vec{c}_2/\vec{x}_2) \rightarrow \cdots$$
$$\rightarrow \Box \forall \vec{x}_n [\Delta_1(\vec{c}_n/\vec{x}_n) \rightarrow \bot] \cdots]],$$

178

where, for all k $(1 \le k \le n)$, $\vec{c}_k$ is the list $c_{k1}, \ldots, c_{kj_k}$ $(0 \le j_k < \omega)$ of all the constants of $\mathcal{L}_{v_k} \setminus \mathcal{L}_{v_{k-1}}$ occurring in wffs of Δ (for $k = 1$, $\mathcal{L}_{k-1} = \mathcal{L}$).

For Δ infinite, Δ is said to be coherent iff all its finite subdiagrams are coherent. As one can expect, the following lemma holds.

Lemma 6.4. *Let Δ be a finite and coherent diagram. Then:*
(a) $\Delta \cup \{\langle w, \top \rangle\}$ *is a coherent diagram for any $w \in Q^+$;*
(b) *for any $w \in Q^+$ and $A \in \mathcal{L}_w$, if $\Delta \cup \{\langle w, A \rangle\}$ is not a coherent diagram then $\Delta \cup \{\langle w, \neg A \rangle\}$ is a coherent diagram; and*
(c) *for $w \in \mathrm{Supp}(\Delta)$, if $\langle w, \exists x_i A(x_i) \rangle \in \Delta$ then $\Delta \cup \{\langle w, A(x_i/d) \rangle\}$ is a coherent diagram for some constant $d \in C_w$.*

The next lemma, which depends on Axiom 3, is the crucial step in the construction (see [1] for the proof).

Lemma 6.5. *Let Δ be a finite coherent diagram and let $w \in \mathrm{Supp}(\Delta)$. If $\langle w, \Diamond A \rangle \in \Delta$ then there is a rational number s, $w \le s$, such that $\Delta \cup \{\langle s, A \rangle\}$ is a coherent diagram.*

Theorem 6.6. *$Q - S4.3$ is characterized by the extended frame $\langle Q^+, \le, D \rangle$ with nested domains.*

Analogously, it can be proved that $Q - K4.3.D.X$ is characterized by the extended frame $\langle Q^+, <, D \rangle$ with nested domains and that $BF - K4.3.D.X$ is characterized by the extended frame $\langle Q^+, <, D \rangle$ with constant domains.

7. BEYOND COMPLETENESS

Throughout this section, let a quantified modal logic $L \supseteq Q - K$ be fixed. Given a wff A, by $\mathcal{L}(A)$ we denote the language determined by the individual constants and predicates occurring in A. We wonder if L satisfies the following theorem.

Interpolation theorem. *If $\vdash_L A \to B$ then there is a wff $C \in \mathcal{L}(A) \cap \mathcal{L}(B)$ such that $\vdash_L A \to C$ and $\vdash_L C \to B$.*

An *L-theory* T (or a theory in the logic L) is a set of sentences, called the specific axioms of T. We say that A is a *theorem* of T iff $T \vdash_L A$. If not otherwise stated, by a theory we intend a modal theory in the logic L. A *model* for a theory T is a quintuple $\langle W, R, D, \mathfrak{I}, w \rangle$, where $\langle W, R, D \rangle$ is an extended frame for L, $w \in W$, and $\langle W, R, D, \mathfrak{I} \rangle$ is a model for the

179

language $\mathcal{L}$ such that every wff of T is true at w. We give two definitions of implicit definability of an n-ary predicate P in a theory T. Let T_0 be the theory obtained by substituting a new n-ary predicate P_0 for any occurrence of P in the axioms of T, and let $T \cup T_0$ denote the L-theory whose language and whose specific axioms are the language and specific axioms of T plus the language and specific axioms of T_0.

Definition 7.1.

(a) P is said to be *(a)-implicitly definable* in T iff

$$T \cup T_0 \vdash_L \forall x_1 \cdots \forall x_n (P(x_1, \ldots, x_n) \to P_0(x_1, \ldots, x_n)).^4$$

(b) P is said to be *(b)-implicitly definable* iff, for any two models of T, $\mathfrak{M}_1 = \langle W, R, D, \mathfrak{I}, w \rangle$ and $\mathfrak{M}_2 = \langle W, R, D, \mathfrak{J}, w \rangle$ of T; if $\mathfrak{I}(Q) = \mathfrak{J}(Q)$ for any predicate Q different from P, then $\mathfrak{I}_w(P) = \mathfrak{J}_w(P)$.

Note that Definition 7.1(a) implies 7.1(b) and is equivalent to it in case L is strongly complete. (We say that L is *strongly complete* when every consistent L-theory has a model.)

Definition 7.2. An n-ary predicate P is explicitly definable in T iff there is a wff B, not containing the predicate P, such that

$$T \vdash_L \forall x_1 \cdots \forall x_n (P(x_1, \ldots, x_n) \leftrightarrow B).$$

A logic L is said to possess the (a)-definability property (the (b)-definability property) according to the following theorem.

Definability theorem (Beth). *An n-ary predicate P is (a)-implicitly definable ((b)-implicitly definable) in an L-theory T iff P is explicitly definable.*

As for the classical case, the interpolation theorem implies the (a)-definability theorem. Beth's definability theorem states a sort of "expressive completeness" of a given modal logic, and it would be desirable that this theorem hold for as many modal theories as possible.

We first report on a negative result due to Fine [5].

Theorem 7.3 (Fine). *$BF - S5$ does not possess the interpolation property.*

The method used by Fine consists of considering the theory T containing as axioms

$$p \to \Diamond(\forall x(P(x) \to \Box(p \to \neg P(x)))$$

and

$$\neg p \to \Box \exists x (P(x) \wedge \Box(\neg p \to P(x))).$$

180

Clearly, p is (b)-implicitly definable in T because every model $\langle W, R, D,$ $\mathcal{I}, w \rangle$ of T is such that $\models_w p$ iff, for some v related to w, $\mathcal{I}_v(F) \cap \mathcal{I}_w(F) = \emptyset$. In order to show that p is not explicitly definable in T, Fine exhibits two models $\mathfrak{M}_1 = \langle W_1, R_1, D_1, \mathcal{I}_1, w_1 \rangle$ and $\mathfrak{M}_2 = \langle W_2, R_2, D_2, \mathcal{I}_2, w_2 \rangle$ of T such that in their distinguished worlds w_1, w_2 the same sentences not containing p are true, but p is false in $\mathfrak{M}_1$ at w_1 and true in $\mathfrak{M}_2$ at w_2. The (b)-definability property consequently fails and, since $BF - S5$ is strongly complete, the (a)-definability property fails too; thus the interpolation property does not hold.

Fine extends his theorem to all the logics L such that $BF - K \subseteq L \subseteq BF - S5$. The situation seems to be better for logics not containing the Barcan formula; for example, systems like $Q - K$, $Q - T$, $Q - S4$ possess the interpolation property (see [7] and [26]). However, one should be careful with interpolation theorems in modal logics; many negative results arise also in the propositional case ([24] and [25] give an almost complete picture of the situation for normal propositional logics stronger than $S4$).

The model theory for modal logic that was developed by using classical model theory as a guide (and so trying to prove for quantified modal logic the counterparts of theorems valid for classical logic) has not had great success up to now; such a purely mechanical translation of results from classical logic appears to be either trivial or impossible (e.g., even very simple theories are undecidable; see [8]). What seems more promising is the development of an original model theory for modal logics, one that takes account of the intrinsic suggestions of the modal context. An interesting example of this approach is Fine ([3], [4], [6]).

8. INCOMPLETENESS RESULTS

The completeness results of the previous sections should not suggest that most usual modal systems are K-complete. We shall see that, on the contrary, K-incompleteness is very common in quantified modal logic. How can we prove that a system L is *not* K-complete? The usual procedure consists of introducing a different semantics, and using it to show that some formula is valid in all the extended Kripke frames for L but not true in a model for L in the new semantics. Consequently, that formula is not a theorem of the system and the completeness theorem fails. The traditional alternative semantics is the "algebraic" semantics; although it is possible to obtain some incompleteness theorems by means of this, it is better to introduce some non-Kripkean "possible world" semantics, such as the "functor" semantics of [13] and [12] or the "Kripke bundles" semantics of [28]. We describe now the main features of the former.

181

Given two possible worlds v and w, in Kripke semantics only two possibilities are given: either they are accessible or not. In the functor semantics, we consider the possibility of having many ways to move from v to w. These ways can be formally described as arrows of a category, whose objects still must be thought of as possible worlds. The domain function is replaced by a set-valued functor. Let us recall the elementary notions involved: a (small) *category* $\mathbf{C}$ is a pair consisting of two sets, the set of objects $v, w, \ldots$ and the set of arrows $k, l, \ldots$ of $\mathbf{C}$. To each arrow are assigned two objects, its *domain* and its *codomain;* the notation $k: v \to w$ expresses that k is an arrow whose domain is v and whose codomain is w. Moreover, with each object v is associated an arrow 1_v (the identity of v) having domain and codomain equal to v. Finally, with each pair of arrows $k: v \to w$ and $l: w \to z$ (notice that the codomain of k is the same as the domain of l, in which case k and l are *composable*) is associated an arrow $kl: v \to z$ such that the following *unity* and *associativity* conditions hold:

$$1_v k = k = k1_w \quad \text{for } k: v \to w;$$

$$(kl)m = k(lm) \quad \text{for } k: v \to w, \ l: w \to z, \ m: z \to t.$$

Notice that any preorder $\langle W, \leq \rangle$ is a category whose objects are the elements of W and whose set of arrows of domain v and codomain w is a singleton (in case $v \leq w$) or the empty set (in case $v \nleq w$).[5]

Given a category $\mathbf{C}$, a set-valued functor D is a map associating with each object v a set D_v and with each arrow $k: v \to w$ a function $D_k: D_v \to D_w$, in such a way that $D_{1_v} = 1_{D_v}$ and $D_{kl} = D_k D_l$ (for every object v and for every pair of composable arrows k, l). Notice that a Kripke extended frame $\langle W, \leq, D \rangle$ on a preorder is a set-valued functor; in fact, if $v \leq w$ then the function assigned to the unique arrow having domain v and codomain w is the inclusion between D_v and D_w (recall the nested domain condition of Section 1). The concept of a set-valued functor reveals a trivialization implicit in Kripke semantics: An individual living in a possible world is identified with an individual living in another accessible one by means of the nested domains condition, which refers to an inclusion map.[6] In the context of a set-valued functor D this is no longer possible, owing to the presence of many arrows; for the sake of clarity, it is better to assume the D_v's to be pairwise disjoint and leave the transition maps D_k to play the role of telling us what an individual becomes after the "possible transformation" k. There is, however, an important feature that this new semantics still shares with Kripke semantics: Every individual in a world w has exactly one counterpart along any possible transformation $k: w \to v$ in v; that is, no splitting or death is admitted. This makes nonproblematic the use of w-assignments, which (we emphasize) are infinitary assignments.

We are now ready for the main definition. A *functor model* $\mathfrak{M}$ is a triple $\langle \mathbf{C}, D, \mathfrak{I} \rangle$, where

(1) $\mathbf{C}$ is a category;
(2) D is a set-valued functor (we need the usual restriction $D_v \neq \emptyset$ for every v, but see Section 9); and
(3) $\mathfrak{I}$ is a function that associates with every predicate letter P_j^n and with every object v a subset $\mathfrak{I}_v(P_j^n)$ of $(D_v)^n$.

Given a functor model $\mathfrak{M} = \langle \mathbf{C}, D, \mathfrak{I} \rangle$, a v-assignment μ, and a formula A, we define *the truth of A in $\mathfrak{M}$ at v under the v-assignment μ*, $\mathfrak{M}^\mu \vDash_v A$, in the following way:

$\mathfrak{M}^\mu \vDash_v P_j^n(x_{i_1}, \ldots, x_{i_n})$ iff $\langle \mu(x_{i_1}), \ldots, \mu(x_{i_n}) \rangle \in \mathfrak{I}_v(P_j^n)$;

$\mathfrak{M}^\mu \nvDash_v \bot$;

$\mathfrak{M}^\mu \vDash_v A_1 \to A_2$ iff [if $\mathfrak{M}^\mu \vDash_v A_1$ then $\mathfrak{M}^\mu \vDash_v A_2$];

$\mathfrak{M}^\mu \vDash_v \forall x_i A$ iff for every $a \in D_v$, $\mathfrak{M}^{\mu(x_i/a)} \vDash_v A$; and

$\mathfrak{M}^\mu \vDash_v \Box A$ iff for every w and for every $k: v \to w$, $\mathfrak{M}^{\mu k} \vDash_w A$, where μk is the w-assignment obtained by composing μ and $D_k: D_v \to D_w$.

This kind of semantics has been widely used in the context of intuitionistic logic (see e.g. [23] for the completeness theorem).

Given a functor model $\mathfrak{M}$ and a formula A, we say that *A is true in $\mathfrak{M}$* ($\mathfrak{M} \vDash A$) iff, for every object v of $\mathbf{C}$ and for every v-assignment μ, $\mu \vDash_v A$. Given a category $\mathbf{C}$, a set-valued functor D, and a formula A, we say that *A is valid on $\langle \mathbf{C}, D \rangle$* ($\langle \mathbf{C}, D \rangle \vDash A$) iff $\langle \mathbf{C}, D, \mathfrak{I} \rangle \vDash A$ for every interpretation $\mathfrak{I}$; finally, given a category $\mathbf{C}$ and a formula A, the notation $\mathbf{C} \vDash A$ means that $\langle \mathbf{C}, D \rangle \vDash A$ for every set-valued functor D. If L is a quantified modal logic, $\mathbf{C} \vDash L$ means that $\mathbf{C} \vDash A$ for every $A \in L$. Notice, however, that if $\mathbf{C}$ is a category then the set of formulas valid on $\mathbf{C}$ contains all the formulas provable in $Q - S4$, is closed under all the rules of $Q - S4$ but *not under uniform substitution,* and so is not a logic. This can be easily seen when $\mathbf{C}$ has only one object; in that case, modalities collapse when applied to sentences but do not collapse when applied to arbitrary formulas (i.e., $A \to \Box A$ holds for sentences only).

In the following we discuss the case of first-order modal systems that are minimum quantified extensions of modal propositional logics (always stronger than $S4$ in this section); we recall from Section 2 that if L is one such logic then its minimum quantified extension is indicated by $Q - L$. In order to state the main lemma, we need to introduce the concept of the *frame representation* $\mathfrak{F}(\mathbf{C})$ of a category $\mathbf{C}$. $\mathfrak{F}(\mathbf{C})$ is the preorder $\langle W_{\mathbf{C}}, \leq_{\mathbf{C}} \rangle$, where $W_{\mathbf{C}}$ is the set of arrows of $\mathbf{C}$ and $k_1 \leq_{\mathbf{C}} k_2$ holds for k_1: $v_1 \to w_1$ and $k_2: v_2 \to w_2$ iff [$v_1 = v_2$ and there exists $l: w_1 \to w_2$ such that $k_1 l = k_2$]. See Figure 1.

183

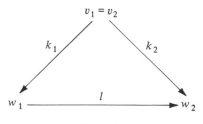

Figure 1

Lemma 8.1. *Let L be any modal propositional logic and let* **C** *be any category; then* $\mathfrak{F}(\mathbf{C})$ *is a frame for L iff* $\mathbf{C} \models Q - L$.

Theorem 8.2. *Let L be a propositional logic such that* $L \not\supseteq S5$ *and such that* $Q - L$ *is K-complete. Then* $L \subseteq S4.3$ *and, in case* $L \neq S4.3$, *L is of unbounded width.*[7]

We cannot give the proof of the theorem here (see [12]), but we can explain the main idea by outlining the proof of a considerably weaker statement. Suppose that L satisfies the hypotheses of the theorem. Because L is complete and L does not extend $S5$, $L \not\vdash \Diamond A \rightarrow \Box \Diamond A$ and so there is a frame for L that is not symmetric. Such a frame can be mapped by a p-morphism onto the frame[8]

$$ \mathcal{C}^{\bullet} \longrightarrow \mathcal{C}^{\bullet} \, , $$

which is consequently a frame for L. However, that frame is the frame representation of the category **M** having only one object v and, in addition to the identity arrow 1_v, an idempotent arrow – that is, an arrow m: $v \rightarrow v$ such that $mm = m$. According to Lemma 8.1, $\mathbf{M} \models Q - L$. Thus, every first-order formula A such that $\mathbf{M} \not\models A$ is not provable in $Q - L$; this is because our new semantics is obviously sound with respect to $(Q - L)$-derivability. The formula ML^* of Section 3 is of this kind. Since we supposed $Q - L$ to be Kripke complete, that formula must have a counterexample in a Kripke extended frame $\langle W, \leq, D \rangle$ for $Q - L$. A direct check shows that $\langle W, \leq \rangle$ must contain an infinite linearly (in particular, antisymmetric) ordered subset; consequently all the frames

$$ \mathcal{C}^{\bullet} \longrightarrow \mathcal{C}^{\bullet} $$
$$ \mathcal{C}^{\bullet} \longrightarrow \mathcal{C}^{\bullet} \longrightarrow \mathcal{C}^{\bullet} $$
$$ \cdots $$
$$ \mathcal{C}^{\bullet} \longrightarrow \mathcal{C}^{\bullet} \cdots \mathcal{C}^{\bullet} \longrightarrow \mathcal{C}^{\bullet} \longrightarrow \mathcal{C}^{\bullet} $$
$$ \cdots $$

are p-morphic images of it. Now all these frames are frames for L, since $\langle W, \leq \rangle$ is. These frames characterize a well-known modal propositional system, the system $S4.3.Grz$, so we have proved that *if $Q-L$ is Kripke complete and $L \not\supseteq S5$, then $L \subseteq S4.3.Grz$.*

We do not yet have a clear picture of which logics of the kind $Q-L$ are Kripke complete; our impression is that in any case the completeness results should be very rare. The situation seems to be better if we take into consideration quantified extensions with the Barcan formula BF. However, there is an example of a propositional logic L (namely $S4.2$) such that $Q-L$ is K-complete (see Section 5) but $BF.Q-L$ is not (see [27]). The same result can be obtained by using the functor semantics of this section; see [11].

In some cases, it is possible to restate K-completeness simply by adding some new and simple schemata (see e.g. [1] and [2]); in fact, given any predicate modal logic L, there exists the smallest logic L_0 that is an extension of L and is K-complete: L_0 is the logic characterizing the class of Kripke extended frames that are frames for L. However, this mere fact does not give any information on L_0; we would like to know if there is some *recursive* set of axioms for L_0.

This observation leads to a slightly new and important problem. Suppose we are interested in a particular class of Kripke extended frames (e.g., those based on preorders that are finite or well-ordered, etc.) and we ask about a (*recursive*) *axiomatization* of their modal logic. Quite often these questions are mathematically very difficult to answer, and one might suspect there is no solution. How can we prove this? Clearly, if there were a solution then the set of formulas valid in this class of frames would be recursively enumerable, and one could try to show this is not the case by selecting some set known to be of higher complexity and by interpreting it in the set of valid formulas. To this end, one could use some classical theorems, like Trachtenbrot's and Tarski's theorems, and so provide suitable examples of nonrecursively enumerable sets (also non-arithmetical in Tarski's case). In the late 1960s it was proved in this way (the main idea is due to D. Scott) that the modal logic of the frame of natural numbers is not recursively axiomatizable. More recently, Skvortsov [28] applied this strategy to the parallel case of intermediate logics, thus obtaining very strong and beautiful results.

Although much work remains to be done in this field (completeness and axiomatizability), we can no longer expect many positive results; thus it seems preferable to investigate some alternatives to Kripke semantics. We have given one example in this section, the functor semantics. However, the generalization can be made in a more radical direction, which seems to be suggested by both mathematical and philosophical motivations.

9. Beyond Kripke semantics

In the previous section, we generalized Kripke extended frames by introducing categories instead of preorders. Here we concentrate again on preorders, but we modify the domain function. In this way a new semantics is obtained, the semantics of *relational universes,* which is only a particular case (as we shall see) of a *topological* semantics that takes into consideration topological spaces varying over a topological space. The approach followed here is developed in [13], [14], and [10]; we provide the main information about it.

Let $\langle W, R \rangle$ be any frame. A relational universe D on this frame consists:

(i) for every $v \in W$, of a (possibly empty) set D_v; and
(ii) for every $v, w \in W$ such that vRw, of a relation $D_{vw} \subseteq D_v \times D_w$.

If we view the elements of the preorder as possible worlds and the sets D_v as the sets of the individuals living in them, then we may read the relations D_{vw} as "counterpart" relations. For instance, if vRw and $\langle a, b \rangle \in D_{vw}$ (we abbreviate this fact by aDb), then in a temporal reading this may mean that "a becomes b" after the time from v to w is passed or that "b is a future counterpart of a."[9] So an individual is subject to changes (splitting and death included) when passing from a possible world to a related one. As a consequence, when we evaluate a boxed formula – say, $\Box P(x)$ – at a world w, we consider only the accessible worlds v where some counterpart of the individual assigned to x exists. Consequently, *we dispense with infinitary assignments;* that is, we use only finitary assignments of variable length according to the free variables actually occurring in the formula to be evaluated. Although not formally needed, we take the D_v's to be pairwise disjoint.

For the sake of simplicity, we deal in this section with logical calculi in which the $S4$ axioms are assumed. Hence we must add the following additional requirements to the definition of a relational universe D on a frame $\langle W, R \rangle$:

(iii) R is reflexive and transitive;[10]
(iv) the "lax functor" conditions hold; that is, for every v, w, z such that $vRwRz$:

$$(l) \qquad 1_{D_v} \subseteq D_{vv}, \quad D_{vw} \circ D_{wz} \subseteq D_{vz},$$

there 1_{D_v} indicates the identity relation (i.e., the set of identical pairs $\langle a, a \rangle \in D_v \times D_v$) and $D_{vw} \circ D_{wz}$ the composite relation (i.e., the set of pairs $\{\langle a, c \rangle \in D_v \times D_z \mid \text{there is } b \in D_w \text{ such that } \langle a, b \rangle \in D_{vw} \text{ and } \langle b, c \rangle \in D_{wz}\}$).[11]

If a relational universe D is given, a model is of course a map associating with every n-ary predicate letter and with every possible world v a subset of $(D_v)^n$; truth clauses for classical connectives and quantifiers are the standard ones. As to the necessity operator, we should say that an n-tuple of individuals necessarily enjoys the property A iff all the n-tuples of their respective counterparts enjoy it. This amounts to setting

$$(\Box_R) \quad \langle a_1, \ldots, a_n \rangle \vDash_v \Box A(x_1, \ldots, x_n) \quad \begin{array}{l} \text{iff for every } w, b_1, \ldots, b_n \in D_w \\ \text{s.t. } vRw, a_1Db_1, \ldots, a_nDb_n, \\ \langle b_1, \ldots, b_n \rangle \vDash_w A(x_1, \ldots, x_n). \end{array}$$

There are two main problems at this point. First, the condition $(\Box_R)$ is in fact defective: We supposed that $x_1, \ldots, x_n$ are exactly the variables occurring free in A and evaluated the formula with respect to an n-tuple of elements of D_v. But it is clear that, when passing to subformulas, the free variables actually occurring could not only increase (as happens for quantified formulas) but also decrease; consequently, the n-tuple considered would happen to be "too long." We need to take care of this in some way. In fact, the addition of extra (possibly vanishing) elements to an n-tuple is not harmless, and could actually restrict the n-tuples of counterparts to be taken into consideration when evaluating a boxed formula (this is why we preferred to avoid the use of infinitary assignments).[12] Despite the fact that this problem is a serious one, we ignore it in the present informal discussion because it will completely disappear after the changes we shall make in the language.

Second, in the semantics just introduced, some well-known logical laws (provable in the quantified extension of any normal modal propositional system) fail; for instance:

$$\exists x_i \Box A \to \Box \exists x_i A;$$

$$x_1 = x_2 \to \Box(x_1 = x_2).$$

In order to avoid the provability of similar formulas, it is usually argued that classical quantificational and identity laws should not be applied to modal contexts. The remedies proposed to these unpleasant features of quantified modal logic sometimes produce unclear and complicated axiomatizations (see [9] for further information). The strategy we follow here is rather different: In the next section (which is independent from the final Section 11), we generalize the present semantics to the topological one in order to provide a clear mathematical and conceptual explanation of these difficulties. On the basis of the analysis so performed, in Section 11 we define a modal language and a modal calculus for which the semantics just outlined is the appropriate one.

187

Let T be any topological space. A *topological universe over T* is a pair consisting of another topological space X and of a continuous map π_X: $X \to T$. For every point $p \in T$ let us indicate with X_p the fiber over p, that is, the set of points $a \in X$ such that $\pi_X(a) = p$. A model in this context is a map associating with every n-ary predicate letter a subset of the n-times fibered product $X \times_T \cdots \times_T X$.[13] Again using finite assignments, we can define (by induction) for every formula $A(x_1, \ldots, x_n)$, for every $p \in T$, and for every $a_1, \ldots, a_n \in X_p$ the notion of truth of A, relative to the finite assignment $a_1, \ldots, a_n$, in such a way that the box operator represents the *interior in the fibered product topology.* We set:

$\langle a_1, \ldots, a_n \rangle \vDash \Box A(x_1, \ldots, x_n)$ iff for every neighborhood $J, I_1, \ldots, I_n$ of $p, a_2, \ldots, a_n$, respectively, and

($\Box_T$) for every $q \in J$ and $b_1 \in I_1 \cap X_q$, $\ldots, b_n \in I_n \cap X_q$, we have that $\langle b_1, \ldots, b_n \rangle \vDash A$.

This semantics really generalizes relational universes. In fact, every preorder is a topological space with (one of the two) topologies induced by the preorder relation, and to any relational universe D over a preorder $\langle W, R \rangle$ is naturally associated a topological universe over the topological space corresponding to $\langle W, R \rangle$, whose fiber over v is precisely the set D_v. After performing this construction, it is immediately seen that the truth clause ($\Box_R$) for the box operator of Section 9 reduces to ($\Box_T$).

Once a model is given on a topological universe X based over some topological space T, we can associate with every formula $A(x_1, \ldots, x_n)$ a subset of $X \times_T \cdots \times_T X$, namely the set $[A(x_1, \ldots, x_n)] = \{\langle a_1, \ldots, a_n \rangle \mid \langle a_1, \ldots, a_n \rangle \vDash A\}$. Clearly, the classical propositional connectives correspond to the Boolean operations on the subsets associated to the formulas, and the box operator corresponds (as we just noted) to the operation of taking the interior in the fibered product topology. For the existential quantifier, notice that it corresponds to the direct image along projections; for example, $[\exists x_i A(x_1, \ldots, x_i, \ldots, x_n)]$ is the direct image of $[A(x_1, \ldots, x_n)]$ along the projection associating to the n-tuples $\langle a_1, \ldots, a_i, \ldots, a_n \rangle$ the $(n-1)$-tuples $\langle a_1, \ldots, a_{i-1}, a_{i+1}, \ldots, a_n \rangle$. For this analysis to be complete, we also need this decisive observation: *substitution corresponds to inverse image.* In fact, we have (for instance) that $[A(x_1, x_1)] = \Delta^{-1}[A(x_1, x_2)]$, where $\Delta: X \to X \times_T X$ is the diagonal map associating with $a \in X$ the pair $\langle a, a \rangle \in X \times_T X$. It is possible to reconstruct the whole first-order logic in this way, by isolating the properties of Boolean operators and of direct and inverse images (this has been done by Lawvere in

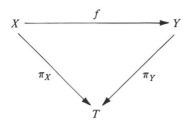

Figure 2

his fundamental papers [20] and [21]). As to inverse image, it can be proved that it commutes with respect to all the classical operators, and for that reason in logic we define substitution directly on atomic formulas. This raises the question: Is it the same in the modal case? The answer is negative. Let X, Y be T-based topological spaces; a T-continuous map is a continuous map $f: X \to Y$ fiber-preserving, that is, a map such that $f\pi_Y = \pi_X$ (see Figure 2). Clearly, the diagonal map and the projections map are T-continuous. However, for a T-continuous (as well as for an ordinary continuous) map $f: X \to Y$, we have only a semicommutativity of inverse image with respect to the interior operator; that is, for every $A \subseteq Y$ we have that

(SC) $$f^{-1} \square_Y A \subseteq \square_X f^{-1} A$$

(here $\square_Y, \square_X$ mean the interior in Y, X, respectively). Let us call a T-continuous map *open* when the converse of (SC) holds for every $A \subseteq Y$. It is easily seen that, in a T-based topological space X, the schema

$$\exists x_i \square A(x_1, \ldots, x_i, \ldots, x_n) \to \square \exists x_i A(x_1, \ldots, x_i, \ldots, x_n)$$

holds iff the projection maps are open, and the formula

$$x_1 = x_2 \to \square(x_1 = x_2)$$

holds iff the diagonal map is open.[14]

In conclusion, we need a reformulation of the logical language that does not allow substitution to be performed directly on the atomic formulas. Otherwise, we implicitly assume the converse of (SC) to hold, which is semantically (and also syntactically, as we shall see) equivalent to *nonvalid modal principles*. This reformulation is immediate if we follow the doctrinal approach (see [13]); it can of course also be performed in a more traditional linguistic setting, as indicated in the next section.

189

We present for the sake of simplicity only one-sorted languages and languages not containing function and constant symbols; for the related analysis and applications, see [14]. The first step consists of a more careful definition of a formula. A formula is a pair $A : n$ consisting of a string of symbols from the alphabet and a natural number (greater than or equal to 0), built in accordance with the following rules:

(1) $P_j^m(x_{i_1}, \ldots, x_{i_m}) : n$ is an (atomic) formula provided $i_1, \ldots, i_m \leq n$;
(2) $\bot : n$ is an (atomic) formula for every $n \geq 0$;
(3) if $A_1 : n$ and $A_2 : n$ are formulas then so is $(A_1 \rightarrow A_2) : n$;
(4) if $A : n+1$ is a formula then so is $(\forall x_{n+1} A) : n$; and
(5) if $A : m$ is a formula then so is $(\Box A)(x_{i_1}, \ldots, x_{i_m}) : n$, provided that $i_1, \ldots, i_m \leq n$.

Thus a formula such as $A : n$ expresses a property of a n-tuple. Moreover, this definition[15] treats modalized formulas as atomic formulas with respect to substitution. We shall call modalized formulas (i.e., formulas of the kind $(\Box A)(x_{i_1}, \ldots, x_{i_m}) : n$), as well as ordinary atomic formulas, *generalized atomic formulas*. Quantifiers bind the variable with greatest index. This choice is mainly due to technical convenience and does not affect the expressive power of the language (we are only less free in using alphabetical variants). $\Box A : n$ stands for $(\Box A)(x_1, \ldots, x_n) : n$. Notice that, for instance, $\Box P(x_1) : 1$ and $\Box P(x_1) : 2$ are different formulas. Such distinctions give a correct answer to the first problem discussed in Section 9 and are crucial in those semantics where inverse image along projections does not commute with respect to modal operators. For example, in relational universes the first formula is true at v relative to a given individual a_1 iff all the counterparts of a_1 have property P, whereas the second formula is true at v relative to a pair $\langle a_1, b_1 \rangle$ iff all the pairs $\langle a_2, b_2 \rangle$ of their respective counterparts are such that a_2 has property P. (The two conditions do not coincide, because the latter is relative to a pair and the former to a single individual; moreover, notice that b_1 can "die.")

The next step is the definition of substitution. It is obvious that any formula $A : m$ is built up in a unique way from generalized atomic subformulas $B_1 : n_1, \ldots, B_k : n_k$ using only the connective $\rightarrow$ and the quantifiers. Suppose that $i_1, \ldots, i_m \leq n$; then the formula $A[x_{i_1}, \ldots, x_{i_m}] : n$ (called the formula *obtained from $A : m$ by substitution with $x_{i_1}, \ldots, x_{i_m} : n$*) is the formula obtained by replacing externally[16] in $B_1 : n_1, \ldots, B_k : n_k$, x_1 by $x_{i_1}, \ldots, x_m$ by x_{i_m}, and the variables $x_{m+1}, x_{m+2}, \ldots$ (which can occur externally only bounded by a quantifier) by $x_{n+1}, x_{n+2}, \ldots$. Thus, for instance, $((\Box P(x_1, x_2))(x_1, x_1) \vee \exists x_3 R(x_2, x_3))[x_3, x_4] : 5$ is equal to

$(\Box P(x_1, x_2))(x_3, x_3) \vee \exists x_6 R(x_4, x_6) : 5$. This definition of substitution (an inductive formulation is also possible, which we leave to the reader) avoids conflicts between free and bound variables and agrees with the "Beck conditions" – in categorical logic, those conditions of commutativity between quantifications and inverse images. Notice that $(\Box A)[x_{i_1}, ..., x_{i_m}] : n = (\Box A)(x_{i_1}, ..., x_{i_m}) : n$.

We are now ready to explain the axioms and the rules for the syntactic calculus. Of course all the propositional tautologies (the reflexivity, transitivity and distributivity schemata of $S4$) will be taken as axioms; modus ponens and necessitation will be included in the list of rules. As to the predicate part of the calculus, we can take any usual axiomatization, being careful to write down only formulas in the sense of the preceding definition. For instance, the *dictum de omni* schema is

$$(\forall x_{n+1} A)[x_1, ..., x_n] \to A : n+1;$$

the generalization rule is

$$\frac{A[x_1, ..., x_n] \to B : n+1}{A \to \forall x_{n+1} B : n;}$$

and the substitutivity schema (in case we have equality) is

$$x_1 = x_2 \wedge A[x_1, x_3, ..., x_n] \to A[x_2, x_3, ..., x_n] : n.$$

(Notice that this last schema is unrestricted, in the sense that it does not distinguish between modalized and nonmodalized formulas.) We also need a substitution rule,[17]

$$\frac{A : m}{A[x_{i_1} ... x_{i_m}] : n,}$$

provided $x_{i_1}, ..., x_{i_m} \le n$. Finally, we have an additional modal axiom, the *semicommutativity* or *continuity* axiom, explaining the link between substitution and modalities:

$$(SC) \qquad (\Box A)[x_{i_1}, ..., x_{i_m}] \to \Box(A[x_{i_1}, ..., x_{i_m}]) : n.$$

This system is sound and complete with respect to both relational and topological semantics. For the sake of clarity, we give here the precise definition of a model for relational semantics. A *model* is a quadruple $\langle W, R, D, \mathcal{I} \rangle$, where $\langle W, R, D \rangle$ is a relational universe and $\mathcal{I}$ is a map associating with every predicate letter P_j^m and with every $v \in W$ a subset of $(D_v)^m$. From this data, we can inductively define $\langle a_1, ..., a_n \rangle \models_v^n A$ for every $v \in W$, for every n-ary formula $A : n$, and for every $a_1, ..., a_n \in D_v$:

$$\langle a_1, ..., a_n \rangle \models_v^n P_j^m(x_{i_1}, ..., x_{i_m}) \quad \text{iff} \quad \langle a_{i_1}, ..., a_{i_m} \rangle \in \mathcal{I}_v(P_j^m);$$

$$\langle a_1, ..., a_n \rangle \models_v^n x_{i_1} = x_{i_2} \qquad\qquad \text{iff} \quad a_{i_1} = a_{i_2};$$

191

$\langle a_1, \ldots, a_n \rangle \not\models_v^n \bot;$

$\langle a_1, \ldots, a_n \rangle \models_v^n A_1 \to A_2$ iff if $\langle a_1, \ldots, a_n \rangle \models_v^n A_1$
                                               then $\langle a_1, \ldots, a_n \rangle \models_v^n A_2;$

$\langle a_1, \ldots, a_n \rangle \models_v^n \forall x_{n+1} A$ iff for all $a_{n+1} \in D_v,$
                                               $\langle a_1, \ldots, a_n, a_{n+1} \rangle \models_v^{n+1} A;$

$\langle a_1, \ldots, a_n \rangle \models_v^n (\Box A)(x_{i_1}, \ldots, x_{i_m})$ iff for all $w, b_1, \ldots, b_m \in D_w$
                                               s.t. $vRw, a_{i_1} Db_1, \ldots, a_{i_m} Db_m,$
                                               $\langle b_1, \ldots, b_m \rangle \models_w^m A.$

Notice that the truth clause for the box operator is in fact a combination of $(\Box_R)$ of Section 9 and of the implicit clause for substitution. The latter is fully expressed in the following metatheorem:

$$\langle a_1, \ldots, a_n \rangle \models_v^n A[x_{i_1}, \ldots, x_{i_m}] \quad \text{iff} \quad \langle a_{i_1}, \ldots, a_{i_m} \rangle \models_v^m A.$$

In the logical calculus just given, we cannot prove

$$\exists x_{n+1} \Box A \to \Box \exists x_{n+1} A : n.$$

We can pass from

$$A \to (\exists x_{n+1} A)[x_1, \ldots, x_n] : n+1$$

to

$$\Box A \to \Box ((\exists x_{n+1} A)[x_1, \ldots, x_n]) : n+1$$

by means of obvious steps, but in order to perform the usual proof we need the schema

(Cpr) $\Box (B[x_1, \ldots, x_n]) \to (\Box B)[x_1, \ldots, x_n] : n+1,$

which is not allowed because it expresses the commutativity of the modal operators with respect to inverse images along projections. Conversely, if we assume the schema $\exists x_{n+1} \Box A \to \Box \exists x_{n+1} A : n$, then (Cpr) is provable. For systems containing Axiom B of Section 3 (and also in tense logic), it can be easily shown that (Cpr) is equivalent to the Barcan formula. A similar (slightly more difficult) analysis shows that the formula $x_1 = x_2 \to \Box(x_1 = x_2) : 2$ is equivalent to the schema

(Cdiag) $\Box (C[x_1, x_1]) \to (\Box C)[x_1, x_1] : 2,$

expressing the commutativity of the modal operator with respect to inverse images along diagonals.[18] As can be expected, the schema (Cpr) axiomatizes the relational universes D such that all the relations D_{vw} are totally defined, while the schema (Cdiag) axiomatizes the relational universes D such that all the relations D_{vw} are partial functions.

1. We have fixed this countable set $\mathbf{C}$ in order to guarantee that W is a set and not a proper class.

2. Notice that $BF.Q-K.B=Q-K.B$, because $Q-K.B \vdash BF$.

3. The method can be adapted to $Q-K$ by deleting (in the canonical subordination frame) all the successor nodes of w in case T_w contains $\square\perp$. This is a simple example showing that the subordination method should be refined, in the sense that it is often better not to fix a frame beforehand but rather to build it during construction of the model.

4. We point out that alternative (perhaps not equivalent) definitions are also possible; for instance, we could permit a box in this formulation.

5. In this section we concentrate only on logics that are extensions of $Q-S4$, but it is also possible to obtain a semantics for $Q-K$ (simply by abolishing identity and composition).

6. The same trivialization is still present if we drop the nested domain condition, as in [19] (see Section 1). In that case we again refer implicitly to an inclusion, which is now not a total (but rather a partial) function.

7. We say that a propositional logic L is of *bounded width* iff there is a natural number n such that, in every frame $\langle W, \leq \rangle$ for L, the following condition is satisfied: for every $w, v_1, \ldots, v_n \in W$, if $wRv_1, \ldots, wRv_n$ then there are i, j $(i \neq j)$ such that $v_i \leq v_j$ or $v_j \leq v_i$.

8. A *p-morphism* f between two preorders $\langle W_1, \leq_1 \rangle$ and $\langle W_2, \leq_2 \rangle$ is an order-preserving map such that, for every $v \in W_1$ and $w \in W_2$, if $f(v) \leq_2 w$ then there exists $v_0 \in W_1$ s.t. $v \leq_1 v_0$ and $f(v_0) = w$. By a standard result, if f is surjective and $\langle W_1, \leq_1 \rangle$ is a frame for a propositional logic L, then so is $\langle W_2, \leq_2 \rangle$.

9. We do not argue that this formalization corresponds to Lewis's suggestions (e.g., the counterpart relation is far from being symmetrical). In our opinion, however, relational universes are quite reasonable from an intuitive point of view, and (last but not least, as we shall see) lead to a clear and simple logical axiomatization. For an attempt to formalize Lewis's ideas, see [16].

10. Again it would be better to use a category instead of a preorder; the generalization is easy and we leave it to the reader.

11. In conditions (l), inclusions can be replaced by equality without losing the completeness theorem.

12. A solution to a similar problem is presented in [8], but it has the unpleasant consequence of altering the underlying propositional modal logic.

13. We recall that this space contains exactly the n-tuples from X living in the same fiber; the topology is the relativization of the product topology.

14. An alternative and very important characterization is the following: π_X is an open map iff the schema $\exists x_i \square A(x_1, \ldots, x_i, \ldots, x_n) \to \square \exists x_i A(x_1, \ldots, x_i, \ldots, x_n)$ holds, and π_X is locally injective iff the formula $x_1 = x_2 \to \square(x_1 = x_2)$ holds; consequently, we have the validity of both just in the case where X is an étalé space over T. The same applies also to the particular case of relational universes, where the projections are open iff the transition relations are totally defined and the diagonal is open iff the transition relations are partial functions.

15. The alternative is to simplify the box clause by saying "if $A:m$ is a formula then so is $(\square A):m$" and to add the following clause: "if $A:m$ is a formula and $i_1, \ldots, i_m \leq n$, then $A(x_{i_1}, \ldots, x_{i_m}):n$ is a formula." This alternative is

clearer from a conceptual point of view (substitution really is a logical operator like connectives and quantifiers); however, it complicates the notation unnecessarily.

16. We call an occurrence of a variable in a formula *external* iff it does not lie within the scope of modal operators.

17. The *dictum de omni* schema of the usual calculi is a combination of the above *dictum de omni* schema and a substitution rule (we need both of them to prove, e.g., $\forall x_2(A[x_1,x_2]) \to A[x_1,x_1] : 1$). We prefer to separate them, because in this way it is evident that the *dictum de omni* schema and the generalization rule simply express the adjointness relation between substitution and universal quantifier.

18. It can be shown that *(Cpr)* and *(Cdiag)* together imply the converse of *(SC)*. Notice that adding them to our calculus is not yet sufficient to obtain a system equivalent to $Q - S4$; we need also an axiom like $\exists x_1 \top : 0$, because the calculus of this section is sound with respect to empty domains. However, if this axiom is added, too, then the equivalence can be proved (see [10] for details).

REFERENCES

[1] Corsi, G. "The Quantified Modal Logic of $\langle Q^+, < \rangle$ and $\langle Q^+, \leq \rangle$ and Some of Their Extensions." Preprint.

[2] Corsi, G., and Ghilardi, S. (1990), "Directed Frames." *Archive for Mathematical Logic* 29: 53–67.

[3] Fine, K. (1978), "Model Theory for Modal Logic, I." *Journal of Philosophical Logic* 7: 125–56.

[4] Fine, K. (1978), "Model Theory for Modal Logic, II." *Journal of Philosophical Logic* 7: 277–306.

[5] Fine, K. (1979), "Failure of the Interpolation Lemma in Quantified Modal Logic." *Journal of Symbolic Logic* 44: 201–6.

[6] Fine, K. (1981), "Model Theory for Modal Logic, III." *Journal of Philosophical Logic* 10: 293–307.

[7] Gabbay, D. (1972), "Craig's Interpolation Theorem for Modal Logics." In W. Hodges (ed.), *Conference in Mathematical Logic, London, 1970.* Amsterdam: North Holland, pp. 111–27.

[8] Gabbay, D. (1976), *Investigations in Modal and Tense Logics with Applications to Problems in Philosophy and Linguistics.* Dordrecht: Reidel.

[9] Garson, J. W. (1984), "Quantification in Modal Logic." In D. Gabbay and F. Günther (eds.), *Handbook of Philosophical Logic,* vol. II. Dordrecht: Reidel, pp. 249–308.

[10] Ghilardi, S. (1990), "Modalità e Categorie." Tesi di dottorato, Università degli Studi di Milano.

[11] Ghilardi, S. (1990), "Presheaf Semantics and Independence Results for Some Nonclassical First-order Logics." *Archive for Mathematical Logic* 29: 125–36.

[12] Ghilardi, S. (1991), "Incompleteness Results in Kripke Semantics." *Journal of Symbolic Logic* 56: 517–38.

[13] Ghilardi, S., and Meloni, G. C. (1988), "Modal and Tense Predicate Logic: Models in Presheaves and Categorical Conceptualization." In F. Borceux (ed.), *Categorical Algebra and Its Applications,* Proceedings (Louvain-La-Neuve, 1987). Lecture Notes in Mathematics, 1348. Berlin: Springer, pp. 130–42.

[14] Ghilardi, S., and Meloni, G. C. (1991), "Philosophical and Mathematical Investigations in First-order Modal Logic." In G. Usberti (ed.), *Problemi Fondazionali Nella Teoria del Significato, Atti del Convegno di Pontignano.* Firenze: Olsckhi, pp. 77–107.

[15] Goldblatt, R. (1987), "Logics of Time and Computation." Lecture Notes 7, Center for the Study of Language and Information, Stanford University.

[16] Hazen, A. (1979), "Counterpart-theoretic Semantics for Quantified Modal Logic." *Journal of Philosophy,* pp. 319–38.

[17] Hughes, G. E., and Cresswell, M. J. (1968), *Introduction to Modal Logic.* London: Methuen.

[18] Hughes, G. E., and Cresswell, M. J. (1984), *A Companion to Modal Logic.* London: Methuen.

[19] Kripke, S. K. (1963), "Semantical Investigations in Modal Logic." *Acta Philosophica Fennica* 16: 83–94.

[20] Lawvere, F. W. (1969), "Adjointness in Foundations." *Dialetica* 23: 281–96.

[21] Lawvere, F. W. (1970), "Equality in Hyperdoctrines and Comprehension Schema as an Adjoint Functor." In *Applications of Categorical Algebra.* Proceedings of Symposia in Pure Mathematics, 24. Providence, RI: American Mathematical Society.

[22] Lemmon, E. J. (1977), *An Introduction to Modal Logic.* Oxford: Basil Blackwell.

[23] Makkai, M., and Reyes, G. E. (1977), "First Order Categorical Logic." Lecture Notes in Mathematics, 611. Berlin: Springer.

[24] Maksimova, L. L. (1979), "Interpolation Theorems in Modal Logics and Amalgamable Varieties of Topological Boolean Algebras." *Algebra i Logika* 18: 556–86.

[25] Maksimova, L. L. (1980), "Interpolation Theorems in Modal Logics. Sufficient Conditions." *Algebra i Logika* 19: 194–213.

[26] Plaza, J. (1989), "The Craig Interpolation Lemma for Quantified Modal Logics (abstract)." *Journal of Symbolic Logic* 54: 669.

[27] Shehtman, V. B., and Skvortsov, D. P. (1989), "Semantics of Non-classical First Order Predicate Logics." In *Proceedings of Heyting 88* (Chaika, Bulgaria, September 1988). Amsterdam: North Holland.

[28] Skvortsov, D. P. (1988), "On Axiomatizability of Some Intermediate Predicate Logic (summary)." *Reports on Mathematical Logic* 22: 115–16.

195

12

Epistemic logic and game theory

Bernard Walliser

For sixty years, theoretical economics paid growing attention to agents' expectations, and then to the beliefs that sustain them, in its study of phenomena such as speculation. Now, agents' beliefs regarding their environment constitute a third explanatory principle for decisions, with constraints on their action possibilities and preferences on the social states. The search for generalization led to the introduction of mental representations into game theory, where they are now explicitly modeled in a set formulation that is generally probabilized. Such representations allow a better account of individual behavior under uncertainty, and of equilibria resulting from the conjunction of actions of several players.

For thirty years, cognitive sciences were mainly interested in agents' knowledge in the study of (for instance) their capacity in problem solving. Knowledge was first explored in terms of static organization and later in terms of dynamic revision – that is, when new information becomes available. The search for formalization has recently led to mental representations being axiomatized in a propositional form by epistemic logic, a variety of modal logic. This allows a better account of an agent's knowledge, of others' knowledge, and of the collective improvement of knowledge resulting from the exchange of information between agents.

Although grounded in distinct formal frameworks, the two trends confronted the same questions and yielded more or less similar (and complete) answers. The trends advanced separately at first, but their meeting was unavoidable; each took advantage of the other's results for integrating them into its own problematics. Their marriage is now almost achieved and has led to equivalent frameworks, except that epistemic logic is content to examine rules of knowledge while game theory links knowledge to principles of action. In this chapter, logic and game theory will be examined and compared; with respect to each aspect, epistemic logic (although last born) will be presented first.

1. BASIC SYNTAX

Epistemic logic (Hughes and Creswell 1985, Halpern and Moses 1985) considers a set of m agents, each agent i being endowed with a unitary "knowledge operator" K_i such that $K_i p$ means "agent i knows proposition p." From finite primitive propositions $\hat{p} \in \mathbf{P}$, let us call $\mathfrak{L}_m(\mathbf{P})$ the smallest set of propositions that contains $\mathbf{P}$ and is closed under the operators $\neg$ (not), $\wedge$ (and), and K_i (i knows). In this set, the propositions that are really known constitute the agent's "knowledge system" $\mathbf{K}_i = \{p \text{ s.t. } K_i p\}$. In general, a proposition p may be considered true ($K_i p$), false ($K_i \neg p$), or undecidable ($\neg K_i p \equiv \neg K_i \neg p$). However, the system is contradictory if it contains both p and $\neg p$; in this case, the system is denoted $\bar{\mathbf{K}}$.

As usual, a proposition p is said to be *provable* (denoted $\vdash p$) if it is a theorem deduced (by the modeler) from primitive propositions through a system of axioms and rules; the proposition is *consistent* if $\neg p$ is not provable. In such a syntactic framework, five axioms and two rules are generally considered:

A1: all tautologies of propositional calculus;
A2: deductive closure axiom: $[K_i p \wedge K_i(p \rightarrow q)] \rightarrow K_i q$
    (an agent knows all logical consequences of his knowledge);
A3: truth axiom: $K_i p \rightarrow p$
    (what an agent knows is true);
A4: positive introspection axiom: $K_i p \rightarrow K_i K_i p$
    (an agent knows what he knows);
A5: negative introspection axiom: $\neg K_i p \rightarrow K_i \neg K_i p$
    (an agent knows what he does not know);
R1: modus ponens: from $\vdash p$ and $\vdash p \rightarrow q$, deduce $\vdash q$
    (the modeler deduces all logical consequences of her assumptions);
R2: generalization rule: from $\vdash p$, deduce $\vdash K_i p$
    (the agent is able to reproduce the reasoning of the modeler).

These axioms have been discussed at length in terms of their realism (Vardi 1986); they will be weakened in what follows. A2 and R2 assume that the agent is able to reason perfectly (logical omniscience), although the modeler herself does not always know all the consequences of her knowledge. A3 claims that the agent cannot have false knowledge, in the sense that his knowledge is always compatible with that of the modeler (although not necessarily with reality). A4 and A5 postulate that an agent knows the extent of his knowledge and ignorance, which means that he considers all propositions that might be considered by the modeler. In any case, if the deduction rules are admitted then sequential acceptance of axioms leads to nested formal systems:

$$R1 + R2 + A1 + A2 + A3 + A4 + A5$$
$$\downarrow \quad \downarrow \quad \downarrow \quad \downarrow$$
$$K_m \quad T_m \quad S4_m \quad S5_m$$

Game theory. In game theory, a set of players is considered; players are either passive (nature, denoted 0) taking random states a_0, or active (denoted i) choosing actions a_i by a deliberation process. The selected action a_i^* stems from a rationality principle defined on players' preferences, beliefs, and constraints, the last defining a set A_i of available actions. Preferences were for a long time strongly axiomatized, and are expressed by a utility function $U_i(a_i, a_{-i}, a_0)$ associated with an optimizing rationality principle. However, beliefs R_i (which, unlike preferences, may evolve in a dynamic game) are still defined less formally, and may be inspired by the preceding axiomatization.

Each player i is assumed to know the structure of the game (number of players, general rules, etc.); i is also assumed to know his own "determiners" A_i, U_i, and R_i, and his own past actions a_i^-. Each player may have an *exogenous* uncertainty about future states of nature, a_0 (which are independent of his own actions). Each player may also have some uncertainty regarding past actions of the other players, a_{-i}^-, which are then treated as actions of nature (imperfect information). Each player has an essential *endogenous* uncertainty about future actions of the active players, a_{-i} (depending on his own); if a player considers the others to have a decision process identical to his own then his action may be linked to his knowledge of others' determiners A_{-i}, U_{-i}, and R_{-i}, which are themselves uncertain (incomplete information). Moreover, each player simulates the others' behavior by a formal calculus, and not a deductive reasoning, in accordance with a poorly developed modal logic of predicates.

2. Kripke semantics

In the most popular semantics associated with the basic axioms, the modeler translates the actor's knowledge (or rather uncertainty) by considering that the actor perceives several possible worlds. More precisely, a Kripke structure $\mathcal{S}$ is defined by the following elements:

(1) a set S of possible worlds s;
(2) a truth function v giving the truth value of each primitive proposition in each world, $v(s, \hat{p}) \in \{0, 1\}$; and
(3) a binary accessibility relation $\mathcal{K}_i$ between worlds: $(s, t) \in \mathcal{K}_i$ iff at real world s, agent i considers world t to be possible (where "iff" is shorthand for "if and only if").

199

The set of worlds accessible from world s is denoted $H_i(s)$; all accessible sets form a set $\mathcal{H}_i = \{H_i(s) \mid s \in S\}$.

A proposition p is said to be *valid at a world s* in a Kripke structure (denoted $(\mathcal{S}, s) \models p$) if it stems from the following conditions, classical for the first three and characterizing the knowledge operator for the fourth:

B1: $(\mathcal{S}, s) \models \hat{p}$ iff $v(s, \hat{p}) = 1$;
B2: $(\mathcal{S}, s) \models \neg p$ iff $(\mathcal{S}, s) \not\models p$;
B3: $(\mathcal{S}, s) \models p \wedge q$ iff $(\mathcal{S}, s) \models p$ and $(\mathcal{S}, s) \models q$;
B4: $(\mathcal{S}, s) \models K_i p$ iff $(\mathcal{S}, t) \models p$, $\forall t \in H_i(s)$
(agent i knows p at world s if and only if p is valid at all worlds accessible from s).

A proposition p is said to be *valid in* $\mathcal{S}$ (denoted $\mathcal{S} \models p$) if it is valid at each world and *satisfiable in* $\mathcal{S}$ if $\neg p$ is not valid in $\mathcal{S}$. The following theorems are the counterparts of the first two axioms and the two rules of the axiomatic system, which are automatically verified in all Kripke structures:

(1) all tautologies are valid (cf. A1);
(2) $\forall p, q \in \mathcal{L}_m(\mathbf{P})$, $\models [K_i p \wedge K_i(p \to q)] \to K_i q$ (cf. A2);
(3) $\forall p, q \in \mathcal{L}_m(\mathbf{P})$, if $\models p$ and $\models p \to q$ then $\models q$ (cf. R1);
(4) $\forall p \in \mathcal{L}_m(\mathbf{P})$, if $\models p$ then $\models K_i p$ (cf. R2).

To obtain the counterpart of the remaining axioms, the binary relation $\mathcal{H}_i$ must exhibit usual properties:

(1) reflexive: $(s, s) \in \mathcal{H}_i$ (cf. A3)
(the real world is considered by the agent as possible);
(2) transitive: $[(s, s) \in \mathcal{H}_i$ and $(t, u) \in \mathcal{H}_i] \Rightarrow (s, u) \in \mathcal{H}_i$ (cf. A4)
(each world indirectly accessible is also directly accessible);
(3) Euclidean: $[(s, t) \in \mathcal{H}_i$ and $(s, u) \in \mathcal{H}_i] \Rightarrow (t, u) \in \mathcal{H}_i$ (cf. A5)
(two worlds accessible from a third one are mutually accessible).

It is demonstrated that K_m (resp., T_m, $S4_m$, $S5_m$) is both a sound (each provable proposition is also valid) and complete (each valid proposition is provable) axiomatization of a Kripke structure if $\mathcal{H}_i$ is a binary relation (resp., reflexive, reflexive and transitive, reflexive and transitive and Euclidean).

The "natural set" of worlds is $\{0, 1\}^{\mathbf{P}}$, each world corresponding to a combination of truth values of the primitive propositions. In any Kripke structure, to each proposition p it is possible to associate its field $|p|$, that is, the set of worlds where p is valid:

$$|p| = \{s \in S \mid (\mathcal{S}, s) \models p\}.$$

These fields obey the following rules, so that the set of fields forms an algebra $(S, \mathcal{E})$:

$$|\top| = S, \qquad |\neg p| = {}^c|p|,$$
$$|\bot| = \emptyset, \qquad |p \wedge q| = |p| \cap |q|,$$

with $p \vdash q$ iff $|p| \subseteq |q|$. Moreover, agent i knows p at world s if $H_i(s) \subseteq |p|$, so that

$$|K_i p| = \{s \in S \mid H_i(s) \subseteq |p|\}.$$

The knowledge set $\mathbf{K}_i$ of an agent i then has an image $\mathcal{K}_i$, to be understood as the set of worlds where i knows all propositions of $\mathbf{K}_i$:

$$\mathcal{K}_i = \bigcap_{p \in \mathbf{K}_i} |K_i p| = \left\{ s \in S \mid H_i(s) \subseteq \bigcap_{p \in \mathbf{K}_i} |p| \right\}.$$

In order to study the properties of the knowledge operator K_i, it is possible to reason on subsets rather than on propositions, with the corresponding operator $\hat{K}_i$ bijectively related to the accessibility relation $\mathcal{K}_i$:

$$\hat{K}_i E = \{s \in S \mid H_i(s) \subseteq E\},$$
$$H_i(s) = \bigcap \{E \in \mathcal{E} \mid s \in \hat{K}_i E\}.$$

An event G_i is a truism (or public event) for agent i if it is known as soon as it happens: $G_i \subseteq \hat{K}_i G_i$; then agent i knows E in s if there exists G_i such that $s \in G_i \subseteq E$. The operator $\hat{K}_i$ obeys axioms that are the counterpart of the basic axioms (Bacharach 1985, Binmore and Brandenburger 1988):

$$\hat{K}_i \emptyset = \emptyset, \qquad \hat{K}_i E \subseteq E, \tag{cf. A3}$$

$$\hat{K}_i({}^c E) = {}^c \hat{K}_i E, \qquad \hat{K}_i E \subseteq \hat{K}_i \hat{K}_i E, \tag{cf. A4}$$

$$\hat{K}_i(E \cap F) = \hat{K}_i E \cap \hat{K}_i F, \qquad {}^c \hat{K}_i E \subseteq \hat{K}_i({}^c \hat{K}_i E). \tag{cf. A5}$$

For example, with the axiomatic system $S5_m$:

(1) $\mathcal{K}_i$ is an equivalence relation;
(2) $\mathcal{K}_i$ is a partition on S, $H_i(s) \cap H_i(t) = \emptyset \Rightarrow H_i(s) = H_i(t)$;
(3) $\hat{K}_i$ associates to E the elements of $\mathcal{K}_i$ contained in E; and
(4) a truism is a union of subsets of the partition.

Game theory. In game theory, the uncertainty of a player is generally analyzed in the semantic framework rather than in the syntactic one, mainly because it is easier to work with set theory than with propositional calculus. The uncertainty of agent i concerns the value of a variable when, in a dynamic game, i cannot recognize some others' past actions a_{-i} and so considers an "information set," regrouping those actions in the game tree. Uncertainty concerns the specification of a function when others' preferences U_{-i} are ignored, but it is usually considered possible to parametrize such functions with underlying explanatory variables (Harsanyi 1976).

The uncertainty is then placed on a parameter – the type $c_i \in C_i$ of a player, which is treated as an exogenous variable as before. Uncertainty also involves some more complex characteristics, harder to reduce, such as the others' beliefs or rationality.

More generally, one considers an uncertainty space common to all players and comprising all relevant characteristics of the game. This space is formalized by an algebra $(S, \mathcal{E})$, where elements of S are called *states* and elements of $\mathcal{E}$ *events*. On the uncertainty space, Aumann (1976) considers a partition $\bar{\mathcal{K}}_i$ such that if the state s happens then player i knows that the real state is in the subset $H_i(s)$; moreover, he claims (as usual) that player i knows event E in state s if $H_i(s) \subseteq E$. In fact, Aumann implicitly views the accessibility relation as an equivalence relation, with the player satisfying the strongest axiomatic system $S5_m$, and suggests a theoretical justification. First he assumes that $\mathcal{K}_i$ is known by the player; if not, it is still possible for the modeler to extend the uncertainty space to endogenize that uncertainty until $\mathcal{K}_i$ is known by construction. Second, Aumann claims that if $\mathcal{K}_i$ is not an equivalence relation and if player i observes $H_i(s)$ in state s, then i knows that the real state has $H_i(s)$ as an image and belongs to $H_i'(s) = H_i^{-1} H_i(s)$, $\mathcal{K}_i'$ being now an equivalence relation.

3. BOUNDED KNOWLEDGE

The first weakening consists of giving up axiom A5; $\bar{\mathcal{K}}_i$ is then a *basis* for S – that is, a set that covers S and for which

$$H_i(s) \cap H_i(t) \neq \emptyset \Rightarrow H_i(s) \cap H_i(t) \text{ is a union of elements of } \bar{\mathcal{K}}_i.$$

A subclass for such a basis is obtained when $\bar{\mathcal{K}}_i$ is nested (Geanakoplos 1989); that is,

$$H_i(s) \cap H_i(t) \neq \emptyset \Rightarrow H_i(s) \subseteq H_i(t) \text{ or } H_i(t) \subseteq H_i(s).$$

A sufficient condition on $\mathcal{K}_i$ (necessary with $\mathcal{K}_i$ reflexive) is that it be semi-anti-Euclidean:

$$[(s, u) \in \mathcal{K}_i \text{ and } (t, u) \in \mathcal{K}_i] \Rightarrow (s, t) \in \mathcal{K}_i \text{ or } (t, s) \in \mathcal{K}_i.$$

But this property is not a weakening of the Euclidean property. It may be interpreted syntactically as a natural order on primitive propositions and hence on all propositions.

The second weakening consists of giving up axiom A3, or better, replacing it with a weaker one:

A3': $\neg K_i \bot$

    (an agent does not believe in contradictory propositions).

The knowledge operator is then called a belief operator B_i such that

$$B_i p = \neg K_i \neg p,$$

and the axiomatic system is $KD45_m$. In the Kripke structure, $\mathcal{K}_i$ is no longer reflexive, becoming instead serial:

$$\forall s, \exists t \text{ such that } (s, t) \in \mathcal{K}_i.$$

The set $\bar{\mathcal{K}}_i$ then forms a partition; however, on the restricted set S' of worlds where $\mathcal{K}_i$ is reflexive,

$$S' = \{s \in S \mid s \in H_i(s)\}.$$

The third weakening, more radical, consists of substituting for axiom A2 and rule R2 a much weaker rule:

R2': from $\vdash (p \leftrightarrow q)$, infer $\vdash (K_i p \leftrightarrow K_i q)$
    (an agent simultaneously knows all equivalent propositions).

One can consider an associated "explicit knowledge" operator N_i (as opposed to the implicit K_i) obtained by deductive closure:

$$N_i p \wedge N_i (p \rightarrow q) \rightarrow K_i q.$$

On the semantic side, the Kripke structure must be replaced by a weaker one, the Scott structure $\mathcal{C}$ (Chellas 1980). In each world s, agent i now considers a family $\mathfrak{N}_i(s)$ of subsets of worlds; the validity of a proposition is given by conditions B1–B3 and:

B4'': $(\mathcal{C}, s) \models K_i p$ iff $|p| \in \mathfrak{N}_i(s)$
    (agent i knows proposition p at s if its field is a considered subset).

A Kripke structure appears as a special case when

$$\mathfrak{N}_i(s) = \{E \mid H_i(s) \subseteq E\}.$$

To obtain axiomatic system K_m, A2' must be completed by the following axioms:

A21: $N_i(\top)$
    (an agent knows each tautology);
A22: $N_i p \wedge N_i q \rightarrow N_i (p \wedge q)$
    (knowing two propositions is knowing their combination);
A23: $N_i(p \wedge q) \rightarrow N_i p \wedge N_i q$
    (knowing the combination of two propositions is knowing each of them).

These axioms have, as a semantic counterpart, the following properties:

(1) $S \in \mathfrak{N}_i(s)$ (cf. A21);
(2) stability by intersection: if $E, F \in \mathfrak{N}_i(s)$ then $E \cap F \in \mathfrak{N}_i(s)$ (cf. A22);
(3) stability by augmentation: if $E \in \mathfrak{N}_i(s)$ and $E \subseteq F$ then $F \in \mathfrak{N}_i(s)$ (cf. A23).

Moreover, axioms A3 to A5 have a direct interpretation:

(1) if $E \in \mathfrak{N}_i(s)$ then $s \in E$ (cf. A3);
(2) if $EP\mathfrak{N}_i(s)$ then $\{t \mid E \in \mathfrak{N}_i(t)\} \in \mathfrak{N}_i(s)$ (cf. A4);
(3) if $E \notin \mathfrak{N}_i(s)$ then $\{t \mid E \notin \mathfrak{N}_i(t)\} \in \mathfrak{N}_i(s)$ (cf. A5).

Game theory. In game theory, axiom A5 (like axiom A4) is assumed to be empirically satisfied if the game is a "small world," the players considering all characteristics that are relevant and being able to assess if they know them or not. However, if the game is too complex, some factors that are put into the uncertainty space by the modeler are neglected or ignored by the players – for instance, the others' knowledge. Axiom A3 is also assumed to be satisfied if the players consider the true value of each variable as being possible; in probabilistic knowledge (see Section 4), this simply means that players assign a nonzero probability to the true value. However, if the players have sufficiently precise knowledge then A3 may be rendered false (e.g., by poorly observing past actions or considering wrong preferences for others).

The weakening of axiom A2 is a very old idea in economics, an agent having a "bounded rationality" (Simon 1986) due to constraints of gathering and processing information. This idea (Walliser 1989) was first applied to instrumental rationality adequation between available means and pursued objectives. Optimizing behavior was replaced by ϵ-optimizing (action giving optimal utility within ϵ) or satisficing (first action exceeding aspiration levels on partial goals). It is now adapted to cognitive rationality (adequation between available information and adopted beliefs). The reasoning of the player (considered sometimes as an automaton) rests only on a finite number of internal states, his memory being in particular limited; or implies costs of calculus, leading to a trade-off with the aimed utility; or even is submitted to upper bounds for crossed beliefs (see Section 5).

4. PROBABILISTIC KNOWLEDGE

In order to qualify agents' uncertainty, many attempts were made to give a probability structure to logical propositions, that is, to define $\Pr_i(p)$ as the probability given by agent i to proposition p. Some authors (Fagin and Halpern 1988, Fagin, Halpern, and Meggido 1988) go a bit further and define a new type of logical proposition $K_i^\alpha p$, which means "for agent i, proposition p holds with probability at least α": $\Pr_i(p) \geq \alpha$. They give an axiom system K_m^α (formed by axiom A1 and rule R1), six technical axioms allowing linear combinations of probabilized propositions ($\Sigma_k a_k \Pr_i(p_k) \geq \alpha$), and four axioms classically related to probability reasoning:

P1: positivity: $\Pr_i(p) \geq 0$;
P2: certainty: $\Pr_i(\top) = 1$;
P3: additivity: $\Pr_i(p \wedge q) + \Pr_i(p \wedge \neg q) = \Pr_i(p)$; and
P4: distributivity: $\Pr_i(p) = \Pr_i(q)$ if $p \leftrightarrow q$.

On the semantic side, a probabilistic structure $\mathcal{P}$ is defined by four elements, including:

(1) a set S of possible worlds s;
(2) a value function: $v(s, \hat{p})$; and
(3) a probabilized σ-algebra $(H_{is}, \mathcal{E}_{is}, \Pi_{is})$ defined on each state s, where $\mathcal{E}_{is}$ are the measurable subsets for the probability distribution Π_{is}.

The proposition $K_i^\alpha p$ is then valid in s if the following fourth condition is satisfied:

B4″: $(\mathcal{P}, s) \vDash K_i^\alpha p$ iff $\Pi_{is}(H_i(s) \cap |p|) \geq \alpha$
   (an agent knows p with probability α if the perceived field of p has probability α).

It has been shown that the axiomatic system K_m^α is sound and complete for the probability structure $\mathcal{P}$.

Probabilization refines the accessibility relation, in that the propositions are not ranked in three classes but rather on a continuous scale; a proposition known by an agent has probability 1: $K_i p \to K_i^1 p$. It seems then possible to introduce a probabilistic knowledge system $\mathbf{K}_i^\alpha = \{p$ s.t. $K_i^\alpha p\}$, but the usual rules of the probability calculus show that this set is not deductively closed:

$$K_i^\alpha p \wedge K_i^\alpha (p \to q) \not\to K_i^\alpha q \quad \text{if } \alpha \neq 0, 1.$$

On another point of view, the probability distribution Π_{is} is sometimes restricted as follows:

U: uniformity: $\Pi_i(s) = H_i(t)$ and $\Pi_{is} = \Pi_{it}$, $\forall(s, t)$
   (the probability distribution of an agent does not depend on the real world);
O: objectivity: $H_i(s) = H_j(s)$ and $\Pi_{is} = \Pi_{js}$, $\forall(i, j)$
   (the probability distribution is the same for all agents).

In the semantic framework (here uniformity is assumed, for simplicity), the Choquet capacity Q_i, a weaker notion than probability, is defined by

$$Q_i(\emptyset) = 0,$$
$$Q_i(S) = 1,$$
$$E \subseteq F \Rightarrow Q_i(E) \le Q_i(F).$$

The Choquet capacity can alternatively be defined by its Möbius inverse m_i, also called *mass function,* defined on $\mathcal{P}(S)$:

$$m_i(E) = \sum_{F \subseteq E} (-1)^{|E \setminus F|} Q_i(F) \quad \text{satisfies } m_i(\emptyset) = 0 \text{ and } \sum_{E \subseteq S} m_i(E) = 1.$$

The mass function is always positive if the capacity satisfies the property of "monotonicity at order k": for each k,

$$Q_i\left(\bigcup_{j=1}^{k} E_j\right) \ge \sum_{J \subseteq [1, \ldots, k] \setminus \emptyset} (-1)^{|J|+1} Q\left(\bigcap_{j \in J} E_j\right).$$

Under this condition, it is possible to define two notions in duality (Shafer 1976):

(1) credibility: $\mathrm{Cr}_i(E) = Q_i(E) = \sum_{F \subseteq E} m_i(F)$;
(2) plausibility: $\mathrm{Pl}_i(E) = 1 - Q_i(\neg E) = \sum_{E \cap F \ne \emptyset} m_i(F)$.

The *focal set* of the mass function is the set of subsets where the mass is strictly positive: $\mathcal{C}_i = \{E \mid m_i(E) > 0\}$. It may coincide with the set of subsets $\hat{E}$ corresponding to the primitive propositions $\hat{p}$. If E is then the field of p, the *credibility* of E is the sum of the masses of the arguments $\hat{p}$ in favor of p, and the *plausibility* the sum of the masses of the arguments $\hat{p}$ not contradicting p. Two specific cases are of interest:

(1) $\mathcal{C}_i$ is composed of strictly nested subsets: $E_1^i \subset \cdots \subset E_k^i \subset \cdots \subset E_n^i$.
Credibility is then called *necessity* and plausibility is called *possibility* with the generic properties

$$\mathrm{Ne}_i(E \cap F) = \min(\mathrm{Ne}_i(E), \mathrm{Ne}_i(F)),$$
$$\mathrm{Po}_i(E \cup F) = \max(\mathrm{Po}_i(E), \mathrm{Po}_i(F)).$$

(2) $\mathcal{C}_i$ is composed of all singletons: $\{s\}$.
Credibility and plausibility are equal, and form a probability with the generic property

$$\mathrm{Pr}_i(E \cup F) + \mathrm{Pr}_i(E \cap F) = \mathrm{Pr}_i(E) + \mathrm{Pr}_i(F).$$

Game theory. In game theory, Aumann considers that players have a prior probability distribution $\Pi_i^-(t)$ on the finite states; when combined with a message taking the form of a partition $\mathcal{H}_i$, they obtain a posterior

206

distribution $\Pi_{is}^+(t)$ depending on the real state s by applying Bayesian rules (see Section 8). The prior distribution is assumed to satisfy uniformity condition U; the posterior distribution also satisfies condition U, but only for each element of the partition $\Pi_{is}^+ = \Pi_{it}^+$ if $t \in H_i(s)$. The prior distribution is frequently required to satisfy objectivity condition O, according to the "Harsanyi doctrine"; $\Pi_i^-(t) = \Pi_j^-(t)$. The Harsanyi doctrine states that similar agents have no reason to have different priors, as their difference of probability assessment can only stem from different private information (translated into different partitions).

Mertens and Zamir (1985) or Tan and Werlang (1986) define a probabilized uncertainty space $(S, \mathcal{E}_i, \Pi_i)$, where a player knows that event E happens if $\Pi_i(E) = 1$. They implicitly suppose that the uniformity condition U is satisfied but not the objectivity condition O (as is usual in Bayesian decision theory), since Π_i is a subjective probability distribution directly defined or stemming from Savage's axioms. Previously, Shackle (1961) theorized that agents assess not the probability of an event but rather the "potential surprise" $\mathrm{Sp}_i(E)$ when the event occurs. In modern terms, it can be shown that Shackle's potential surprise is nothing more than a possibility measure: $\mathrm{Sp}_i(E) = 1 - \mathrm{Po}_i(E) = \mathrm{Ne}_i(\neg E)$. More recently, by weakening Savage's axioms, Gilboa (1987) showed the existence of a subjective credibility distribution.

5. Levels of knowledge

The notion of crossed knowledge between agents is easy to formalize in a syntactic framework, through cumulative operators such as $K_i K_j K_i p$ (agent i knows that j knows that i knows that p). This notion is harder to formalize in a Kripke structure because the number of needed worlds may rapidly grow with the number of levels when knowledge does not evolve regularly; this is why Fagin, Halpern, and Vardi (1988) proposed a third semantic interpretation, the "knowledge structure," grounded on piled-up worlds. At level 0, a function f_0 defines the propositions considered as true by the modeler. At level k, the function $f_k(i)$ defines, in a set W_k of possible k-worlds, the k-worlds w_k^n that agent i really considers as possible; a function $g_k(i)$ defines, in the same set W_k, the k-worlds that agent i is assumed *by another agent* to consider as possible. A k-world is then formed of possible knowledge of all agents at all inferior levels: $w_k^n = \langle g_0, ..., g_{k-1} \rangle$, where $g_l = \langle g_l(1), ..., g_l(i), ... \rangle$.

The $(k+1)$-world actually considered by all agents is denoted $\hat{f}_{k+1} = \langle f_0, ..., f_k \rangle$. It obeys interlevel conditions, one being a consistency condition and the others being counterparts of the basic axioms:

K0: extension axiom: $\langle g_0, \ldots, g_{k-2} \rangle \in f_{k-1}(i)$ iff $\exists g_{k-1}$ s.t.
$\langle g_0, \ldots, g_{k-1} \rangle \in f_k(i)$
(an agent always builds $(k+1)$-knowledge by extending k-worlds and conversely);

K3: factuality axiom: $\langle f_0, \ldots, f_{k-1} \rangle \in f_k(i)$ (cf. A3)
(an agent considers, in his k-worlds, the real k-world to be possible); and

K4-5: self-knowledge axiom: if $\langle g_0, \ldots, g_{k-1} \rangle \in f_k(i)$ then
$g_{k-1}(i) = f_{k-1}(i)$ (cf. A4-5)
(an agent know his own $(k-1)$-worlds in the real k-world).

If the three conditions are satisfied between all successive levels k and $k-1$, conditions K0 and K4-5 are satisfied between any levels k and l ($l < k$); moreover, the sufficiency of condition K0 stems from the other conditions.

A traditional example is the following:

level 0: p is true;

level 1: Alice does not know p, but Bob knows: $\neg K_A p$, $K_B p$;

level 2: Alice knows that Bob knows p or $\neg p$ (without knowing p herself), and Bob does not know whether Alice knows p:
$K_A(K_B p \wedge K_B \neg p)$, $\neg K_B K_A p$.

The counterpart is the knowledge structure where conditions K0, K3, and K4-5 are satisfied:

level 0: $f_0 = p$;

level 1: $f_1(A) = \langle p, \neg p \rangle$
$f_1(B) = p$;

level 2: $f_2(A) = \langle w^1, w^2 \rangle$ with $w^1 = \langle p, g_1(A) = \langle p, \neg p \rangle, g_1(B) = p \rangle$
$f_2(B) = \langle w^1, w^3 \rangle$ $\qquad w^2 = \langle \neg p, g_1(A) = \langle p, \neg p \rangle, g_1(B) = \neg p \rangle$
$w^3 = \langle p, g_1(A) = p, g_1(B) = p \rangle$

(but other worlds may be considered; e.g., if Alice thinks that Bob can have false knowledge).

Let us define the *depth* of proposition p (denoted δp) as the highest level of crossed knowledge between agents in that proposition:

$$\delta \hat{p} = 0, \quad \delta p \vee q = \max(\delta p, \delta q),$$

$$\delta \neg p = \delta p, \quad \delta K_i p = 1 + \delta p.$$

At the $(k+1)$-world $\hat{f}_{k+1}$, p is then said to be *valid* (denoted $\hat{f}_{k+1} \models p$) under the following conditions:

$$\hat{f}_{k+1} \vDash \hat{p} \quad \text{iff } \hat{p} \in f_0;$$

$$\hat{f}_{k+1} \vDash \neg p \quad \text{iff } \hat{f}_{k+1} \nvDash p;$$

$$\hat{f}_{k+1} \vDash p \vee q \quad \text{iff } \hat{f}_{k+1} \vDash p \text{ and } \hat{f}_{k+1} \vDash q;$$

$$\hat{f}_{k+1} \vDash K_i p \quad \text{iff } \langle g_0, ..., g_{k-1} \rangle \vDash p \; \forall \langle g_0, ..., g_{k-1} \rangle \in f_k(i)$$

(an agent knows p in a $(k+1)$-world if p is valid in each possible k-world).

The axioms A1 and A2 and rules R1 and R2 are automatically satisfied by this notion of validity. Moreover, if a proposition of depth k is valid in a $(k+1)$-world then it is valid in each superior l-world $(l > k)$.

Formally, a knowledge structure appears as a very peculiar Kripke structure: heterogeneous worlds gathered in successive levels, with the real universe consisting of one world at each level. The accessibility relation of agent i joins a world of level k to the world of inferior levels that it considers as possible. This relation forms a lattice that has specific properties corresponding to conditions K0, K3, and K4-5, and defines the validity of $K_i p$. The most interesting point is that each world comprises not only worlds of the next inferior level but rather *all* (cumulated) inferior ones; it would not be possible otherwise to express all subtleties of knowledge. In the standard example, it is necessary to consider simultaneously three successive levels in order to formalize the following statement: Alice knows either that p is true and Bob knows it, or that p is false and Bob does not know it.

Game theory. In game theory, crossed knowledge was first introduced in a binary form, concerning the information of two players about an exogenous event. At level 1, information is said to be ignored, personal, or shared if (respectively) zero, one, or both players know it; at level 2, information (shared, for example) is said to be secret, private, or public if (respectively) zero, one, or both players know that the other knows it. Crossed knowledge was also suggested, in probabilistic terms, with respect to the assessment of an event by the player himself (Ellsberg 1962). At level 1, the player defines the "occurrence" of an event by a probability distribution; at level 2, the "weight" of this preceding evaluation is defined by a probability distribution on the preceding one. Of course, in orthodox Bayesian terms the two assessments are collapsed into a unique one.

In Aumann's partition framework, at level 1 in a state s, player i knows that event E happened if $H_i(s) \subseteq E$; at level 2, player i knows that j knows that event E happened if $H_i(H_j(s)) \subseteq E$, where $H_i(A) = \bigcup_{s \in A} H_i(s)$. In fact, player i is assumed to know the partition of player j and constructs a combined accessibility relation $\mathcal{K}_{ij}$ such that

$$(s, t) \in \mathfrak{IC}_{ij} \quad \text{iff} \quad \exists u \text{ such that } (s, u) \in \mathfrak{IC}_i \text{ and } (u, t) \in \mathfrak{IC}_j.$$

A first restriction of this framework is that each level is defined only on the preceding one; K0 being satisfied, the relation $\mathfrak{IC}_{ij}$ is an equivalence relation if $\mathfrak{IC}_i$ and $\mathfrak{IC}_j$ are, which corresponds to conditions K3 and K4–5. The second restriction is that each player has a real uncertainty only at level 1, but is certain of others' uncertainty at all superior levels; for this framework to stand on a more general basis, it would be necessary to consider second-level partitions of basic ones.

In the probabilistic framework, at level 1 each player has a probability distribution t_i^1 on the states $t_i^1 \in \Delta_i(S)$; at level 2, a probability distribution t_i^2 is defined on others' probability distributions $t_i^2 \in \Delta_i \Delta_j(S)$; and so on. Condition K0 is satisfied by construction (and K3 as well) if all inferior distributions are considered at the superior level. For Tan and Werlang (1986), t_i^k belongs to $T_i^k = \Delta(S \times T_{-i}^{k-1})$. This construction defines each level only with respect to the preceding one, except for allowing a correlation with the 0-level; it must satisfy a minimal consistency requirement, analogous to K4–5: $\text{marg}_{S_0} t_i^k = t_i^{k-1}$ (the same event is evaluated similarly at all levels by a given agent). For Mertens and Zamir (1985), t_i^k belongs to $T_i^k = \Delta(S \times T_i^1 \times T_{-i}^1 \times \cdots \times T_i^{k-1} \times T_{-i}^{k-1})$. This construction defines each level with respect to all the preceding ones, and permits correlations between successive levels; it is also constrained to satisfy a consistency requirement similar to K4–5:

$$\text{marg}_{T_i^{k-1}} t_i^k = t_i^{k-1}.$$

6. KNOWLEDGE DISTRIBUTION

When moving up through the crossed levels of knowledge to infinity, a *complete knowledge structure* f is a sequence $f = \langle f_0, \ldots, f_k, \ldots \rangle$; a proposition p is said to be *valid in* f (denoted $f \models p$) if it is valid in the $(k+1)$-world $\hat{f}_{k+1}$, where $k = \delta p$. It can easily be shown that $S5_m$ is a sound and complete axiomatization of a complete knowledge structure under conditions K0, K3, and K4–5. One can even show that it is possible to associate a family of knowledge structures $f_s(S)$ to each Kripke structure S, the same propositions being valid in (S, s) and f_s; this means that each world corresponds now to a state of crossed knowledge, just as the transition between worlds does to the uncertainty about crossed knowledge. Finally, the probabilization of a Kripke structure may be extended to a knowledge structure; the syntactic counterpart creates no problem since $K_j^\beta K_i^\alpha p \equiv \Pr_j(\Pr_i(p) \geq \alpha) \geq \beta$ is well defined.

210

For a group of agents, it is convenient to define syntactically specific knowledge operators (Halpern and Moses 1988), corresponding to knowledge "always better allocated":

Dp: distributed knowledge: Dp iff $\exists\{q_i\}$ s.t. $K_1 q_1 \wedge \cdots \wedge K_m q_m \to p$
(p may be known by associating the knowledge of all agents);
Sp: scattered knowledge: $Sp = K_1 p \vee K_2 p \vee \cdots \vee K_m p$
(p is known by at least one agent);
Ep: expanded (or mutual) knowledge: $Ep = K_1 p \wedge K_2 p \wedge \cdots \wedge K_m p$
(p is known by all agents);
Cp: common knowledge: $Cp = Ep \wedge E^2 p \wedge \cdots \wedge E^k p \wedge \cdots$
(everybody knows p, knows that the others know, …).

If all K_i satisfy A2, the same is true for D, E, and C but not S; if all K_i satisfy A3 then $Xp \to p$ is satisfied by D, S, E, and C; if all K_i satisfy A4 then $Xp \to X^2 p$ is satisfied by S or C but not by D or E; if all K_i satisfy A5 then $\neg Xp \to X \neg Xp$ is satisfied by C only.

The common knowledge operator (Shin 1988) is here defined as a hierarchical framework, which may be problematic to the extent it involves an infinite conjunction of propositions. It may also be defined in a circular framework, with the following axioms:

C2: closure axiom: $(Cp \wedge C(p \to q)) \to Cq$
(all logical consequences of common knowledge are common knowledge);
C3–4: fixed point axiom: $Cp \equiv E(p \wedge Cp)$
(a proposition is common knowledge if everybody knows that it is and if it is true);
RC1: induction rule: from $\vdash p \to Ep$, deduce $\vdash p \to Cp$
(if a proposition is inferred to be known by all as soon as it is true, it is common knowledge).

In a synthetic framework, it may be shown (Barwise 1988) that the circular view entails the hierarchical one, but not conversely. A third view, developed in an original semantic framework (Barwise 1988), is grounded on the idea that an event is common knowledge if it is mutually manifest (each agent observes it in the presence of the others).

In the standard Kripke structure, if axiom A3 is satisfied then $|E^k p| \subseteq |E^{k-1} p|$, and the field of crossed knowledge of p decreases with reductions in the level. If the set S is finite, there must then exist a level k at which the crossed knowledge becomes common knowledge: $Cp \equiv E^k p$. In a rich and complex environment, it may be difficult to acquire common knowledge in either the hierarchical approach (although a recurrence, once initiated, may be continued quite automatically) or in the circular one (even if a fixed point may be easily discovered). It is then possible to introduce an "almost common knowledge" operator C^ϵ (from the operator $K_i^\epsilon p$, "i almost knows p") or a "probabilistic common knowledge" operator C^α (from the operator $K_i^\alpha p$ and a fixed-point approach).

Game theory. In a framework close to game theory, following some ideas of Schelling (1960), Lewis (1965) introduced qualitatively the hierarchical notion of common knowledge about some structural features of a social situation. More precisely, he defined a *convention* as a regularity R in the behavior of members of a population P, considered in a recurrent situation S, by four conditions that are common knowledge:

(1) each member conforms to R;
(2) each member thinks that the others conform to R;
(3) each member prefers general conformity to R to less general conformity; and
(4) there exists another regularity satisfying the three preceding conditions.

In fact, Lewis restricts himself to a coordination game, and the preceding conditions then define a coordination equilibrium, a notion that is notably stronger than a Nash equilibrium (see Section 10).

In Aumann's partition framework, an event E is said to be *common knowledge* at state s if $H(s) \subseteq E$, where $H(s)$ is an element containing s of partition $\mathcal{H} = \mathcal{H}_i \vee \mathcal{H}_j$; $\mathcal{H}$ is the finest common coarsening of partitions $\mathcal{H}_i$ and $\mathcal{H}_j$ and is obtained as follows:

$$t \in H(s) \leftrightarrow \exists\, u_0 = s, u_1, \ldots, u_{n-1}, u_n = t \text{ s.t. } u_n \in H_i(u_{n-1}) \text{ or } H_j(u_{n-1}).$$

This hierarchical definition can be shown to be directly equivalent to the circular one, by reasoning on the fields associated with the propositions (Milgrom 1981, Bacharach 1985). But the definition also assumes that the partitions $\mathcal{H}_i$ themselves are common knowledge, in a qualitative sense (they are not events); Aumann argues (not very convincingly) that if this is not so then the set of states is incomplete and must be augmented, the final set then having partitions by construction. As for his probabilistic extension, Aumann also assumes that the priors Π_i^- are common knowledge for the same reasons; it is then compatible with the partition definition as soon as partition $\mathcal{H}_i \vee \mathcal{H}_j$, the coarsest common refinement of partitions $\mathcal{H}_i$ and $\mathcal{H}_j$, has all subsets with nonzero probability.

In a general probabilistic framework, it is also possible to complete the recursion to infinity, a hierarchy of knowledge defining the (cognitive) "type" of a player: $t_i = (t_i^1, t_i^2, \ldots) \in T_i$. An event E is then common knowledge if it is given a probability of 1 by all plyers at all levels. This hierarchy is sufficient to summarize a player's knowledge, without necessitating knowledge about the others' type, because there exists a canonical homeomorphism between T_i and $\Delta(S \times T_{-i})$; it also satisfies the consistency conditions (cf. A4–5) previously considered, and the closure condition (cf. A3) stating that each agent considers as possible the others' type (t_i assigns a probability of 1 to T_j). Using this approach, Brandenburger and Dekel (1985) show that Aumann's assumption of common knowledge of partitions can be justified by extending S to $S \times \bar{T}$ (where $\bar{T}$ is a maximal closed set of hierarchical knowledge), under the assumption of common knowledge of the consistency property. In the same vein, Brandenburger and Dekel (1987b) show that it is possible to abandon the hypothesis of nonzero probability of the subsets of $\bar{\mathfrak{K}}_i \vee \bar{\mathfrak{K}}_j$ by reformulating the posterior probabilities. Finally, Stinchcombe (1988) as well as Monderer and Samet (1989) introduce "almost common knowledge" in game theory.

7. COMPARISON OF KNOWLEDGE

In the syntactic framework, a knowledge system is more informative (denoted $\succsim_i$) than another if it contains more propositions and is consistent:

$$\mathbf{K}_i' \succsim_i \mathbf{K}_i \Leftrightarrow \mathbf{K}_i' \supseteq \mathbf{K}_i;$$

in a given Kripke structure, the field $\mathfrak{K}_i'$ is smaller than $\mathfrak{K}_i$. In the semantic framework, a Kripke structure is more informative than another if the agent considers fewer possible worlds in each world:

$$S' \succsim_i S \Leftrightarrow H_i'(s) \supseteq H_i(s), \quad \forall s.$$

Still in the semantic framework, a knowledge structure is more informative than another at level k if the agent considers fewer possible k-worlds:

$$f' \succsim_i^k f \Leftrightarrow f_k'(i) \supseteq f_k(i).$$

However, if condition K4–5 is satisfied then a knowledge structure is more informative than another only if they are equivalent; this is because an agent knows exactly in his $(k+1)$-world the k-worlds he knows.

In a knowledge structure, it is possible to extend a given $(k+1)$-world $\hat{f}_{k+1} = \langle f_0, \ldots, f_k \rangle$ into a $(k+2)$-world $\hat{f}_{k+2}$ according to axioms K0 through K4–5, with two polar cases. In the extension without information, agent i learns nothing about the inferior level, except about himself:

$$\tilde{f}_{k+1}^-(i) = \{\langle g_0, \ldots, g_k \rangle \text{ s.t. } g_k(i) = f_k(i)\}.$$

In the extension with perfect information, agent i learns exactly what happens at the inferior level about the others, but this must not induce new knowledge for himself at that inferior level:

213

$$\tilde{f}_{k+1}^+(i) = \{\langle g_0, \ldots, g_k \rangle \text{ s.t. } g_k(j) = f_k(j) \cup f_k(i)\}.$$

For the standard example of Bob and Alice, starting from level 1:

$$\tilde{f}_2^-(A) = \{w^1, w^2, w^4, w^5\}$$
$$\text{with } w^4 = \langle p, g_1(A) = (p, \neg p), g_1(B) = (p, \neg p)\rangle$$
$$\tilde{f}_2^-(B) = \{w^1, w^3\} \qquad w^5 = \langle \neg p, g_1(A) = (p, \neg p), g_1(B) = (p, \neg p)\rangle$$
$$\tilde{f}_2^+(A) = \{w^1, w^2\}$$
$$\tilde{f}_2^+(B) = \{w^1\}.$$

When considering Choquet capacities on the world space, a credibility function is said to be more informative than another if its values are higher on all subsets:

$$\text{Cr}_i' \succeq_i \text{Cr}_i \Leftrightarrow \text{Cr}_i'(E) \geq \text{Cr}_i(E), \ \forall E.$$

In the specific case of a probability function, this condition is too weak to compare nonidentical distributions:

$$\text{Pr}_i' \succeq_i \text{Pr}_i \Leftrightarrow \Pi_i'(s) = \Pi_i(s), \ \forall s.$$

The preceding order characterizes the precision permitted at a second level in the assessment of the uncertainty of the worlds. It must be distinguished from orders characterizing at the first level the uncertainty of the worlds themselves, for example the "dissonance" D (Yager 1983):

$$D(\text{Cr}_i) = - \sum_{E \in \mathcal{C}_i} m_i(E) \log_2 \text{Pl}_i(E).$$

In the specific case of probabilities, dissonance yields entropy:

$$I(\text{Pr}_i) = - \sum_{s \in S} \Pi_i(s) \log_2 \Pi_i(s).$$

Game theory. In game theory, in the partition framework, in accordance with the Kripke structure definition a partition $\mathcal{K}_i'$ is said to be more informative than another $\mathcal{K}_i$ if it is finer: $H_i'(s) \subseteq H_i(s)$, $\forall s$. If A3 is satisfied then all partitions on the state space form a lattice, the coarsest corresponding to the situation where no state is distinguished, $\bar{H}_i(s) = S$, and the finest to the situation where all states are distinguished: $\underline{H}_i(s) = \{s\}$. Moreover, if at the first level the partitions $\mathcal{K}_i'$ are finer than $\mathcal{K}_i$ for all agents, then the partitions corresponding to higher levels of crossed knowledge are likely ordered, and the partition $\mathcal{K}'$ corresponding to common knowledge is finer than $\mathcal{K}$.

In the probabilistic framework, probability distributions may be compared on the first level of knowledge as well as on higher ones. All distributions form a partial order, the coarsest corresponding to the uniform distribution and the finest to the Dirac measure. In decision theory (see Section 9), the states may reflect exogenous factors a_0 acting on the results of an action a_i or directly on the combined results r themselves, generally expressed in monetary terms (i.e., as a real variable). Calling F_i and F_i' the repartition functions of the probability distributions Π_i and Π_i', two definitions of stochastic dominance are given:

first-order: $\Pi_i' \gtrsim \Pi_i$ iff $F_i'(r) \leq F_i(r)$;
second-order: $\Pi_i' \gtrsim \Pi_i$ iff

$$\int_{-\infty}^{R} F_i'(r) \leq \int_{-\infty}^{R} F_i(r)\, dr, \quad \forall R.$$

8. UPDATING OF KNOWLEDGE

In a syntactic framework, an agent may receive a message in the shape of a new proposition q generally assumed to be true, and updates his old knowledge system $\mathbf{K}_i$ in a revised one $\mathbf{K}_i^*(q)$. If the message is compatible with the preceding knowledge, it is simply added to it, together with all combined consequences:

$$\mathbf{K}_i^*(p) = \mathbf{K}_i^+(p) = \{p \mid \mathbf{K}_i \cup q \vdash p\}.$$

If the message is not compatible with the preceding knowledge then the knowledge revision is not so well defined; some old propositions (which together imply $\neg p$) must be removed and replaced by q. This procedure is similar to Duhem–Quine's problem in epistemology (Duhem 1906), which asks what assumptions must be removed when a testable consequence of a set of assumptions is falsified by experience. A frequently stated conservation principle suggests that the revision be made by reorganizing the knowledge system in the least disruptive way possible.

In a given Kripke structure, when the message is compatible with the preceding knowledge, the possible worlds shrink to the subset $H_i'(s) = H_i(s) \cap |q|$; otherwise, $H_i(s) \cap |q|$ is empty and $H_i'(s)$ is indeterminate (except that it must be taken in $|q|$ and be nonempty). More generally, a "message structure" announces a new proposition conditional on the actual world, and takes the form of a new accessibility relation $\mathfrak{M}$; the old accessibility relation $\mathfrak{K}_i$ is then revised into a new one $\mathfrak{K}_i' = \mathfrak{K}_i \vee \mathfrak{M}$, with $H_i'(s) = H_i(s) \cap M(s)$ (but some subsets may be empty and duly replaced). When several messages are coming successively, a learning process is induced that may or may not converge toward a perfect knowledge

215

of the real state; only on the condition that no message contradicts preceding knowledge is the updating independent from the order of the messages and the knowledge considered to be progressively more informative.

To formalize the revision of knowledge, Gärdenfors (1988) developed a syntactic system of axioms in order to construct $\mathbf{K}_i^*(q)$:

D1: $\mathbf{K}_i^*(q)$ is a knowledge system;
D2: $q \in \mathbf{K}_i^*(q)$;
D3: $\mathbf{K}_i^*(q) \subseteq \mathbf{K}_i^+(q)$;
D4: if $\neg q \in \mathbf{K}_i$ then $\mathbf{K}_i^+(q) \subseteq \mathbf{K}_i^*(q)$;
D5: $\mathbf{K}_i^*(q) = \bar{\mathbf{K}}$ if $\vdash \neg q$;
D6: if $\vdash q \leftrightarrow r$ then $\mathbf{K}_i^*(q) = \mathbf{K}_i^*(r)$;
D7: $\mathbf{K}_i^*(q \vee r) \subseteq (\mathbf{K}_i^*(q))^+(r)$; and
D8: if $\neg r \notin \mathbf{K}_i^*(q)$ then $(\mathbf{K}_i^*(q))^+(r) \subseteq \mathbf{K}_i^*(q \wedge r)$.

There are many revision procedures satisfying these axioms; in the semantic framework, each corresponds to a specific choice of $H_i'(s)$:

(1) if $H_i(s) \cap |q| \neq \emptyset$ then $H_i'(s) = H_i(s) \cap |q|$;
(2) if $H_i(s) \cap |q| = \emptyset$ then $H_i'(s) = X_i \subseteq |q|$ and $H_i'(s) \neq \emptyset$.

According to the conservation principle, X_i may be chosen as the nearest world to $H_i(s)$ only if a metric is defined on the world space.

The choice of X_i may in any case be specified by defining an "epistemic entrenchment" index on propositions (Gärdenfors 1988, Gärdenfors and Makinson 1988) such that, if $\mathbf{K}_i$ contains p and if agent i learns that q, then i keeps p if $p \wedge q$ is more entrenched than $p \wedge \neg q$. The order of entrenchment is given by:

E1: transitivity: $p \leq_i q$ and $q \leq_i r \Rightarrow p \leq_i r$;
E2: dominance: if $p \vdash q$ then $p \leq_i q$;
E3: conjunctivity: $p \geq_i p \vee q$ or $q \geq_i p \vee q$;
E4: minimality: $\neg p \notin \mathbf{K}_i$ iff $p \geq_i q, \forall q$;
E5: maximality: if $q \geq_i p \ \forall q$ then $\mathbf{K}_i \vdash p$.

In fact, as axiom E3 clearly shows (Dubois and Prade 1989), epistemic entrenchment is an ordinal possibility distribution on the propositions (or the worlds). Thus it is possible to define the focal set associated with the distribution as $E_1^i = H_i(s) \subset E_2^i \subset E_3^i \subset \cdots$ (Spohn 1988); X_i is then defined by $E_k^i \cap |q|$, where k is the smallest number h such that $E_h^i \cap |q| \neq 0$.

In a probabilistic framework, when a proposition q is learned, the usual updating rule is the Bayesian rule transforming a prior probability Pr_i^- into a posterior probability Pr_i^+, a procedure that is here indeterminate if $\mathrm{Pr}_i^-(q) = 0$:

$$\mathrm{Pr}_i^+(p) = \mathrm{Pr}_i(p/q) = \mathrm{Pr}_i^-(p \wedge q)/\mathrm{Pr}_i^-(q).$$

216

In the world space, if q is learned then the posterior probabilities of all worlds in $|q|$ are homothetically augmented while the worlds out of $|q|$ are affected with a zero probability. In a Dempster–Shafer framework, when a proposition q is learned several updating rules may be proposed, especially Dempster's rule (Dempster 1967), all generalizing the Bayes rule; they are generally obtained by revising directly the elements of the focal set through distinct methods (Walliser 1992). Finally, again in a probabilistic framework, if the message is no longer certain (q is known only with probability $Pr_i^+(q)$), several updating methods are available, especially Jeffrey's rule (Jeffrey 1965); they are obtained by considering the message as a specific Dempster–Shafer distribution.

Game theory. In game theory, in Aumann's framework, a message $\mathfrak{M}$ gives the announcement m_k: "the real state is in subset H_k," where $H_k = H(s)$ is the subset of the partition $\mathfrak{K}$ containing the real state s:

$$\Pi(m_k/s) = \begin{cases} 1 & \text{if } s \in H_k, \\ 0 & \text{otherwise.} \end{cases}$$

Before receiving the message, the prior probability of receiving each m_k defines a capacity on the state space:

$$m_i(H_k) = \sum_{t \in H^{-1}(H_k)} \Pi_i^-(t).$$

After receiving the message, the posterior probability of each state t when the message is m_k (or the real state is s) is

$$\Pi_i^+(t/m_k) = \Pi_i^+(t/s) = \begin{cases} \Pi_i^-(t) / \sum_{t \in H_i(s)} \Pi_i^-(t) & \text{if } t \in H_i(s), \\ 0 & \text{otherwise.} \end{cases}$$

Here, a message $\mathfrak{M}'$ is more informative than a message $\mathfrak{M}$ if partition $\mathfrak{K}'$ is finer than $\mathfrak{K}$. In fact, the whole construction above can easily be generalized for any accessibility relation $\mathfrak{K}$.

In the probabilistic framework, a message $\mathfrak{M}$ was first conceived as an announcement m, selected in a given set of alternative occurrences ($m \in \mathfrak{M}$), its probability conditional to the real state s being known: $\Pi(m/s)$. When receiving the message m, the prior distribution $\Pi_i^-(s)$ is changed into a posterior distribution $\Pi_i^+(s/m)$ according to the Bayesian rule:

$$\Pi_i^+(s/m) = \frac{\Pi_i^-(s)\check{\Pi}(m/s)}{\sum_s \Pi_i^-(s)\Pi(m/s)}.$$

A message $\mathfrak{M}'$ is then said to be more informative than a message $\mathfrak{M}$ if $\mathfrak{M}$ is a "mixture" of $\mathfrak{M}'$:

$$\exists \gamma_{mm'} \text{ s.t. } \Pi(m/s) = \sum_{m' \in M} \gamma_{mm'} \Pi(m'/s).$$

The less informative message is the trivial message $\mathfrak{M}_0$ with only one occurrence, $\Pi(m_0/s) = 1$ $\forall s$; the most informative message is the precise message $\mathfrak{M}_s$ that indicates the correct state: $\Pi(m_s/s) = 1$ $\forall s$. Aumann's framework is just a special case where $\Pi_i(m_k/s) = 1$ if $s \in H_k$ and 0 otherwise.

If a player i can observe player j's posterior probability on an event, then i may try to reveal the information j received (which is implicitly embedded) and use that information in his own assessment. Aumann (1976) showed that, if two players have the same priors and intersecting partitions $\mathfrak{K}_i$ and $\mathfrak{K}_j$, and if their posteriors are common knowledge, then these posteriors must be identical because i and j cannot "agree to disagree." A player may even globally reveal others' information merely by observing only a mixture of all posteriors (McKelvey and Page 1986); two players can also progressively reveal their information by a rapidly converging learning process (Geanakoplos and Polemarchakis 1982). In fact, the impossibility of agreeing to disagree holds for weaker accessibility relations than partitions – for instance, with axioms A3 and A4 only (Samet 1987) or with axiom A3 and an axiom of "balanced" relations (Geanakoplos 1989).

9. KNOWLEDGE AND DECISION

At the individual level, epistemic logic is concerned with the properties of an agent's knowledge and is not associated with a praxeologic study of that agent's action principles. In contrast, decision theory studies the deliberation process by which the beliefs and preferences of an agent are computed, according to rationality assumptions, in order to select a given action. However, the deliberation process can be considered to take place entirely in the "mental sphere" of the agent, who knows his determiners and consciously chooses an "intended action" that is afterward faithfully implemented into the real sphere. Epistemic logic is then able to formalize it through linked propositions such as: "agent i considers preferences U_i," "agent i acts in a Bayesian rational fashion," or "agent i intends to play action a_i."

An agent generally takes material actions to achieve general goals, and these actions give him some utility conditional on his environment; a very special case is when he bets on a given event, his utility depending only on the event occurring or not. An agent also takes informational actions to know the environment better, and their utility then depends on the material actions in which the information is used. Information may be

218

obtained (at a higher or lower cost) by direct observation, by a purchase from a specialized agency, or by revelation from the observed actions of others. Of course, some actions are mixed; the agent may learn some information as a by-product of a material action. An agent can even move away from optimal material action in order to explore the environment and acquire more information.

Remarks. In *decision theory,* player i considers his uncertainty space to be the states of nature $S = A_0$, and his utility is defined on $A_i \times A_0$. In the classical probabilistic approach, i chooses an optimal action a_i^* maximizing expected utility:

$$a_i^* = \arg \max_{a_i} \sum_{s \in S} \Pi_i(s) u_i(a_i, s).$$

If the combination of an action and a state yields a monetary result r, then it is possible to construct a probability distribution directly on the results, conditional on an action $\Pi_i(r/a_i)$, with the following expected utility of the action (also called lottery):

$$U(a_i) = \sum_r \Pi_i(r/a_i) u_i(r).$$

For risk-averse player, Rothschild and Stiglitz (1970) have shown that if a lottery a_i^* dominates a lottery a_i at first (second) order, then its expected utility is higher for a nondecreasing (concave) utility function.

In the partition approach, or (more generally) in the accessibility relation approach, the player chooses an optimal decision function $f_i(s)$, depending on the real state s, through his information (Bayesian rational behavior):

$$f_i(s) = \arg \max_{a_i} \sum_{s \in S} \Pi_i^+(t/s) u_i(a_i, t)$$

$$= \arg \max_{a_i} \sum_{s \in H_i(s)} \Pi_i^-(t) u_i(a_i, t).$$

Of course, the decision function must be $\mathcal{K}_i$-measurable, giving the same action for all states having similar images:

$$H_i(s) = H_i(t) \Rightarrow f_i(s) = f_i(t).$$

Technically, by a redefinition of the states, it is always possible to associate a decision problem grounded on a nonpartition $\mathcal{K}_i$ with a decision problem grounded on a partition $\mathcal{K}_i'$, giving the analogous decision function; but the priors are subsequently changed and depend on the accessibility relation (Brandenburger, Dekel, and Geanakoplos 1989).

In the probabilistic approach, using the optimal action after receiving the message $\mathfrak{M}$ always gives player i an expected utility:

$$U_i(\mathfrak{M}) = \sum_m \Pi(m) \max_{a_i} \sum_t \Pi_i^+(t/m) u_i(a_i, t).$$

For a risk-averse player, Blackwell (1951) has shown that when a message $\mathfrak{M}'$ is more informative than a message $\mathfrak{M}$, its expected utility is higher:

$$\mathfrak{M}' \geq_i \mathfrak{M} \Rightarrow U_i(\mathfrak{M}') \geq U_i(\mathfrak{M}).$$

When the message is constituted by a partition, it is possible to show directly that if partition $\mathfrak{K}'$ is more informative than $\mathfrak{K}$, its expected utility is higher:

$$\mathfrak{K}' \geq_i \mathfrak{K} \Rightarrow U_i(\mathfrak{K}') \geq U_i(\mathfrak{K}).$$

This result holds (Geanakoplos 1989) when considering weaker accessibility relations that satisfy axioms A3 and A4 as well as nestedness; these axioms are, moreover, minimal to assure the result.

10. Knowledge and equilibrium

At the collective level, epistemic logic is mainly interested in the distribution of knowledge between agents, rather than in the coordination of their actions. Game theory studies equilibrium notions where the actions of all players are made compatible, essentially through institutional devices (such as the "Nash auctioneer"). However, in their mental sphere, all agents have expectations of others' actions resulting from knowledge of their determiners, and an equilibrium may be obtained by an ex ante coordination of their beliefs. Epistemic logic is then able to formalize the agents' reasonings, leading to compatibility of their intended actions; a special case concerns a pure exchange of information between agents, leading from distributed knowledge to its progressive homogenization and converging possibly to common knowledge.

In the "problem of the hats," k individuals are wearing a black or a white hat, but each one can only observe the others' hats and not his own; the possible worlds consist then of all combinations of hat colors, even if each agent considers only two of them. At the beginning of play, an observer says: "At least one individual wears a black hat"; at successive dates, each individual must say whether he knows the color of his hat or not. If the black hats are exactly k, it is possible to show (by a recurrence on k starting from the observer's remark) that the answer will be negative during $(k-1)$ periods and positive for period k, simultaneously for all individuals. In fact, the private knowledge of each individual climbs one level of crossed knowledge at each answer, and the mutual knowledge of level k (but not of level $k-1$) happens to be a common knowledge.

In the "problem of the generals," two generals wish to make a coordinated attack on an enemy at the right moment chosen by the first one, the attack being launched if each knows that the other will attack simultaneously. At the right moment, the first one sends a message that has a probability ϵ of being lost; if the second one receives it, he sends a countermessage – with the same probability of being lost – to confirm the reception, and so on. The possible worlds correspond then to exactly k messages arriving, when k varies, and each general considers a partition formed of subsets of two successive worlds. It can be shown that the attack never occurs, for each general keeps some doubts about the other's attitude. In fact, the private knowledge of each general also climbs one level at each message, but the mutual knowledge at level k never becomes common knowledge, which is needed.

Game theory. In game theory, from the point of view of the modeler, Nash equilibrium is the basic equilibrium, each action of a player being the best response to the others'; each player can play a mixed action – that is, a probability distribution $\Pi_i(a_i)$ on his pure actions that can be interpreted as the others' probabilistic assessment of his own actions:

$$\Pi_i^*(a_i) = \arg\max_{\Pi_i} \sum_{a_i, a_j} \Pi_i(a_i)\Pi_j(a_j)u_i(a_i, a_j).$$

A rationalizable equilibrium considers more generally that each action of a player is a best response to the others' actions, itself considered as a best response, and so on; a sophisticated equilibrium is even weaker and obtains when each agent eliminates his dominated actions, considers that the other one does the same afterward, and so on. A correlated equilibrium also enlarges the Nash equilibrium, but by considering that there exists an uncertainty space common to the players, each agent choosing a decision function which is a best response to that of the others. The correlated equilibrium is objective or subjective depending on whether the players have the same priors:

$$f_i(s) = \arg\max_{a_i} \sum_{t \in S} \Pi_i^+(t/s)u_i(a_i, f_j(t))$$

$$= \arg\max_{a_i} \sum_{t \in H_i(s)} \Pi_i^-(t)u_i(a_i, f_j(t)).$$

Taking now the point of view of the players, all these equilibria have recently received a Bayesian justification by considering that players play as if against an uncertain nature that actually consists of the others' actions (Bernheim 1986, Aumann 1987, Brandenburger and Dekel 1987a, Tan and Werlang 1988). More precisely, each player is considered to be

Bayesian rational, and takes for an uncertainty space the crossed knowledge $S = \Pi_i T_i^\infty$, where T_i^∞ is itself constructed on the basic space of actions $\Pi_i A_i$; moreover, each player is assumed to know his own action. Three more assumptions can be made:

(1) it is common knowledge that the players are Bayesian rational (F1);
(2) it is common knowledge that the players' priors are different (F2) or identical ($\neg$F2); and
(3) it is common knowledge that the players play with correlation (F3) or independently ($\neg$F3).

Asserting F1, the other assumptions lead to the following equilibria:

(1) F2 + F3: sophisticated equilibrium = subjective correlated equilibrium;
(2) F2 + $\neg$F3: rationalizable equilibrium;
(3) $\neg$F2 + F3: objective correlated equilibrium; or
(4) $\neg$F2 + $\neg$F3: Nash equilibrium.

Considering now dynamic games, the main equilibrium from the modeler's point of view is the Bayesian perfect equilibrium that combines cognitive and instrumental rationality: At each information set, the player updates his beliefs considering past observations and optimizes his action by expecting the others' action in a backward induction procedure. In this framework, the "problem of the generals" can be formalized by assessing the players' relevant utilities regarding whether they attack or not (Rubinstein 1989); Bayesian equilibrium implies that they don't attack, but the coordinated attack becomes possible if the players have bounded rationality (such as ϵ-rationality) or bounded crossed knowledge. One can also study the problem of two agents likely to transact on a contingent good when they have identical utility functions (implying the same risk aversion) and priors on the states of nature. If their private information forms a partition then no transaction will take place (their intended action reveals their information), but this result may disappear with weaker information structures (Milgrom and Stokey 1982, Geanakoplos 1989).

Taking again the point of view of the players, for a simple game without uncertainty it is possible to stress the backward induction paradox (Reny 1985, Binmore 1987/88, Bicchieri 1988). At the beginning of the game, a player defines the equilibrium path by calculating the rational action of each player at each node of the tree, but this path goes through some nodes and not through others. When a player finds himself at a node out of equilibrium, this is not compatible with his beliefs, which now must be updated in the difficult case of a contradiction, leading to an indefinite action that destroys the equilibrium. The paradox indeed disappears when rationality is no longer common knowledge; but when keeping these assumptions, one must conclude that the perfect equilibrium cannot be implemented by pure reasoning of the players.

REFERENCES

Aumann, R. J. (1976), "Agreeing to Disagree." *Annals of Statistics* 6(4): 1236–9.
Aumann, R. J. (1987), "Correlated Equilibrium as an Expression of Bayesian Rationality." *Econometrica* 55(1): 1–18.
Bacharach, M. (1985), "Some Extensions of a Claim of Aumann in an Axiomatic Model of Knowledge." *Journal of Economic Theory* 37: 167–90.
Barwise, J. (1988), "Three Views about Common Knowledge." In M. Y. Vardi (ed.), *Theoretical Aspects of Reasoning about Knowledge*. San Mateo, CA: Morgan Kaufmann.
Bernheim, B. D. (1986), "Axiomatic Characterizations of Rational Choice in Strategic Environments." *Scandinavian Journal of Economics* 88(3): 473–88.
Bicchieri, C. (1988), "Common Knowledge and Backward Induction: A Solution to the Paradox." In M. Y. Vardi (ed.), *Theoretical Aspects of Reasoning about Knowledge*. San Mateo, CA: Morgan Kaufmann.
Binmore, K. G. (1987/88), "Modeling Rational Players." *Economics and Philosophy* 3: 179–214; 4: 9–59.
Binmore, K. G., and Brandenburger, A. (1988), "Common Knowledge and Game Theory." Discussion Paper, London School of Economics.
Blackwell, D. (1951), "Comparison of Experiments." In J. Neyman (ed.), *Proceedings of the Second Berkeley Symposium on Mathematical Statistics and Probability*. Berkeley: University of California Press.
Brandenburger, A., and Dekel, E. (1985), "Hierarchies of Beliefs and Common Knowledge." Research Paper 841, Department of Economics, Harvard University.
Brandenburger, A., and Dekel, E. (1987a), "Rationalizability and Correlated Equilibria." *Econometrica* 55(6): 1391–402.
Brandenburger, A., and Dekel, E. (1987b), "Common Knowledge with Probability 1." *Journal of Mathematical Economics* 16: 237–45.
Brandenburger, A., Dekel, E., and Geanakoplos, J. (1989), "Correlated Equilibrium with Generalized Information Structures." Mimeo, Department of Economics, Yale University.
Chellas, B. (1980), *Modal Logic*. Cambridge: Cambridge University Press.
Dempster, A. P. (1967), "Upper and Lower Probabilities Induced by a Multivalued Mapping." *Annals of Mathematics and Statistics* 38: 325–39.
Dubois, D., and Prade, H. (1989), "Epistemic Entrenchment, Partial Inconsistency and Abnormality in Possibilistic Knowledge." *Busefal* 39: 66–74.
Duhem, P. (1906), *La théorie physique, son objet et sa structure*. Paris: Chevalier et Rivière, Vrin.
Ellsberg, D. (1962), "Risk, Ambiguity and the Savage Axioms." *Quarterly Journal of Economics* 75: 643–95.
Fagin, R., and Halpern, J. Y. (1988), "Uncertainty, Belief and Probability." Mimeo, IBM Almaden Research Center, San Jose, CA.
Fagin, R., Halpern, J. Y., and Meggido, N. (1988), "A Logic for Reasoning about Probabilities." Mimeo, IBM Almaden Research Center, San Jose, CA.
Fagin, R., Halpern, J. Y., and Vardi, M. Y. (1988), "A Model-Theoretic Analysis of Knowledge." Mimeo, IBM Almaden Research Center, San Jose, CA.
Gärdenfors, P. (1988), *Knowledge in Flux*. Cambridge, MA: MIT Press.
Gärdenfors, P., and Makinson, D. (1988), "Revision of Knowledge Systems Using Epistemic Entrenchment." In M. Y. Vardi (ed.), *Theoretical Aspects of Reasoning about Knowledge*. San Mateo, CA: Morgan Kaufmann.

223

Geanakoplos, J. (1989), "Game Theory without Partitions, and Applications to Speculation and Consensus." Cowles Commission Discussion Paper, Yale University.

Geanakoplos, J., and Polemarchakis, H. (1982), "We Can't Disagree Forever." *Journal of Economic Theory* 28(1): 192–200.

Gilboa, I. (1987), "Expected Utility with Purely Subjective Nonadditive Probabilities." *Journal of Mathematical Economics* 16: 65–88.

Halpern, J. Y., and Moses, Y. (1985), "A Guide to the Modal Logics of Knowledge and Belief." Mimeo, IBM Almaden Research Center, San Jose, CA.

Halpern, J. Y., and Moses, Y. (1988), "Knowledge and Common Knowledge in a Distributed Environment." Mimeo, IBM Almaden Research Center, San Jose, CA.

Harsanyi, J. C. (1976), "Advances in Understanding Rational Behavior." In J. Harsanyi (ed.), *Essays on Ethics, Social Behavior and Scientific Explanation*. Dordrecht: Reidel.

Hughes, G., and Cresswell, M. (1985), *An Introduction to Modal Logic*. London: Methuen.

Jeffrey, R. (1965), *The Logic of Decision*. Chicago: University of Chicago Press.

Lewis, D. K. (1965), *Convention, a Philosophical Study*. Cambridge, MA: Harvard University Press.

McKelvey, R., and Page, T. (1986), "Common Knowledge, Consensus and Aggregate Information." *Econometrica* 54: 106–27.

Mertens, J. F., and Zamir, S. (1985), "Formulation of Bayesian Analysis for Games with Incomplete Information." *International Journal of Game Theory* 14(1): 1–29.

Milgrom, P. (1981), "An Axiomatic Characterization of Common Knowledge." *Econometrica* 49(1): 219–22.

Milgrom, P., and Stokey, N. (1982), "Information, Trade and Common Knowledge." *Journal of Economic Theory* 26: 17–27.

Monderer, D., and Samet, D. (1989), "Approximating Common Knowledge with Common Beliefs." *Games and Economic Behavior* 1: 170–90.

Reny, P. J. (1985), "Backward Induction and Common Knowledge in Games with Perfect Information." Mimeo, Department of Economics, University of Western Ontario.

Rothschild, M., and Stiglitz, J. (1970), "Increasing Risk, a Definition." *Journal of Economic Theory* 2: 315–29.

Rubinstein, A. (1989), "The Electronic Mail Game: Strategic Behavior under Almost Common Knowledge." *American Economic Review* 79: 385–91.

Samet, D. (1987), "Ignoring Ignorance and Agreeing to Disagree." Mimeo, Department of Economics, Northwestern University.

Schelling, T. C. (1960), *The Strategy of Conflict*. Cambridge, MA: Harvard University Press.

Shackle, G. L. (1961), *Decision, Order and Time in Human Affairs*. Cambridge: Cambridge University Press.

Shafer, C. (1976), *A Mathematical Theory of Evidence*. Princeton, NJ: Princeton University Press.

Shin, H. S. (1988), "Logical Structure of Common Knowledge." Mimeo, Nuffield College, Oxford University.

Simon, H. A. (1986). "From Substantive to Procedural Rationality." In S. Latsis (ed.), *Method and Appraisal in Economics*. Cambridge: Cambridge University Press.

Spohn, W. (1988), "Ordinal Conditional Functions: A Dynamic Theory of Epistemic States." In W. Harper and B. Skyrms (eds.), *Causation in Decision, Belief Change and Statistics*. Dordrecht: Kluwer.

Stinchcombe, M. (1988), "Approximate Common Knowledge." Mimeo, University of California, San Diego.

Tan, T. C., and Werlang, S. R. (1986), "On Aumann's Notion of Common Knowledge, an Alternative Approach." Mimeo, Department of Economics, University of Chicago.

Tan, T. C., and Werlang, S. R. (1988), "The Bayesian Foundations of Solution Concepts of Games." *Journal of Economic Theory* 45: 370–91.

Vardi, M. Y. (1986), "On Epistemic Logic and Logical Omniscience." In J. Y. Halpern (ed.), *Theoretical Aspects of Reasoning about Knowledge.* San Mateo, CA: Morgan Kaufmann.

Walliser, B. (1989), "Instrumental Rationality and Cognitive Rationality." *Theory and Decision* 27: 7–36.

Walliser, B. (1992), "Belief Revision and Decision under Complex Uncertainty." In P. Bourgine and B. Walliser (eds.), *Economics and Cognitive Science.* London: Pergamon.

Yager, R. (1983), "Entropy and Specificity in a Mathematical Theory of Evidence." *International Journal of General Systems* 1(1): 25–38.

13

Abstract notions of simultaneous equilibrium and their uses

VITTORIOEMANUELE FERRANTE

1. INTRODUCTION

This chapter discusses the notions of information and rationality of agents that are implicit in interpretations of formal definitions of simultaneous equilibrium in normal-form games. The examination of the nature of such notions as information and rationality aims in particular at ascertaining the possibility of regarding the definitions of equilibrium as well-defined instances of a theory of rational decision; in other words, we inquire into whether it is possible to characterize an equilibrium point of a normal-form game as the outcome of a rational decision process, that is, as a *rational equilibrium*.

The models that allow for some definitions of equilibrium can be classified as "topological" or "Boolean algebraic." The former class allows for the definition of a Nash equilibrium, and the models are often interpreted as representing multiperson decision making "under conditions of certainty." The latter class includes "neo-Bayesian" models and the notion of a "correlated equilibrium," and are often construed as representing decision making "under uncertainty."

2. NORMATIVE USE OF A NASH EQUILIBRIUM

Let A^i and A^j be the two sets of actions (or strategies) available to the players of a two-person, normal-form game. The Cartesian product $\Omega := A^i \times A^j$ will be the preference-relevant set of states of things, or outcomes of the game. Assume the two players to have well-defined preference orderings of the points of Ω. There may be a point of Ω such that no unilateral change in either coordinate would lead to a point that is preferred by the corresponding player; the former point will then be a Nash equilibrium.

Mathematical existence of such a point is sometimes used to give a normative interpretation of the model, either to independently prescribe

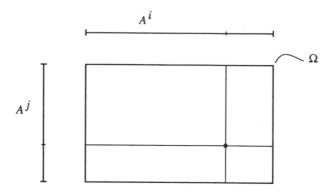

Figure 1

each player's implementation of the strategy corresponding to his own coordinate of the equilibrium point, or to forecast such an outcome as the result of choices of "rational" players. Mathematical existence is thus given the interpretation of *necessary* existence of such an equilibrium: Player i, it is sometimes argued, should (or will) choose his component strategy of the equilibrium outcome because he can legitimately expect that j will do the same, on the grounds of their "common knowledge" that the resulting outcome is indeed an equilibrium. (See Figure 1.)

Mathematical existence of multiple Nash equilibria is then somewhat at odds with their all being necessary as well as alternative to each other. Moreover, the prescription is apparently hard to justify as the outcome of a well-defined decision procedure, which would lead each player to the choice of his best option on the grounds of the circumstances of the choice itself. In the context of the game, in fact, such circumstances will be the choice of the other player; requiring the equilibrium choice to be the result of a well-defined decision process will thus imply either that each player has direct knowledge of the strategy chosen by the other, or that each is able to calculate that strategy. Direct knowledge requires determination of its object – that is, predetermination of the strategy selected by the other player – and thus requires a sequential implementation of the strategies, thereby contradicting the "simultaneous" nature of the normal form of the game. Alternatively, calculation of the other player's rational choice does not seem to be well founded, since it requires more than a finite number of steps. This is the well-known logical circularity in the notion of a rational equilibrium, which has been a recurrent theme in game theory since its beginnings: i's calculation of j's choice includes also i's calculation of j's own calculation, which i should

228

reproduce inside his own; but since j also performs an identical calculation, the latter calculation will have to contain i's, and this will not be able to contain j's as its part.[1]

Notice that an assumption of common knowledge of the equilibrium point is not sufficient to prescribe the implementation of such point (although the point is assumed to be unique), unless we assume also that the point *should* indeed be played: Knowing that you know that I know . . . the equilibrium point does not imply knowing that you will actually play your equilibrium strategy. Consequently, it is not clear how to characterize the knowledge that this model assumes the agents to possess, when the model is meant to represent a multiperson decision process under conditions of certainty.

3. BOOLEAN MODELS OF DECISION THEORY

A solution to the problem of the construction of a rational equilibrium has been sought in the application of models of decision theory to a game (Aumann 1987). These models are essentially "vertical," in the sense that knowledge of predetermined circumstances[2] is taken to lead to a choice, and decisions are functions from circumstances to choices.

A single-agent Boolean decision theory can be defined as follows. The decision maker is confronted with well-defined circumstances, represented by one generic point of a set S, which is assumed to be comprehensive of all possible alternative specifications of the world. A Boolean algebra of events is defined by a field $\mathcal{F}$ of subsets of S, S itself being the universal element of $\mathcal{F}$. A set-theoretically atomic field $\mathcal{F}^i$, which is a subset of $\mathcal{F}$, represents the information structure of i. The atoms of $\mathcal{F}^i$ are i's information partition $\mathcal{P}^i$, and i's decision function, $\delta^i : \mathcal{F}^i \to A^i$, relates i's information to his choice.[3] Only maximal pieces of information are actually used: These are the atoms of $\mathcal{F}^i$ (i.e., the elements of the information partition). The occurrence of a state of things $s \in S$ will imply the occurrence of the element $P(s) \in \mathcal{P}^i$, which includes $\{s\}$; $\delta^i(P(s))$ will thus be i's choice.

In this one-person model, the information partition is sometimes construed as the result of reading instruments of measurement of the "outside" world. The metric scale of the instruments generally allows for knowledge of the circumstances with a varying degree of refinement; typically, the graded bar placed beside the mercury column of a thermometer allows specification of only the segment – that is, the element of the partition – within which lies the level of the mercury and thus the temperature. This model leaves no room for a notion of "incorrect information." In other words, no formalization can be given of the idea that the actual circumstances are not a point of the subset (of S) that gathers all the points i believes to be possible on the grounds of his knowledge. This implies no loss of generality in the model with one person, since there will be no one else's information that can be in conflict with i's. The issue becomes relevant in models with more persons, when the agents' behavior may arise from a disagreement over the circumstances of the choice.

4. ALGEBRAIC MODELS OF KNOWLEDGE AND BELIEF

A formal notion of incorrect information can be constructed by distinguishing the event that the decision maker believes to be true from the event that the decision maker holds such belief, and by allowing the two events not to coincide.

Bacharach (1985) defines a so-called modal operator K_i on the algebra of events $\mathcal{F}$, interpreted as a knowledge operator: If $E \in \mathcal{F}$ is an event then $K_i E \in \mathcal{F}$ is the (possibly null) event that i knows that E is true. Leaving aside some technical details, Bacharach's axiomatization of K_i is essentially that of an algebraic model for the modal logic known as $S5$:

K1: $\bigcap_{\alpha \in A} K_i E_\alpha = K_i \bigcap_{\alpha \in A} E_\alpha$ (axiom of completeness);
K2: $\{s\} \subseteq -K_i - \{s\}$ for all $s \in S$ (axiom of veridicality);
K3: $-K_i E = K_i - K_i E$ (axiom of "wisdom").

K1 grants completeness to the "epistemic" events $K_i E$. By K2, the decision maker believes the actual (or true) state of things to be possible; equivalently, each event the decision maker has knowledge of is actually true.[4] By K3, if the decision maker knows something, then he knows that (and vice versa). By K1 and K3, events of the kind $K_i E$ define a set-theoretically atomic field $\mathcal{F}^i$ that is a subfield of $\mathcal{F}$ and whose atoms $K_i e$ are thus a partition of S. The information partition is obtained by considering the smallest elements e^* of the inverse images of such atoms (the knowledge operator is here viewed as a function), and by showing that $e^* = K_i e^* = K_i e$. In other words, the event representing i's maximal information in Bacharach's system coincides with the event that such information is held by i.[5]

A weaker operator B_i can be defined on the algebra of events (Ferrante 1988) and construed as a belief operator:

B1: $\bigcap_{\alpha \in A} B_i E_\alpha = B_i \bigcap_{\alpha \in A} E_\alpha$;
B2: for all $s \in S$, there exists an s' such that $\{s'\} \subseteq -B_i - \{s\}$;
B3: $-B_i E = B_i(-B_i \cup -E)$;
B4: if $B_i e^1$ and $B_i e^2$ are two atoms of the subfield made of the images of B_i in the field, then $e^1 \cap e^2 = \emptyset$.

B1 is the same as K1. B2 is a clear weakening of K2, and makes it possible that what the decision maker believes is true. By B3, if the decision maker does not believe that E is true, then he will believe that if he believed that E were true, E would not be true:[6] If what the decision maker believes were always true – as is the case with K2 – then B3 would turn out to be equivalent to K3. B4 can be read as an axiom of independence: If the $B_i e$ are disjoint segments of the analogical part of the instrument of measurement (the mercury column of the thermometer), then the coded part (the graded scale adjacent the column) will not change as a function of the state (the level of the mercury).[7]

The smallest elements e^* of the inverse images of the atoms $B_i e$ of the subfield composed of "doxastic" events, which are of the form $B_i E$ ($E \in \mathfrak{F}$), constitute again a partition – the doxastic partition – that will not necessarily coincide with the partition composed of the atoms $B_i e$. The doxastic partition will not necessarily be veridical.

Bacharach's epistemic information partition is, in fact, veridical. Veridicality of knowledge is the essence of the so-called no-trade theorems; these follow from the fact that if two (or more) decision makers are characterized by an identical decision function and by veridical information partitions (or by other veridical information structures), and if their choices are common knowledge, then such choices will not differ. In applications of decision theory to economics, this makes it impossible to have transactions based on differing opinions – if the prior is common and known (Aumann 1976) – owing to the fact that the transaction itself is necessarily common knowledge, and the information that might lead to differing ex post probability evaluations is thus made common to all. The axiomatic system B1–B4 defines information partitions that are not necessarily veridical. Such a notion of incorrect information allows people to "agree to disagree"; in other words, they can enter into bets that are based on a difference of opinion that cannot be reduced to differing priors.[8]

Epistemic and doxastic partitions do not depend on the state. If (cf. Aumann 1976, 1987) the generic state s is viewed as containing the description of an information partition – that is, a list of lists of states whose occurrence the decision maker is unable to distinguish – then such a description would not change with the states in S. Things could be thought of as being otherwise: If s contained a list of points of a set $P^s(s)$ of states that the decision maker in s would not distinguish from s itself, and a list of points of a set $P^s(s')$ of states that the decision maker in s would not have distinguished from s' if s' had been the actual state of things, then s' could in turn contain (among other things) a list of points of a set $P^{s'}(s')$ of states that the decision maker in s' would not distinguish from s' itself; but $P^{s'}(s')$ might not coincide with $P^s(s')$; and so forth.

On the other hand, it would not be possible to describe all information partitions in the states of S, short of restricting the set of allowed partitions. This is because the cardinality of the set of all partitions of any set S is greater than that of S itself: For each subset of S, a different partition can be defined by the subset and its complement in S. In other (not quite rigorous) terms, there are at least as many partitions as subsets of S, and the set of these subsets has a greater cardinality than S itself. Consequently, if i's information structure were to be considered as part of the state of things, and if all structures (or even just all partitions) that the decision maker may have were to be so described, then a set-theoretical antinomy would result.

5. Models of decision theory in games

A Boolean model can be constructed to represent i's decision problem in a context with many decision makers (as in a normal-form game) by identifying S and A^j, the set of strategies of i's opponent. In the example of Figure 2, A^j is parted into two subsets, and a choice in A^i is paired with each. Agent j's choice could be represented symmetrically. Incorrect information will consist here of j's choosing a strategy lying out of the subset of A^j that i deems to contain j's choice. (Notice that implicit in the model is the idea that the information partition could have been different from the actual one; this idea also underlies the notion of an incorrect partition, which echoes a counterfactual existence of an alternative and correct partition.) However, no formal notion of an ex ante equilibrium can be defined here, since the two decision functions are defined on different spaces. Only some ex post notion of consistency can be obtained by assuming that each player's actual choice be one of those the other player deems to be possible, or at least one of those that determine a state of things whereby the latter player (say i) believes the strategy of j to be one that i had considered possible in choosing his own (whether or not such beliefs are correct). Furthermore, each decision process takes the choice of the other player as predetermined, which is in some friction with the game's presumed simultaneity.

A notion of ex ante equilibrium is constructed in a second Boolean model, which is usually employed to formalize multiperson decision theory. The set of circumstances S is here identified with $\Omega := A^i \times A^j$, and i's information partition of S, $\mathcal{P}^i$, is the natural one that is made of the subsets of Ω whose projections onto A^i are points. An ex ante equilibrium is then defined as a pair of decision functions, defined on such information partitions, whose images are the very points that are the projections of the elements of the information partitions. In Figure 3, an element of the

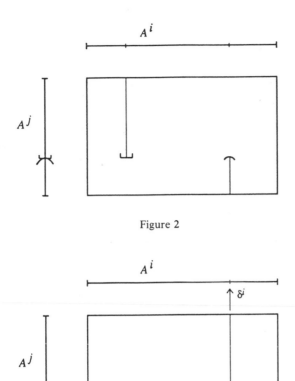

A^i

A^j

Figure 2

A^i

A^j

$\uparrow \delta^i$

Figure 3

domain of δ^i and its image are represented. Leaving aside for the moment the question of which mechanism or procedure may determine the occurrence of the generic state of things $\omega \in \Omega$, this occurrence will imply the occurrence of the two events (respectively, two elements of the two information partitions) whose intersection is the singleton containing ω itself: Knowledge of the occurrence of the elements of the respective information partition will make it optimal for each player to choose precisely the strategy that composes ω in the respective coordinate, by the ex ante equilibrium decision function just defined. This is the case for all $\omega \in \Omega$.

In the neo-Bayesian specification of this model, such as for Aumann's (1987) correlated equilibrium, player i's (say) process of information acquisition consists – in Harsanyi's (1967/68) terms – of learning one's own

233

"type." In other words, it consists of learning that the possible outcomes are the points of a subset P_i of Ω, $P_i \in \mathcal{P}^i$. These points are pairs of elements: The first component of each pair is a constant element of A^i, and the second component will vary over all points in A^j. Both the points in A^i and the points in A^j will first include i's and j's actions (respectively); they will then include both i's and j's ex post probability evaluations of such pairs of actions. The definition of the type that the player comes to learn will further consist of higher-order probability distributions, which are defined on one's own and the opponent's just-described probability distributions. This has the result of enriching the construction of the points (i.e., the types) of A^i and A^j, and hence of Ω, by placing beside the description of actions the description of the probabilistic beliefs that render the respective actions optimal – given cardinal utilities – as well as the relative infinite hierarchy of beliefs whose levels will be connected by appropriate consistency restrictions (Mertens and Zamir 1985, Tan and Werlang 1985). If a "signal" Γ, a random variable taking values in Ω, is such that the values of Γ that are constant in the first component (and which are the points of a P_i) include probability evaluations that make i's expected utility – relative to the action described in the first component of the pair of actions that initially forms the points of P_i – over the a posteriori distribution (which is also contained in the points of P_i) not smaller than that which can be calculated over the distributions contained in all other elements of i's information partition; and if this also happens to Γ for a subset P_j, an element of j's information partition; then the a priori distribution of Γ is a "correlated equilibrium."[9]

A player's information is here of a special nature, since it is not concerned with the observation of external circumstances but rather with the specification of his own type: The signal the player receives is precisely the set of worlds made possible by his action, and are also characterized by the beliefs that make the action optimal in a Bayesian sense. There is no room here for the idea that the information partition is incorrect, nor for the possibility that the partition could have been different. The possibility of an irrational choice of an action that is not optimal with respect to the related hierarchy of beliefs would require learning a type that is not defined in the game (all points of A^i and A^j are ex post optimal combinations of actions and beliefs; i.e., they all describe Bayesian types). On the other hand, the idea that a player's choice could be different from what actually is – that is, both the choice of an action and of the beliefs that make the action optimal – is nonsense, short of thinking of a space of signals that *describe* rather than *are* the states; only in this case can a player be thought of as reading such a description incorrectly. But in this case, a Bayesian decision based on such a (possibly incorrect) reading will require assuming a similar Bayesian rationality in the other player, thereby falling back into the definition of a correlated equilibrium mentioned in note 9, together with its logical circularities.[10]

Aumann's correlated equilibrium could be read as an application of Jeffrey's (1965) "ratifiability" decision criterion in the game between decision maker and Nature. There also, the decision maker derives a posteriori probability evaluations of the states of Nature by "conditioning upon his own choice": The criterion is relevant in the case of externalities, that is, when Nature's choice changes with the decision maker's. On the other hand, this model assumes that the decision maker knows and exploits a relation that is assumed to exist between his own actions and the probability distributions of Nature's "actions." In an industrial economics classification, Jeffrey's is a sequential game where the decision maker is the leader and Nature is the follower.

There is in fact no signal in Jeffrey's construction of the decision process, if a signal is taken to be the sign of an otherwise well-determined state of things. We believe Aumann's (1987) signal incorrectly identifies the sign and its reference – one might say it neglects Quine's (1953) inverted commas, which distinguish things from their names. It is for this reason that there is no room for variations ("mistakes") of names with respect to things in Aumann's construction. On the other hand, Aumann needs true signals – and thus the determination of their objects – to construct a decision-theoretic structure; but these objects are the result of the decision and therefore cannot be predetermined.

However, the idea of an information mistake seems to be necessary to allow for a "rational" disagreement, that is, for the existence of transactions based on opinions whose differences result from possibly incorrect instruments of measurement. But this does not seem to be compatible with the Bayesian theory of player types on which is based Aumann's theory of correlated equilibrium, a theory that he suggested be used to construct a notion of rational equilibrium that is free of logical circularities.

6. DESCRIPTIVE USE OF NASH EQUILIBRIA

If normative interpretations of simultaneous equilibria encounter logical problems, or imply a distortion of the notion of information that leads to a decision, then the Nash equilibrium may be given a descriptive function.

There are indeed objects in the world whose existence requires the joint implementation of actions of many agents; among these are economic objects such as transactions and contracts. In the construction of such objects, the acts of the parties are viewed symmetrically, that is, as simultaneous acts: Either both parties will be legally committed to their ex ante

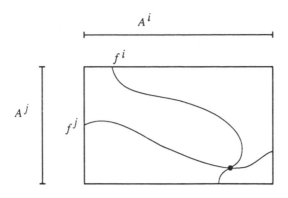

Figure 4

offers, or they will both maintain the right not to accept the offer of the counterpart. This is one reason for the economist's need to construct models of simultaneous equilibria. If an explanation of these objects is a justification for their actual existence, this will require the construction of a model where possible outcomes other than the actual one are described, and where necessary conditions for actual existence are given.

If A^i and A^j are both compact then so is $\Omega := A^i \times A^j$. The set Ω is naturally parted into a class $\mathcal{P}^j$ of subsets whose projections onto Ω are the points of A^j, that is, into subsets of states characterized by one and the same action chosen by j. The axiom of choice will then allow i to choose a point in each element of $\mathcal{P}^j$, and to construct a "choice function" f^i; a similar argument will allow for the construction of f^j. Well-known continuity conditions will grant the existence of points where the two functions intersect. These points will be Nash equilibria.

The choice functions can be construed as commitment schedules of counterfactual replies to the opponent's actions: A "net supply function" (i.e., a price–quantity schedule) may be regarded as such a choice function.[11] (See Figure 4.)

The economic interpretation we have just given to the choice function might seem to require the existence of markets that are contingent on all the possible choices of an opponent. Things may be seen to be otherwise if we take a notion of "existence" that reflects the ontology of the economic object. Economic events such as transactions and contracts may in fact come into existence only by means of concurring actions. But the transaction (i.e., the market) will exist only at the crossing point of the net supply schedules, while incompatible intentions – such as those that are not an equilibrium – will remain vain; that is, they will not exist.

236

Within this framework, mathematical existence of Nash equilibria will grant only that such equilibrium outcomes are possible, and will exclude the possibility of existence of all nonequilibrium outcomes. That such equilibria may number more than one will then come as no surprise, since they characterize only possible existence, rather than necessary existence. A state of things that exists contingently must be a possible outcome; but it will not have to be a necessary, or "rational," outcome.

NOTES

1. "What should [a] player . . . then do? . . . *Game theory does not attempt to prescribe what he should do!*" (Luce and Raiffa 1957, p. 62; italics in the original) On the other hand, it is nowadays sometimes held that: "Game theory is a *normative theory:* its aim is to prescribe what a player should do in a game." (Van Damme 1987, p. 1; italics in the original)
2. He who believes that neo-Bayesian decision theory employs subjective probabilities may not want to require determination of the events on which beliefs are defined. In that case, a justification of the beliefs that are to be *prescribed* should be provided; but then it is difficult to think of such a justification in the absence of a well-defined object of belief. Alternatively, an observed behavior could be tautologically identified with the beliefs that justify it as a Bayesian optimum choice; such beliefs would thus be deduced from the behavior, rather than implying it. On the other hand, the issue here is the normative use of models of equilibrium; even the Nash equilibrium causes no problems as a descriptive notion (see Section 6).
3. A neo-Bayesian model specifies such a decision function as a combination of a "cardinal" utility function and a probabilistic measure on the field of events.
4. That is, K2 is equivalent to: $K_i E \subseteq E$, $E \in \mathfrak{F}$.
5. e^* is thus a fixed point of K_i. Further iterated applications of the knowledge operator, as required by the algebraic definition of common knowledge, are again fixed. This guarantees that, every time two people know something, this is also common knowledge between them; in other words, there is no (extensional) distinction between knowing *that* you know and knowing *what* you know, because every time the former is the case then so is the latter, and vice versa. In the approach here presented, the resulting nature of the points of S – namely, whether their epistemic features are natural circumstances or whether such features remain external to the specification of the state of things – remains unclear. In this sense, the approach does not overcome criticisms that can be raised (essentially for the same reasons) against Aumann's (1976) algebraic definition of common knowledge.
6. A transcription of B3 in the calculus of proposition is $\neg b_i p \equiv b_i (b_i p \rightarrow \neg p)$, where proposition p describes E and the rest of the notation is obvious.
7. One might prefer the temperature (the state of things) to be capable, for instance, of distorting the graded scale.
8. Information structures that are not partitions have been studied recently (see, e.g., Geanakoplos 1989). The need to have information structures that are partitions in our context lies in their application to normal-form games, where the strategies available to a player define a natural partition of the set of outcomes.

237

9. In Aumann's (1974) earlier definition of a correlated equilibrium, the space of the signal is not identified with the space of the outcomes of the game, and probabilistic beliefs are not part of the outcomes because they do not concern the actions of the players. The Bayesian rationality in the optimal choice of a player requires then (in this definition) the presupposition of a similar Bayesian rationality on behalf of the opponent; but this assumption is not justified in any way other than circularly, just as in the case of the Nash equilibrium.

10. In the vein of the issues mentioned in note 2, an external observation of the players' behavior could impute to them some ex post beliefs of all orders (i.e., their types) – concerning the worlds made possible by the behavior itself – in such a way as to make that behavior justifiable within a Bayesian framework. But if the player's observation of his own type is to be a criterion for his own decision, then the matter will be quite different.

11. This choice function is not a *chosen* function: The latter choice of an element in the space of all functions could be justified only by some assumption regarding the function chosen (by the opponent) in the larger game whose strategy spaces are such spaces of functions. Such a reading of our construction would still be normative; but we are interested only in the definition of compatibility conditions of actions.

REFERENCES

Aumann, R. (1974), "Subjectivity and Correlation in Randomized Strategies." *Journal of Mathematical Economics* 1: 67–9.

Aumann, R. (1976), "Agreeing to Disagree." *Annals of Statistics* 4: 1236–9.

Aumann, R. (1987), "Correlated Equilibrium as an Expression of Bayesian Rationality." *Econometrica* 55: 1–18.

Bacharach, M. (1985), "Some Extensions of a Claim of Aumann in an Axiomatic Model of Knowledge." *Journal of Economic Theory* 37: 167–90.

Ferrante, V. (1988), "Theories of Information Partitions, and Applications to Economics." Studi e Discussioni No. 49, Dipartimento di Scienze Economiche, Università di Firenze.

Geanakoplos, J. (1989), "Game Theory without Partitions, and Applications to Speculation and Consensus." Cowles Foundation Discussion Paper No. 914, Yale University.

Harsanyi, J. (1967/68), "Games with Incomplete Information Played by 'Bayesian' Players," Parts I, II, III. *Management Science* 14: 159–82, 320–34, 486–502.

Jeffrey, R. (1965), *The Logic of Decision.* Chicago: University of Chicago Press.

Luce, D., and Raiffa, H. (1957), *Games and Decisions.* New York: Wiley.

Mertens, J.-F., and Zamir, S. (1985), "Formulation of Bayesian Analysis for Games with Incomplete Information." *International Journal of Game Theory* 14: 1–29.

Quine, W. V. O. (1953), *From a Logical Point of View.* Cambridge, MA: Harvard University Press.

Tan, T., and Werlang, S. (1985), "On Aumann's Notion of Common Knowledge – An Alternative Approach." Mimeo, University of Chicago Business School and Princeton University.

Van Damme, E. (1987), *Stability and Perfection of Nash Equilibria.* Berlin: Springer.

14

Representing facts

KRISTER SEGERBERG

Without the concept of action there would be no social science. Much work can be (and is) done without a detailed examination of this concept, but ultimately there is a need for a full conceptual analysis of action. Even though philosophers are already addressing that task, work of a more formal nature is still in its infancy. In this chapter some suggestions toward a formal analysis of action are offered, in the belief that philosophical work of this kind ought eventually to result in formal modelings.

We take our lead from Georg Henrik von Wright, to whom we should like to attribute the insight contained in the slogan "to act is to bring about a fact."[1] As a first step, then, one ought to analyze the notion of a fact; this is attempted in the first five sections, the bulk of this chapter. The last two sections contain some suggestions – sketchy, to be sure – for how a theory of action, in the spirit of von Wright, might be based on our analysis of facts.[2]

1. FACTS ACCORDING TO VON WRIGHT

When a (contingent) proposition is true there corresponds to it a *fact* in the world. It is a well-known view that truth 'consists' in a correspondence between proposition and fact.

This observation was made by von Wright in his book *Norm and Action* ([14], p. 25), and he went on to distinguish three types of fact: states-of-affairs, events, and processes. His use of the term "fact" may not be in complete agreement with normal usage, but it is clear that facts (in his sense of the word) are of crucial importance, not least for the philosophy of action.

It is obvious to von Wright – and indeed to many of us – that, even though events and processes share some similarities, they are different in kind. "Like processes, events are facts which *happen*. But unlike the happening of processes, the happening of events is a *taking place* and not

239

a *going on.*" ([14], p. 26) States-of-affairs, on the other hand, do not happen. Perhaps we might say that they *obtain.* As a common term for "obtaining," "taking place," and "going on," we might use "occurring": facts *occur.*

The present chapter is a study in ontology rather than in logic, so we will not introduce an object language in any formal sense. Nevertheless, it might be helpful from time to time to imagine a language in which certain propositions about the occurrence of facts can be represented. For example, if **S**, **E**, and **P** are terms in the object language denoting (respectively) a state-of-affairs, an event, and a process, then the object language should contain formulas equivalent to the following three central types of statements:

(1) **S obtains**
(2) **E has taken place**
(3) **P is going on**

(where, following Montague, boldface type is used to distinguish this object language from natural language). No doubt one will want the object language to provide for many temporal and model variations of these, but we will not go into details here.

2. State spaces

The development of dynamic logic lends renewed interest to the analysis given in Scott's oft-quoted paper "Advice on Modal Logic" [8], from which the concept of the two spaces defined in the present section is borrowed. We shall assume as a primitive concept that of a given set of states or, more briefly, an (atemporal) *state space.* The intuitive idea is that the world is always in one particular state or other, and that there are only so many states in which it can be. This notion of "state" must not be confused with that of "state-of-affairs." There is a temptation to drop the suffix "-of-affairs" from "state-of-affairs," but it must be resisted whenever confusion is likely to occur. If a distinction is needed, the states of the state space might be referred to as *total states.*

We take our states to be primitive. But in any application one would have to explain how they are to be viewed. This is rarely considered by philosophers, presumably because such a task is likely to be forbidding in any really interesting application. A philosopher may perhaps be allowed the privilege of taking it for granted that such an explanation can be given, at least in principle. Nevertheless, in order to guide the reader's intuition we will outline some possibilities here.

A philosopher who does consider this problem is Carnap (e.g., in [2]). Here a formal language is used in order to obtain his famous *state-descriptions*. Let us call an interpreted language the *auxiliary language* in order to distinguish it from our metalanguage, as well as from the formal object language intimated in Section 1. The problem of explaining the states of the state space would be solved whenever adequate state-descriptions are available in the auxiliary language.

(1) Rolling a pair of dice: Suppose we have two identifiable dice (one red and one green, say). Then a suitable state space would be the set of all ordered pairs (m, n), where m gives the number of dots on the red die and n the number of dots on the green die. Each time the dice are rolled, one of these states will be realized. Notice that this state space is finite.

(2) Turing machines: In the case of any given Turing machine, the state space is given by the set of total states, each of which consists of an inner state, a string of symbols (intuitively, what is written on the tape), and the position of the scanning head. This state space is usually infinite.

(3) Computers: This case is similar to the preceding one. Any instantaneous description of the total state of the computer describes an element of the state space. This state space is infinite unless some restriction is introduced that would make it finite.

(4) Chess: Here a suitable state space is the set of all possible configurations on the chess board. (Such a configuration might be given by a function assigning to each square on the chess board either a certain piece or else a null value, indicating that the square is empty; there are some natural restrictions here.) This state space is finite.

(5) Astronomy: A schoolteacher wishing to show students how the planets move around the sun, how eclipses occur, and so on may build a simplified model of our planetary system (using wire, Styrofoam, etc.). A small number of parameters will suffice to determine the relative position of all participating celestial bodies.

(6) Meteorology: Somebody wishing to study the weather in a certain part of the world would be interested in certain magnitudes or parameters (temperature, air pressure, direction and strength of winds, humidity, cloud cover, etc.). If there are n parameters, a state could correspond to an appropriate n-tuple of real numbers.

(7) The stock market: Somebody wishing to study the behavior of a certain market might wish to fix on certain parameters (the selling and buying courses of certain stocks, the number of stocks traded, the exchange rate of certain currencies, etc.). Not all of these parameters

241

need to be measurable – estimates might be required. (Rules of thumb could be used, for example, to assess factors such as the risk of the deutsche mark weakening against the American dollar, or intangibles such as the public confidence in a certain government.) In the end, as in the previous example, a state could correspond to an appropriate n-tuple of some kind.

Needless to say, these examples are meant only as indications of what can be done. In each case, there would be many possible ways of proceeding. In the first four cases, one would expect the resulting analyses to be equivalent, in a certain sense. In the last two examples, different analysts may well come up with different analyses, which we might describe as different views of the world.

In order to study dynamics, some notion of time is necessary. In the dice, Turing machine, and chess examples just described, time may be said to be implicit in the state space – isomorphic to the set of natural numbers – since the rolling of dice, the running of a Turing machine, and the playing of a chess game all proceed discretely. However, the running of a real computer can be represented in physical time, and one might conceivably wish to model the playing of a chess game in physical space and time. In any case, some concept of time is needed, and here we shall assume a time structure T to be given. (It is often natural to cast T as (some interval of) the set $\mathbb{R}$ of real numbers or else as (some initial segment of) the set $\mathbb{N}$ of natural numbers. Note, however, that $\mathbb{R}$ for T may not be the best choice even when the application is empirical rather than abstract. For example, what time structure to choose in the case of the stock-market analysis may be dictated by the way data are reported; there is also the matter of perspective (day-to-day, week-to-week, etc.).)

Given a state space S and a time structure T, we define the corresponding *temporal state space* as $S \times T$. The temporal state space is the space with respect to which we would wish to evaluate formulas such as (1)–(3) of Section 1. Let us say that a function h is a (*partial*) *history* if dom h (the *domain* of h) is an interval of t and the range of h is included in s. For any time $t \in T$, $h(t)$ is the value of h at t and denotes the state of the world with respect to h at t.

For the moment, let us regard h as giving the past history of some development. If $t \in$ dom h, it makes sense to ask whether a particular state-of-affairs obtains at t. Thus, if **S** denotes a state-of-affairs, we should like to give necessary and sufficient conditions for the sentence ⌜**S obtains**⌝ to be true with respect to h and t (relative to some modeling).

Similarly, it makes sense to ask whether a certain event, as seen from the vantage point of t, has taken place. Thus, if **E** denotes an event, we should like to give necessary and sufficient conditions for the sentence ⌜**E has taken place**⌝ to be true with respect to h and t (relative to some modeling).

Finally, it makes sense to ask whether at t a certain process is going on. Thus, if **P** denotes a process, we should like to give necessary and sufficient conditions for the sentence ⌜**P is going on**⌝ to be true with respect to h and t (relative to some modeling).

This way of putting the questions will suggest the further development of this analysis.

3. STATES-OF-AFFAIRS

Given appropriate assumptions on the relevant state space S, here are some examples of states-of-affairs:

that the sum of the dots on the dice is at least 7;
that the symbol scanned is 1;
that the White King is in check;
that the sun, the moon, and the earth are on a straight line;
that the temperature in Auckland is 21° C;
that the buyers' quote for Brierley is 215.

Given h and t, it is clear that the answer to the question of whether a state-of-affairs obtains is given by $h(t)$: no other part of h is relevant. This suggests that, if **S** denotes a state-of-affairs, then $\|\mathbf{S}\|$ – the *meaning* or *intension* of **S** in S – should be a subset of S. The required truth condition is immediate:

⌜**S obtains**⌝ is true with respect to h and t if and only if $h(t) \in \|\mathbf{S}\|$.

Two comments are in order here. One is that, according to this theory, what counts as a state-of-affairs depends on the given state space S. Thus it might perhaps be natural to say, at a certain stage of a particular chess game, that the White King's not having been moved so far is a state-of-affairs, just as virginity may be regarded as a state-of-affairs. It would be possible to find a state space with respect to which such a view is correct, but such a state space would be different from the one suggested in our example.

Our second comment concerns the relationship between propositions and states-of-affairs: there is a close relationship between a proposition of the type ⌜**S obtains**⌝ (a subset of the temporal state space $S \times T$) and the state-of-affairs $\|\mathbf{S}\|$ (a subset of the atemporal state space S). In this theory, however, propositions and states-of-affairs are different notions.

4. EVENTS

Some examples of events include:

getting a sum of at least 7;
a scanned symbol being erased;
the White King moving;
a solar eclipse;
the temperature at Auckland rising by three degrees;
2,400 Brierley shares being traded.

The astronomical example stands out: We have a name for that kind of event. Eclipses are spectacular; perhaps that is why we have names for them. The other events in the list are individuated by description. Notice that there is a certain ambiguity in the English rendering of these events. If an empirical Turing machine were built (i.e., one operating in the real world), then erasing a symbol would take place in a certain way and would take a certain time, and so the phrase "a scanned symbol being erased" might be taken to refer to a process rather than an event. Similarly, if in the play of a chess game White has lost all pieces except the King, then White would be confined to moving that King in all remaining moves of the game; in such a situation, "the White King moving" might refer to something that probably should be regarded as a process – a "going on," in von Wright's terminology. This is not the sense that is intended here. In each case it is events to which we call attention: the event of getting 7 or more (the result), the event of a scanned symbol being erased (the erasure), the event of the White King moving (the move), the event of the temperature at Auckland rising by three degrees (the rise), the event of 2,400 Brierley shares being traded (the trade).

On von Wright's program, trying to find truth conditions for propositions of the type $\ulcorner$**E has taken place**$\urcorner$ requires that one look for a certain entity that "corresponds" to the proposition. What would such an entity look like? Before answering this question, it is worth considering elementary probability theory.

On the whole, philosophers have balked at the notion that events exist.[3] But statisticians have long been happy to work with the event of a coin landing heads, the event of a coin landing heads three times in a row, the event of a coin landing heads at least three times in five throws, and so forth. On their view, if a suitable outcome space O is agreed upon, an event E is simply a subset of O, and they would say that an outcome x *realizes* E if and only if $x \in E$. This idea is not immediately transferable to our problem, but (as we shall see) with suitable modifications it is.

Let us ask, then, what it would take to "realize" an event such as the White King moving. Intuitively, if the White King is (say) on e1, then the event of his moving would be realized if next (on the completion of White's next move) he is on d1, d2, e2, f1, or f2. In one chess notation, White's move would be recorded by one of the "formulas" Ke1-d1, Ke1-d2, Ke1-e2, Ke1-f1, Ke1-f2. As logicians we might be tempted to represent this by ordered pairs, such as ⟨e1, d1⟩, giving the King's position before and after the move. However, it will turn out to be more useful to represent the move by letting the elements of the ordered pairs be the corresponding total states – that is, configurations on the chess board.

Suppose that x is a certain total state in which the White King is on e1, that there are no pieces on the adjacent squares, that none of Black's pieces bear on the adjacent squares, and that castling is not an option. Furthermore, let us denote by y_1, y_2, y_3, y_4, y_5 the total states resulting if White makes one of the five moves quoted. Let E be the event of the White King moving (in accordance with the rules of chess). What E comes to in total state x is given by the set

$$E(x) = \{\langle x, y_1\rangle, \langle x, y_2\rangle, \langle x, y_3\rangle, \langle x, y_4\rangle, \langle x, y_5\rangle\}.$$

That is to say, each pair $\langle x, y_i\rangle$ (where $i = 1, 2, 3, 4, 5$) is a *possible realization* of E, and $E(x)$ is the *set of all realizations that are possible at x.*

The event of the White King moving is, of course, possible in total states other than x. On the other hand, if x is the initial configuration (at the beginning of a game with all the pieces in their allotted places), then the White King is unable to move, and so $E(x) = \emptyset$. Thus it seems natural to identify E with the set of all possible realizations of E: $E = \bigcup\{E(x): x \in S\}$.

This analysis is readily extended to more complicated moves – such as the event of moving, in exactly three moves, the King's Bishop and the King's Knight out of the way (in either order) and then castle; or the event of the White King moving three steps, not necessarily in consecutive moves.

We may now return to the dice example and see how our account differs slightly from that of the statistician's. Using the state space described in our formulation of the example, the latter would view the event of getting at least 7 as the set of all pairs (m, n) such that $m + n \geq 7$. However, in our theory that set is a state-of-affairs, not an event. Looking more carefully at the situation, we note that the event we are talking about is the event of getting at least 7 the next time the dice are rolled. Before this event the world is in some total state (m, n) (the "prior" state), and after the event in some total state (m', n') (the "posterior" state). It is a curious feature of this particular state space that the prior state in no way contributes to determination of the posterior state. Nevertheless, the correct identification of the event in question, in our terminology, is with the set $\{\langle(m, n), (m', n')\rangle : m' + n' \geq 7\}$.

Against this informal background, let us state the following general definitions. A subset $U \subseteq T$ is *convex* if, for all elements $t, t_1, t_2 \in U$, if $t_1 < t < t_2$ then $t_1, t_2 \in U$ only if $t \in U$. A *T-path* in S is a function p from a convex subset of T to S. Thus, in the special but common case that T is the set $\mathbb{N}$ of natural numbers, a path may be identified with a sequence of states. In the future we will usually omit the reference to T and simply write "path" for "*T-path*." For the set of T-paths in S we write S^*, again omitting the reference to T. We shall usually assume that paths have a first element, which we write $p(0)$; if p has a last element, it will be denoted by $p(\#)$. The empty path $\langle \rangle$ is denoted by $\emptyset$.

An *event* in S is any set of paths. If p is a path, then we say that p *realizes* E if and only if $p \in E$. Thus the set S^* will play the same role for us as the outcome space does for the statistician. The set

$$E(x) = \{p : p(0) = x \,\&\, p \in E\},$$

which might be called the *extension* of E at x, is the set of paths beginning at x that realize E. As in our informal example, E is the union of its possible extensions:

$$E = \bigcup\{E(x) : x \in S\}.$$

Our set-theoretical presentation permits a number of comparisons between events. If two events E and F have the same intension, then they are impossible to distinguish in the modeling and so are identical (in the modeling). "To move a rook" and "to move a rook horizontally or vertically" are in this sense identical in our chess modeling. But two events may have the same extension at some state but not at others. For example, "to move the White King to the second row" and "to check Black" may have the same extension at a certain configuration, but at most configurations they don't and so are different events. Similar remarks apply to inclusion; the following two conditions are not equivalent:

$E(x_0) \subseteq F(x_0)$, where x_0 is a particular state,
$E(x) \subseteq F(x)$, for all states x.

Such observations are germane to the topic of level-generation, introduced by Goldman in [5]. If the latter condition obtains, then any realization of E is also a realization of F, and one might express this by saying that E *generates* F in a very strong sense. If the former conditions obtain, then any realization of E starting at x will also realize F, and this one might express by saying that E *generates* F *at* x. This way of speaking is not quite in agreement with Goldman's own presentation, but the present theory – if supplemented by further concepts in action theory – offers a basis for a formal reconstruction of his ideas.

Similar examples are readily found in any state space. For example, in the dice-rolling example we may consider the events $L_{n,t}$ of getting at least n points in t throws. Then $L_{m,t} \subseteq L_{n,t}$ whenever $m \leq n$ (where $0 \leq m, n \leq 6t$): Any path realizing $L_{m,t}$ also realizes $L_{n,t}$. This example illustrates the customary distinction between "event types" and "event tokens" – another important topic. Event types are events, in our sense; event tokens are paths. Thus an event type with only one possible realization must not be confused with the event token that is identical to that possible realization. For example, the path $p_0 = \langle (6,6), (6,6), (6,6) \rangle$ is an event token, while the event $L_{36,3} = \{p_0\}$ is an event type. It is worth noting that it is usually event types, not event tokens, that can be named in an object language. One might observe ("see") a path develop, but to describe it is to point to events under which it falls.

It may be noted that the usual set-theoretical operations are meaningful for events: the union of two events and the intersection of two events make sense. Even the complement of an event makes formal sense, but it is clear why it is not of much interest: Interesting events tend to be "small," and almost any path would realize the complement of such an event. But occasionally complements do make sense. For example, the event that something happens and the event that nothing happens are both acceptable events.

If h is a history and I is a closed interval in T, then let $h \mid I$ be the restriction of h to I. Somewhat inexactly, we shall identify $h \mid I$ with the path in S to which it obviously corresponds. This convention makes it easy to state the following truth condition for propositions of type (2):

> $\lceil$**E has taken place**$\rceil$ is true with respect to h and t if and only if there is an interval $I \subseteq$ dom h such that $h[I] \in \|\mathbf{E}\|$ and, for all $i \in I$, $i < t$.

This concept of event does not cover all cases that one would want it to cover. One omission is what might be called *events with interruptions*. The event of the White King being moved three steps may take place over several moves, with other pieces being moved in the meantime. *Iterative events* (such as the White King being moved repeatedly) will also require attention. Nevertheless, at the expense of some complication, such events can also be dealt with in this theory.

5. PROCESSES

Raining (rainfall) is a standard example of a process (see [14], p. 25ff). An account of such a process is usually given in context. Thus a condition for it to be raining at a time t would be that there exist some real open interval $(t-\epsilon, t+\epsilon)$ around t, for some $\epsilon > 0$, such that it rains throughout that interval – that is, for each u, if $u \in (t-\epsilon, t+\epsilon)$ then it rains at u. If rainfall is regarded as a homogeneous process then there is in fact an infinite collection of open intervals around t throughout which it rains.

Not all processes are like rainfall in all respects. Some have a definite beginning (the alarm beginning to sound) or a definite end (the train coming to a full stop). Some have a more or less specified duration (raining for several days; raining for more than a week; raining for less than a week). Some are not homogeneous (getting wet). Our problem is to find a sufficiently general semantic representation that allows for this diversity. This is a difficult topic, and among the many tentative and speculative suggestions made in this chapter, the following are the most tentative and speculative.

We begin with some technical definitions. The operation of *concatenation* is taken for granted: If p and q are paths in S such that $p(\#) = q(0)$, then pq is a well-defined path (such that $pq(0) = p(0)$ and $pq(\#) = q(\#)$). For any paths p and q in S we say that p is a *subpath* of q (in symbols, $p \le q$) if there are paths q_0 and q_1, not necessarily nonempty, such that $q = q_0 p q_1$; moreover, p is a *proper* subpath of q if p is a subpath and not identical to q. If c is a set of paths linearly ordered by the subpath relation, then we write glb c for the *greatest lower bound* of c and lub c for the *least upper bound* of c. Formally, $p = \text{glb}\, c$ if (i) $p \le q$ for all $q \in c$ and (ii) if $r \le q$ for all $q \in c$ then $r \le p$. Similarly, $p = \text{lub}\, c$ if (i) $p \ge q$ for all $q \in c$ and (ii) if $r \ge q$ for all $q \in c$ then $r \ge p$. The paths glb c and lub c always exist, but of course they need not be elements of c.

We now define a *complex* as a set c of paths linearly ordered by the subpath relation, subject to the condition that glb $c \ne \emptyset$. The complexes of P are regarded as the possible *realizations* of P. A path in a complex c of P may be said to *exemplify* P. If P is a process and $x \in S$, then we write $P(x)$ for the set of complexes c every path of which goes through x:

$$P(x) = \{c \in P : x \in \text{glb}\, c\}.$$

Notice that P is the set of its possible realizations:

$$P = \bigcup \{P(x) : x \in S\}.$$

It is natural that there should be logical relationships between states-of-affairs and processes. For example, if S is a state-of-affairs then there is the process of *maintaining* S and the process of *suppressing* S:

248

$$\mathbb{P}_{\text{maintain}} S \overset{\text{df}}{=} \{c : \forall p \in c, \forall i \in \text{dom}\, p, p(i) \in S\};$$

$$\mathbb{P}_{\text{suppress}} S \overset{\text{df}}{=} \{c : \forall p \in c, \forall i \in \text{dom}\, p, p(i) \notin S\}.$$

There is an even more intimate relationship between events and processes (which is rather obscured by natural language, where the same verb is sometimes understood as denoting an event and sometimes a process). For example, for every event E there is the process of the *unfolding* of that event:

$$\mathbb{P}_{\text{unfold}} E \overset{\text{df}}{=} \{c : \exists p \in E\; c = \{q : \exists r\; p = qr\}\}.$$

On the other hand, every time a process is run, many events are realized. In particular there is the event of running the process (for a while) – the event of *exemplifying P*:

$$\mathbb{E}P = \{q : \exists c \in P\; q \in c\} = \bigcup P.$$

There is also the event of running it to the end – the event of *finishing P*:

$$\mathbb{E}_{\text{finish}} P = \{q : \exists c \in P(\text{lub}\, c \in c \,\&\, q = \text{lub}\, c)\}.$$

Notice that this event is impossible if P does not contain at least one *terminating* complex; that is, a complex c such that $\text{lub}\, c \in c$ and $\text{lub}\, c$ has a last element.

The following simple observations are obvious, but interesting all the same:[4]

(1) $\mathbb{E}_{\text{finish}} \mathbb{P}_{\text{unfold}} E = E$ for every event E;
(2) $\mathbb{E}_{\text{finish}} P \subseteq \mathbb{E}P$ for every process P;
(3) $E \subseteq \mathbb{E}\mathbb{P}_{\text{unfold}} E$ for every event E.

Let us now provide truth conditions for sentences of type ⌜**P is going on**⌝, where $\|\mathbf{P}\|$ is a process in S. As before, the truth conditions are given with respect to some history h and time t relative to the temporal state space $S \times T$. Suppose first that T is the set of real numbers:

> ⌜**P is going on**⌝ is true with respect to h and t if and only if there is an open interval $I \subseteq \text{dom}\, h$ such that $t \in I$ and, for some $c \in P$, $h|I \in c$.

At first sight it might not appear natural to accept that processes can occur in discrete time. Nevertheless we sometimes express ourselves as if they do. A Turing machine might go into a loop and keep repeating itself, a chess player might be preparing to castle or might be increasing the pressure on Black's Queen's flank, and so forth. Accordingly, if T is the set of natural numbers then we suggest the following definition:

249

⌜**P is going on**⌝ is true with respect to h and t if and only if there is a segment $I = \{i: m < i < n\}$ of natural numbers such that $m+1 < t < n-1$ and, for some $c \in P$, $h \mid I \in c$.

Notice that here it is necessary that $n - m > 3$.

6. EXTENDING THE MODELING

The previous outline (Sections 1–5) has been an attempt to identify the most general logical properties of states-of-affairs, events, and processes.[5] Concentrating on how to represent these properties in set-theoretical terms, our outline omits a number of considerations that are of fundamental importance for our understanding of these concepts.

One main lacuna in our account is that no attention is paid to how events and processes are produced. In many cases causality is indispensable for understanding events and processes, but in others there might be some other source of order. To understand what goes on in a truly chaotic modeling is difficult or perhaps impossible; all understanding involves some ordering, actual or perceived.

Therefore it is important to associate with each state space a law or rule (or set of laws or rules) that regulates what happens in the modeling. For example, one might think of the state space as being the state space of a certain automaton or abstract machine. If S is the state space, let Γ_S be the machine function associated with S; we drop the subscript when the context makes it clear what the state space is. It goes without saying that the same state space can be associated with different machines, so the choice of machine makes a fundamental difference to the modeling.

A machine function in automata theory is a function defined on some subset of the state space, assigning to each total state either a total state (deterministic machines) or a set of total states (nondeterministic machines). For some of our purposes a more general notion may be needed: The domain might be (some subset of) the set of finite paths in the state space, and the value of Γ for the argument p may be a set of possible continuations of p; that is, a set of paths q such that $p(\#) = q(0)$.

Where action is involved, this description is inadequate. The machine function governs what happens in the world. But action is essentially influencing "from outside the world" how the world develops. Von Wright describes it well:

An act *is* not a change in the world. But many acts may quite appropriately be described as the bringing about or *effecting* ('at will') of a change. To act is, in a sense, to *interfere* with 'the course of nature'. ([14], p. 36)

On this view, the world is given by the state space. The wills of the agents are outside the world, yet they are sometimes able to influence what goes on in the world. We might represent this by considering their wills as inputs that serve to determine the machine function. Thus, if there are n agents, the machine function should be seen as a functor $\Gamma(i_0, \ldots, i_{n-1})$ where $i_0, \ldots, i_{n-1}$ are parameters, each i corresponding to the input from an agent. In other words, given arguments $i_0, \ldots, i_{n-1}$, $\Gamma(i_0, \ldots, i_{n-1})$ is a machine function in the sense just suggested. The questions of how to model the agents' inputs, and how these inputs are determined, will not be discussed here. However, for given inputs $i_0, \ldots, i_{n-1}$ and a given state x, $\Gamma(i_0, \ldots, i_{n-1})(x)$ would represent what von Wright calls "the course of nature" and what some would call "the continuation of the *world-line*." The system is *deterministic* if, for all inputs $i_0, \ldots, i_{n-1}$ and all paths p, the set $\Gamma(i_0, \ldots, i_{n-1})(p)$ is a singleton.

There is no agent in the examples of the Turing machine, the computer, the solar system, and the weather. Also, in the example of the dice there is no agent. Of course, there is a device (which could be a person) to keep the dice rolling, but it is not modeled. (This example is unusual in that here Γ is the most general function possible – the constant function that assigns, for any path p, the value $\Gamma(p) = S$. This extreme example exhibits a maximal degree of indeterminacy: Any continuation of any path is possible. This analysis takes no account of probability; a more refined analysis would.) In the stock-exchange example there is an even clearer sense in which there are agents. However, in the modeling outlined in this chapter they are treated as inside the world, and hence are not seen as agents in our sense.

The examples cited were given primarily in order to exemplify events and processes, not actions; this is why they contain no agents. However, in the chess example it would be natural to recognize two agents, W (White) and B (Black). Given a certain past history p, $\Gamma(i_W, i_B)(p)$ would be a singleton set $\{q\}$, where $q = \langle p(\#), s \rangle$ for some $s \in S$; chess is in this sense deterministic. In game theory one might (unrealistically) take White's and Black's inputs to be complete strategies. (Clearly, White's input will not matter when Black is about to move, and vice versa.) This example is one where the arguments of the machine function need to be paths rather than states. For example, according to one rule of chess, a game is declared a draw when for a third time the same configuration appears on the board – this could make a difference for a player who wishes to achieve or avoid a draw.

Thus there are several possible cases, depending on how many agents there are. (1) If there are no agents, or if there are agents but their input is taken as fixed, then one may say that the study of the machine function and what can happen in the world is up to "natural science," so long as we recognize that what is meant by "nature" can vary considerably. (2) If there is only one agent, or if there are more agents than one but the input of all but one agent is kept fixed, then one may study the machine function $\Gamma(i)$ as a functor depending on that agent's input i and what happens in the world as depending on that one agent; this is the approach used in individual decision theory. (3) If there are two agents, or if there are more than two agents but the input of all but two agents is kept fixed, then one may study the machine function as a functor $\Gamma(i, j)$ depending on the input i and j of those two agents; this typifies two-person game theory. (4) Similarly, if there are only n agents (for some $n > 2$), or if there are many agents but the input of all but n agents is kept fixed, then the machine function may be studied as a functor $\Gamma(i_0, \ldots, i_{n-1})$ depending on the input of those n agents; this is done in n-person game theory.

7. AN EXAMPLE FROM THE LOGIC OF ACTION

In order to analyze human action adequately, one would have to introduce further notions such as "intention," "plan," "goal," etc. – all notoriously difficult. Thus our previously defined apparatus, complicated though it is, is not rich enough. On the other hand, it is possible to work out meaningful fragments: relatively uncomplicated theories or models that may serve special, limited purposes. In this section we discuss how notions of "ability" may be modeled.

Typically, the actions of agents are structured in a certain special sense. A random device does one thing after the other, with no discernible pattern to what it is doing; to say that it could have done "the same thing in a different way" seems (at best) artificial. But there are several ways of doing almost anything that a real agent does. For example, I have two standard ways to get to work in the morning: drive or take the ferry. Introducing a technical term, I would like to say that I have two *routines* for getting to work. Each consists of subroutines. Driving to work involves walking to the car, then getting into it, then starting it, then backing out into the street, etc. Taking the ferry means first walking to the ferry, then embarking, then "doing nothing" while the ferry is crossing the harbor, etc. But even these subroutines have subroutines . . . . If one wants to discuss examples like this, it is evidently desirable to agree on some level of analysis for in principle there seems to be no limit beyond which the analysis could not be pushed. (See [9] for a more detailed examination of the concept of "routine.")

Conversely, just as routines have subroutines, so routines can be combined into more complex routines. For example, there are the three so-called *regular operations* +, ;, and *: If α and β are routines, then so are $\alpha + \beta$, $\alpha; \beta$, and α^*, which may be described as follows. The routine

252

$\alpha + \beta$ ("α or β") is performed by either performing α or performing β. The routine $\alpha; \beta$ ("α and then β") is performed by first performing α and then – α completed – performing β. The routine α^* ("α some number of times") is performed by performing α any finite number of times (each performance of α must be completed before the next is begun).

I shall assume, then, that every agent has a *repertoire,* by which I mean a set of available routines. Perhaps what (in Section 6) was called the input of an agent might here be analyzed as the selection of a routine belonging to the agent's repertoire. Let us now make the simplifying assumption that we are dealing with the one-agent case; that is to say, let us focus on a certain agent and keep the input for all other agents (if there are any) fixed. We shall view the running of a routine as a trivial enterprise that in principle can be left to an umpire: Once the agent has made up his mind and selected a certain routine, he will not be consulted again (unless he changes his mind and decides to intercede and is able to do so). Thus our routines are very much like strategies in game theory, and equally implausible if viewed from the point of view of practical life.

Let us now try to form a picture of routine, that is, try to represent it semantically. Suppose that, relative to some state space and with the world in some total state x, the agent selects routine α for running. Things will now happen in the world; in the preceding terminology, the worldline will be continued. The sets $\Gamma(\alpha)$ and $\Gamma(\alpha)(x)$ are events, in our technical sense. There is an immediate temptation to identify the event $\Gamma(\alpha)$ with the routine α and the event $\Gamma(\alpha)(x)$ with the routine α as started at x. Resist that temptation![6] Routines are not events; they are more like programs. Thus, a carrying-out or enactment of a routine may come to a stop, just as a computation according to a program may *halt*. On the other hand, a carrying-out of a routine, like a computation according to a program, may *fail* (if instructions are inadequate or impossible) or *go on indefinitely;* realizations of an event do neither of those things.

For each routine α of the agent we therefore define $H(\alpha)$, the set of halt paths of α; $F(\alpha)$, the set of fail paths of α; and $I(\alpha)$, the set of infinite paths of α. Evidently, the functions H, F, and I must satisfy (among others) certain conditions with respect to the regular operations. Some technical concepts will let us express those conditions. Let A and B be arbitrary sets of paths. Define:

$$A \mid B = \{ pq : p \in A \,\&\, q \in B \,\&\, p(\#) = q(0) \};$$

$$A^0 = \{ \langle x \rangle : \exists p \in A \; x = p(0) \};$$

$$A^{n+1} = A^n \mid A;$$

$$A^* = \{ A^n : n < \omega \}.$$

253

The conditions on H, F, and I are then as follows:

$$H(\alpha + \beta) = H(\alpha) \cup H(\beta),$$

$$H(\alpha; \beta) = H(\alpha) \mid H(\beta),$$

$$H(\alpha^*) = H(\alpha)^*;$$

$$F(\alpha + \beta) = F(\alpha) \cup F(\beta),$$

$$F(\alpha; \beta) = F(\alpha) \cup (H(\alpha) \mid F(\beta)),$$

$$F(\alpha^*) = F(1) \cup (H(\alpha)^* \mid F(\alpha));$$

$$I(\alpha + \beta) = I(\alpha) \cup I(\beta),$$

$$I(\alpha; \beta) = I(\alpha) \cup (H(\alpha) \mid I(\beta)),$$

$$I(\alpha^*) = I(1) \cup (H(\alpha)^* \mid I(\alpha)).$$

Here 1 is a constant term that denotes a particular routine: passivity, the routine of "doing nothing." In machine terms this routine may be thought of as turning on just to turn off:

BEGIN! ; STOP!.

A minimum condition governing this term is that $H(1) \subseteq \{ p : \exists x \, p = \langle x \rangle \}$.

A routine α may thus be semantically represented by a triple $(H(\alpha), F(\alpha), I(\alpha))$. To repeat what was just argued: a routine is distinct from the event of running α; it is also distinct from the process of running α.

We now have a conceptual basis for expressing two interesting concepts of ability, one stronger and one weaker. The stronger one involves control or reliability, in the sense that I can open this window now or that my car (on a given occasion) can do more than 50 kph. This is probably the concept of ability concerning which Moore [7] gave his celebrated – and controversial – analysis. Let R be the repertoire of the agent; let S be any state-of-affairs. We define ∂S as the routine defined as follows:

$$H(\partial S) = \bigcup \{ H(\alpha) : \alpha \in R \, \& \, \forall p \in H(\alpha) \, p(\#) \in S$$
$$\& \, \forall p \in F(\alpha) \cup I(\alpha) \, p(0) \neq x \};$$

$$F(\partial S) = I(\partial S) = \emptyset.$$

The routine ∂S may thus be described as the maximal reliable way for the agent to see to it that S obtains. We way that the agent is *able, in the strong sense,* at x to see to it that S if $H(\partial S)(x) \neq \emptyset$. The weaker concept of ability is more traditional. We say that the agent is *able, in the weak sense,* at x to see to it that S, if there is some routine $\alpha \in R$ and some $p \in H(\alpha)$ such that $p(0) = x$ and $p(\#) \in S$. This is a much weaker concept of ability, but still not as weak as mere logical possibility.

254

With the help of further abstraction, these suggestions may be related to theories in the literature. One abstraction is to keep only the starting and end points of paths and discard all intermediate points; this is the way of propositional dynamic logic (PDL). A model may readily be transformed into one of the more abstract kind as follows:

$$R(\alpha) = \{(x, y) : \exists p \in H(\alpha)\, p(0) = x \,\&\, p(\#) = y\},$$

$$Q(\alpha) = \{(x, \lambda) : \exists p \in F(\alpha) \cup I(\alpha)\, p(0) = x\},$$

where λ is some object not in the state space.[7] The other abstraction is to make the customary identification of a state-of-affairs with the proposition that the state-of-affairs in question obtains.

We propose to use the ordinary object language of PDL, with the following restriction:[8] The program terms are built with the help of the regular operators $+, ;$, and $*$ from expressions δA, where A is a propositional formula and δ is a new operator corresponding to the model-theoretical operation ∂. Moreover, we also admit the symbols $\lozenge$ and OK; for $\lozenge A$ read "it is dynamically possible that A" and for OKα read "α always halts." The semantic conditions of these two operators are as follows:

$\lozenge A$ is true at x if and only if $\exists y((x, y) \in R(\alpha) \Rightarrow A$ is true at $y)$;

OKα is true at x if and only if $Q(\alpha) = \emptyset$.

In this system we may redefine our two concepts of ability as follows:

$$\mathbf{can\text{-}do}_s\, A \stackrel{\mathrm{df}}{=} \mathrm{OK}\alpha \wedge \langle \delta A \rangle \top,$$

$$\mathbf{can\text{-}do}_w\, A \stackrel{\mathrm{df}}{=} \lozenge A,$$

where **can-do**$_s$ reflects strong ability and **can-do**$_w$ reflects weak ability.

NOTES

1. Von Wright never expresses himself in this way, and it is possible that he would find our slogan unacceptable. In [11], a slightly different slogan was flaunted: "To act is to bring about an event." Obviously the two slogans are related, perhaps (given suitable assumptions) even equivalent.
2. A previous effort along similar lines was made in [10].
3. Davidson is one of the best-known exceptions; see [3].
4. *Proof of* (1). Take any $q \in \mathbb{E}_{\mathrm{finish}}\, \mathbb{P}_{\mathrm{unfold}}\, E$. Then there is some c such that $c \in \mathbb{P}_{\mathrm{unfold}}\, E$ and $q = \mathrm{lub}\, c$. Consequently, there is some p such that $p \in E$ and $c = \{q : \exists r\, p = qr\}$. The fact that $q \leq p \leq \mathrm{glb}\, c = q$ implies that $p = q$. Hence $q \in E$. Conversely, take any $p \in E$. Define $c = \{q : \exists r\, p = qr\}$. Since $p(0) \leq q$ for all $q \in c$, it follows that $\mathrm{glb}\, c \neq \emptyset$; therefore $c \in \mathbb{P}_{\mathrm{unfold}}\, E$. It is clear that $p = \mathrm{lub}\, c$ and $p \in c$. Hence $p \in \mathbb{E}_{\mathrm{finish}}\, \mathbb{P}_{\mathrm{unfold}}\, E$. $\square$

   Proof of (2). Take any $q \in \mathbb{E}_{\mathrm{finish}}\, P$. Then there is some $c \in P$ such that $\mathrm{lub}\, c \in c$ and $q = \mathrm{lub}\, c$, and thus $q \in c$. Hence $q \in \mathbb{E}P$. $\square$

   Proof of (3). By combining Theorems 1 and 2. $\square$

5. It is not claimed that the facts studied in this chapter are all the facts there are; there exist "higher-order" facts that are not treated here.
6. Here the author is giving good advice to himself; see [11].
7. There is a similar use of a "limbo" element λ in an unpublished paper by Vaughan Pratt.
8. See [4] for a textbook introduction to PDL or [6] for an extensive introduction. A system closely related to that presented here was treated in [11]. The present exposition has benefited from discussions with T. J. Surendonk and S. K. Thomason and their papers ([12], [13]). For a related yet different analysis of ability, see [1].

REFERENCES

[1] Brown, Mark A. (1988), "On the Logic of Ability." *Journal of Philosophical Logic* 17: 1–26.
[2] Carnap, Rudolf (1947), *Meaning and Necessity: A Study in Semantics and Modal Logic.* Chicago: University of Chicago Press.
[3] Davidson, Donald (1980), "The Logical Form of Action Sentences." In *Essays on Actions and Events.* Oxford: Clarendon Press, pp. 105–48.
[4] Goldblatt, R. I. (1987), "Logics of Time and Computation." CSLI Lecture Notes, vol. 7, Stanford University.
[5] Goldman, Alvin I. (1970), *A Theory of Human Action.* Princeton, NJ: Princeton University Press.
[6] Harel, David (1984), "Dynamic Logic." In D. Gabbay and F. Guenthner (eds.), *Handbook of Philosophical Logic,* vol. 2. Dordrecht: Reidel, pp. 497–604.
[7] Moore, G. E. (1912), *Ethics.* London: Williams & Norgate.
[8] Scott, Dana (1970), "Advice on Modal Logic." In Karel Lambert (ed.), *Philosophical Problems in Logic: Some Recent Developments.* Dordrecht: Reidel, pp. 143–73.
[9] Segerberg, Krister (1985), "Routines." *Synthese* 65: 185–210.
[10] Segerberg, Krister (1988), "Talking about Action." *Studia Logica* 47: 347–52.
[11] Segerberg, Krister (1989), "Bringing It About." *Journal of Philosophical Logic* 18: 327–47.
[12] Surendonk, Timothy J. "More on Action." *Theoria,* to appear.
[13] Thomason, S. K. "Dynamic Logic and the Logic of Ability." To appear in the proceedings of a 1989 conference on model logic in Siena.
[14] Von Wright, Georg Henrik (1963), *Norm and Action: A Logical Inquiry.* London: Routledge & Kegan Paul.

15

Introduction to metamoral

ROBERTO MAGARI

Every "reasonable" ethics is "rational."

The divagations that follow concern "metamoral" problems (the meaning of this term will be clarified) together with some "moral" indications. From a mathematical point of view, the following observations are manifest, although they are explained with more words than necessary (for the benefit of the nonmathematical reader). There are no historical or philological references: The professional historian or philosopher is in a better position to provide them than is the writer, who is a mathematician by profession, and the nonspecialist reader can find better sources by consulting any good history of philosophy. Occasional references to Pascal, Bentham, and others are rather more "historical fiction" than historical, since the author felt it was more important to indicate the sources of certain stimuli than to ascertain historical exactness in the attribution of such and such a thought to Tom or Dick or Harry.

The nonmathematical reader may wish to skip over (at least on a first reading) the mathematical parts.

As a brief clarification of the term "metamoral," note that: I refer to all research aimed at establishing an organized system of prescriptions (i.e., a "moral theory") as "moral research"; to all behavior aimed at influencing one's own or others' behavior, in particular moral theories, as "moral activity"; and finally to all research aimed at ascertaining possible moral theories, the conditions in which they acquire meaning, and their connections to the various cognitive situations with which they must coexist, as "metamoral research." Plainly these three types of activity never occur in pure states, and each of them is dependent on the linguistic method and the cognitive schema in which it is placed.

These concepts have also been discussed in "Sulle Morali Pascaliane," *Rivista di Filosofia* (*Journal of Philosophy*) 75 (1984), pp. 163–83, and in "Morale e Metamorale, un Approccio Probabilistico ai Problemi Morali," CLUEB, Bologna (1986).

In the previous paragraph I used the words "behavior" and "prescription" as undefined terms. A realistic analysis of these terms is not appropriate for the purposes of the research I propose to carry out. I will thus reduce them to their essential principles, leading (in a certain sense) to their impoverishment, only to make them more complex again as we go along.

A first example of what is implicit in the term "behavior" can be illustrated by a chess player's position. After his adversary's tth move the player finds the board in a certain situation $s(t)$ (which should include, because of castling and *en passant,* a part of the history of the game), and reacts with a certain action (move) $a(t)$. For now we may consider behavior, taking the situation just described in an abstract way, as a pair of functions (s, a) whose domain is the set T of natural numbers (which, for convenience and to preserve the images underpinning our schema, we will refer to as "instants in time"). The range (or an appropriate set that contains it) S of s will be called "the set of situations" and A, the range of a, "the set of actions."

Now imagine we are observing a person – or for that matter an area of space, a machine, etc. – without knowing anything about that person's "thoughts" or "experience," much less his "moral criteria." We see this person at instant t (I am assuming a discrete representation of time, thereby avoiding complications that are at present irrelevant) in a certain situation $s(t)$. By $s(t)$ we mean to indicate not only what we know about the person's situation in the strictest sense, but also everything we know about the whole world (or, if you like, about that person's sphere of influence) at instant t. The person under observation carries out a certain action $a(t)$; obviously we must suppose $a(t)$ to be posterior to $s(t)$. For convenience, let's suppose that the person acts between one instant and another of our discrete time, just as in a game of chess a move is executed between one position and another. Let us further imagine the person as the only part of the universe unknown to us: a closed box, into which stimuli enter (relative to $s(t)$) and out of which responses issue ($a(t)$). (The spread of computers, behaviorism, etc. has made these concepts familiar.)

The search for laws that allow us to predict $a(t)$ once $s(t)$ is known (and possibly the person's history in relation to his actions, although that history can be considered part of $s(t)$ by means of a written biography or other records) is a purpose of psychology – at least for as long as it is obliged to work mainly with closed boxes, and if the search is not then reduced to biology and finally to physics (although nothing prevents us from supposing that physics itself concerns "closed boxes," even if it analyzes them in increasingly smaller parts).

However, another type of research is possible. Each of us knows, or thinks he knows – if not by other means than by an introspective analysis of the box he knows best (which is for him, so to speak, less closed, even if in analyzing it he discovers other closed boxes) – that a first analysis is possible. In particular, each believes that he can distinguish in himself a part that seeks not to understand the link between s and a but rather to "prescribe" it.

What does to "prescribe" it mean? – in the first place, to invent or theoretically construct another link different from the one ascertained by the facts. This part of the person (which some would call the super-ego[1] or, to use a traditional term that is loaded with possible misunderstandings, the "moral conscience") has not only a theoretical function but, within certain limits, also contributes to behavior, especially in matters of "moral activity" in the sense indicated at the beginning of this chapter.

For example the "moral conscience" of a professional moralist is generally not sufficient to ensure that his behavior complies with the rules he elaborates, although it does influence his behavior enough to cause him to write books on morals and to compose moral lessons. Perhaps the reader will observe that it is not moral conscience that causes the moralist to write books and compose lessons, but rather (at times) self-love, envy tinged with sadism of other people's pleasure,[2] or – in the best of events – an impulse of love toward others (much more precious when it is love not only for other men but toward all forms of life).

It is up to the psychologist to decide if this is the case, but the fact remains that we are dealing with these books and not others, that the writer's imagination has addressed itself to those subjects and not to others; this allows us to consider a part of his personality as what might be called moral imagination ("imagination" is intended here as a creative function) rather than conscience.

Numerous problems arise for those who accept this analysis or some aspects of it. Useful or not, the fact remains that some of us would like to develop this part, that is, to prescribe rules of behavior or at least to construct them. (Naturally the destruction of rules of behavior is an important and valuable part of this activity.)

Moreover, we also want to influence our own and others' behavior. Undeniably, we have some all-too-efficient tools for this purpose, and psychology is forever offering us others, sometimes terrible: The education of children by potent religious or ideological organizations, advertising and propaganda, and the pressures to which every individual is exposed before relinquishing autonomy provide us with innumerable and repugnant examples.

259

But anyone whose opinions and even sense of values have been changed by contact with people he respects knows that other methods exist, and that the presentation of moral values and schemata can be convincing when just honesty and clarity are employed, without psychological violence. This nonviolent persuasion can be carried out by presenting values and moral schemata and, although clarity of presentation is not everything, it is important.

It is not my intention that a justified uneasiness about teaching by precepts, or that some presumption about the irreducibility of the moral world to formulas, should give rise to misgivings about the analysis we want to carry out. I believe that whatever is "irreducible to formula" in human activity has everything to gain by an examination of what is analyzable. I think the claim that analysis "drys up" and "kills" should be inverted; instead, analysis brings clarity exactly where the emotions look for nourishment in the wrong places – that is, to intellectual errors.

Another and more important objection to the kind of analysis outlined here can arise in those who maintain that we should talk not about metamorals but rather about metabehavior, that is, the science of behavior (or psychology, although perhaps psychology has also other goals). In attempting to clarify the term "prescription" I have already replied in part to this objection; however, I wish to add by way of analogy that for me the relationship between metamorals and psychology is similar to that between symbolic logic (or metalogic) and psychology. Symbolic logic is not concerned with the investigation of effective thought, even if it can contribute something to this. Rather, symbolic logic is (in a certain sense) concerned with proposing prescriptions for thought – or, better still, enriching it – while metalogic is concerned with establishing possible logical calculations, investigating their demonstrative power, and so forth. Symbolic logic and metalogic create new forms, possibilities, and tools.

Returning to the example of the person whose behavior was schematized by way of the functions s and a: let us now introduce a specific analysis of the person, thereby enriching the schema (s, a) and at the same time giving a first (and I think fundamental) example of the "morals" an individual can assume. The mathematical reader can skim this part, which is a presentation of the normal probabilistic utility schema.

Let us connect our person with a function p from the domain $S \times T \times A$ having values in the real interval $[0, 1]$. If s is a "situation," t an instant, and a an "action," the value $p(s, t, a)$ will be called "the probability the person ascribes to the existence of the situation s at the time $t + 1$ if at time t he carries out action a."

Let us also connect our person with a function v from S to the field of real numbers (it is possible to use other ordered fields). If s is a situation then $v(s)$ will be called "the value that the person attributes to situation s." The sentences introduced clarify the meaning of the functions p and v from a commonsense point of view; all this will be made even clearer for the nonmathematical reader if he keeps in mind the example of (say) a roulette player. At instant t, the player's situation can be described in terms of the money he possesses, the past history of the roulette wheel, and so forth. The player has a choice between various actions, and for each one he predicts various possible consequences. Suppose he predicts that if he performs action a then situation s_1 will have a particular probability of coming true, s_2 another probability, and so on. This information is contained in function p. In short, each situation represents an increase or a decrease in the player's capital, presuming for simplicity's sake that the value of these situations is calculated on the basis of the money that might be won in each case.[3]

Here is a well-known (at least since Pascal's time) possible "moral" for our gambler: $a(t)$ – that is, the action that the gambler carries out (or prescribes himself) at time t – is such as to maximize the expression

$$\sum_s p(s, t, a) v(s).$$

We have used a summation operator, presuming for the time being that the sets A and S are finite. Generally, in a first approach to these problems, it is misleading and unnecessarily complicated to deal with the infinite case.

The task of reconstructing the history of this type of moral and establishing its connections with the thought of Pascal, Bentham, and many others is left to the historically inclined reader. From now on, in allusion to the above formula, we will refer to Pascalian behavior, Pascalian morals, etc. Analysis of the type of morals presented here can be hindered by personal attitudes of acceptance or rejection. Keep in mind that we are studying and not accepting or rejecting a morality.

Initially, we might make the mistake of believing that we are dealing with a very special moral schema. Closer examination reveals that if function v is not fixed then the majority of morals can be included, by particular determinations of v, in the indicated schema – especially if (as will be shown) certain finitist hypotheses are abandoned, and in particular if values and probabilities with actual infinites and actual infinitesimals are introduced.

However, the opposite thesis would also be erroneous: that any given behavior and therefore any given moral could be included in the indicated schema by way of an appropriate choice of v. Among other things, it will be observed that function v is not subject to time, which obviously means that any behavior that reacts to identical situations with different actions cannot be included in the schema. (An easy way of sidestepping the problem is to absorb, as it were, the mention of time into the situations.) Nevertheless it's obvious that – with the exception of the fact that nearly every individual's moral schema is continually changed during the course of a lifetime – the majority of morals proposed throughout history have been, so to speak, timeless. It's not that they necessarily lack an explicit notion of time; they may even prescribe different behavior at different times (although generally because it is assumed that at different times the situations will in reality be different). But every time it is pretended that a law is given, it is in reality independent of time. Naturally not all possible morals (nor, perhaps, all those contrived) are of this type. Nor is a morality impossible that forsees in a person the formation of values at present unknown.[4] With regard to the problem of including a given moral in the probability-values schema when neither values nor probabilities are fixed, the reader is referred to Savage's work.[5] I will deal with the situation where the probabilities are fixed. In this case, even allowing for a variation of v with t, the Pascalian schema fails to cover every possibility, as the following example demonstrates.

Example 1. There are three possible actions a_1, a_2, a_3 and two possible situations s_1, s_2. Let us suppose that, according to the person in question, if he carries out action a_1 then s_1 is sure to take place; if he carries out action a_2 then s_2 is sure to take place; and if he carries out action a_3 then either s_1 or s_2 will take place with equal probability. Now, if the value v_1 attributed to s_1 and the value v_2 attributed to s_2 were such as to cause the subject, according to the Pascalian schema, to prescribe himself a_3, then we would have:

$$\tfrac{1}{2}v_1 + \tfrac{1}{2}v_2 > v_1 \quad \text{and} \quad \tfrac{1}{2}v_1 + \tfrac{1}{2}v_2 > v_2,$$

which is impossible.

There are some functions v that, when the Pascalian schema is applied, do not univocally determine the action to be chosen. In the previous example this would occur if $v_1 = v_2$, in which case even action a_3 would be maximized.

If we were to call every function v that (by way of the Pascalian schema) univocally determined the value of $a(t)$ "regular" in relation to certain values of p and t, then the previous example demonstrates that when $a(t)$ is fixed, a regular v that prescribes the given $a(t)$ doesn't always exist. Naturally there is always a nonregular v (e.g., every constant v) that maximizes, in relation to $a(t)$, the summation.

In leaving behind the abstractness of the schema, it should be pointed out that this example is less useful than it would at first seem. Among the results of an action there is generally something that is, as it were, an image or trace of the action; for example, the memory of it. With this in mind, we must admit that for every action there exists at least one possible situation that is not a possible consequence of any other action. Clearly, using this hypothesis, every action can be explained by way of the Pascalian schema. Still, there are very good reasons for working without this hypothesis, as we shall see.

So on the one hand, a regular v that enables us to prescribe a given action doesn't always exist; on the other hand, it is obvious – granted that a person has a Pascalian way of behaving with v independent of time – that we cannot always determine the function v from a restricted number of actions executed; nor, therefore, can we predict future behavior. Nevertheless, within the bounds of our finitist hypothesis, if we are free to choose p in each of our experiments then v can be determined (not univocally but) in such a way as to render future behavior predictable.

The question of how to determine p and v is of enormous importance, not just for our problem but also for the following reason. The problem is raised in epistemic research of finding a working definition for the concept of probability. Among the various theories proposed, some make use of the concept of "subjective probability." Normally, no attempt is made to give the concept a working definition by means of the subject's language; the concept remains undefined and subjected only to criteria and postulates (if anything). Yet, the concept is given (within certain limits) a working definition, with reference to the subject's behavior, by assuming that in substance this behavior is Pascalian and by determining p and v together.[6]

One of the arguments often invoked against assuming a Pascalian moral is that such a moral doesn't make allowance for "inestimable values" or "absolute imperatives" or the like. As we will see, this sort of objection is made invalid by working, when necessary, in a non-Archimedean field. There are also excellent reasons for working in non-Archimedean fields with respect to probabilities, reserving the probability 0 for impossible events.

From now on let's suppose we are working in ordinate fields (non-Archimedean when necessary) that are overfields of the field of real numbers. Specifically, we will suppose that probabilities are taken in a certain field K and values from one of its overfields, L. The nonmathematical reader should suppose for now that $K = L$; the argument may easily be followed by keeping in mind that among the numbers we will use, in addition to ordinary real numbers, there may also be "actual infinitesimals" – positive numbers h such that no whole multiple of h is more than 1 – and "actual infinites" (reciprocals of actual infinitesimals).

Because consideration of time is irrelevant for the properties we wish to examine, and since it will be convenient to make certain hypotheses about p and A, we will refer to a schema that is slightly different from the one used until now. Let S be a finite set of n elements. To indicate the underlying images in the usual way, we will call S "the set of possible situations." Let $P_n(K)$ be the set of n-tuples $(p_1, p_2, \ldots, p_n)$ of elements of K with $0 \le p_i \le 1$ and $\sum_i p_i = 1$; we write P_n for $P_n(K)$ if the mention of K is unnecessary. Let A be a subset of P_n. Then any function of part of the set of nonempty subsets of A to the set itself – let's say f, for whatever X may be, $f(X)$ is included in X – will be called a *moral* in S (relative to K, A). A moral f will be said to be *complete* or (more precisely) *K-complete* if $A = P_n$ and every finite nonempty subset of A belongs to Df (the domain of f).

To clarify the images leading to this definition, let's take our usual example of a chess player. S represents the set of situations that may occur during the course of a game. If we imagine that a player's move – allowing for the fact that his adversary will also move – doesn't lead to a certain situation that is determinable previously but (in his opinion) leads with probability p_1 to the situation s_1, with probability p_2 to the situation s_2, and so on, then for his every possible move we have an n-tuple $(p_1, p_2, \ldots, p_n)$ with which we can abstractly identify the move[7] (and also every move having the same n-tuple): This is the meaning of P_n and A. Obviously, in a game of chess not all the possible "moves" of this type exist; hence A is a proper subset of P_n. In every situation the player has a choice among actions ("moves") from a certain subset X of A, and his "style of play" (or "morality") remains fixed if an element $f(X)$ represents the move the player would choose among those of X.

In the general definition we see that f associates subsets to subsets of X, and not to elements of X. This has been done to include the case where a set of moves or actions is prescribed, among which a choice would be irrelevant. We will say that a moral is *regular* if, for every X different from $\emptyset$, $f(X)$ is a set of only one element – in which case, *par abus de langage,* that element will also be indicated by $f(X)$.

Of course, no chess player has such a precise style. Sometimes, both in chess and other cases, f can be only partially defined, the subject being unable or unwilling to prescribe a reaction to every situation (not everyone is willing to answer questions of the sort "what would you do if . . . ?"). Even though this is of considerable importance, it may be disregarded for our purposes.

Usually a moral f will be called *Pascalian* if and only if there is L, an overfield of K, and a function v from S to L such that however $X \in Df$, $a, b \in f(X)$, $c \notin f(X)$ are taken, we have

$$v(c) < v(a) = v(b),$$

where, for $k = (p_1, p_2, \ldots, p_n)$, we set $v(k) = \Sigma_i \, p_i v(s_i)$.

Note that, by abandoning the hypothesis that A is finite, a new case is introduced in which a function v doesn't necessarily lead to a moral defined for every nonvoid subset of A. Also, a case may arise where there is no maximum for some subset X of A (examples of this are to be found in any treatise on probability or game theory). Normally, however, we are more interested in the opposite problem: that is, whether or not a given moral is Pascalian. To this end, let's first note the following.

Lemma 1. *A moral f is Pascalian if and only if it has a Pascalian "completion."*

(Obviously, for "completions" of a given f we mean every complete g whose suitable restriction is f.) The proof is obvious.

In view of Lemma 1, we can hereafter limit ourselves to the consideration of complete morals. Therefore, let f be a complete moral with reference to a set S of n possible situations, and define a binary relation $<$ on P_n in the following manner:

If $a, b \in P_n$, we say that a is *preferable* to b (written $b < a$ or $a > b$) if and only if at least one subset X of P_n exists such that $a, b \in X$, $a \in f(X)$, and $b \notin f(X)$.

In other words, one action is preferable to another if and only if it is prescribed by f, where the other action is also feasible but not prescribed. Obviously, the relation $<$ is antireflexive (i.e., there is for every a no $a < a$).

We have the following as an obvious consequence.

Lemma 2. *If f is Pascalian then a v from S to an overfield L of K exists such that*

$$a < b \quad \text{if and only if} \quad v(a) < v(b).$$

For convenience, every binary relation on P_n that is antireflexive and satisfies the condition of Lemma 2 will be said to be Pascalian. It is easy to see that our problem is substantially one of determining conditions for an antireflexive binary relation $<$ in P_n to become Pascalian. The next lemma is obvious.

Lemma 3. *If $<$ is Pascalian then it is antisymmetric (it cannot be that both $a < b$ and $b < a$) and transitive (if $a < b$ and $b < c$ then $a < c$).*

Let us define another binary relation incorporating $<$ as follows:

> $a \equiv b$ (a is equivalent to b) if and only if for every x we have
> $x < a$ if and only if $x < b$;
> $x > a$ if and only if $x > b$.

Obviously, $\equiv$ is an equivalence (i.e., reflexive, symmetric, and transitive). It is clear that if $<$ is derived from a regular f then $\equiv$ is the identity; hence we will call every binary relation $<$ *regular* for which $\equiv$ is identical. It is easy to see the following lemma.

Lemma 4. *If $<$ is Pascalian then, however a or b are chosen, one (and only one) of the following is true: $a < b$, $a \equiv b$, or $b < a$.*

In this way we have identified a first group of non-Pascalian morals. It is probable that many people who would not want to assume a Pascalian moral would nevertheless accept the necessity of prescribing a moral for which $<$ satisfies the conditions set out in Lemmas 3 and 4. Let's say then that $<$ is a "near order."

We should note at this point that requiring $<$ to be a near order does not exhaust the requirements that common sense would probably expect of any moral. For example, let $n = 3$ and let the actions $(0, 1, 0)$ be preferable to the actions $(1, 0, 0)$: Common sense would probably consider it "incoherent" for the actions $(\frac{1}{3}, \frac{1}{3}, \frac{1}{3})$ to be preferable to the actions $(\frac{1}{6}, \frac{1}{2}, \frac{1}{3})$. Naturally this doesn't happen in the Pascalian case: Denoting by a_i the n-tuple that carries 1 to the ith place and therefore 0 to the other places, we have the following.

Lemma 5. *If $<$ is Pascalian and $a_r < a_s$, then however two n-tuples (that differ only in the places r and s) are chosen, the one with the largest sth component is preferable.*

We will not bother with the obvious demonstration.

From now on, we will refer to every binary relation on P_n that satisfies the conditions of Lemmas 3, 4, and 5 as *coherent*. As we have already seen, coherence is a necessary condition for Pascalianity; we shall soon see that it is not a sufficient condition.

Given a function v defined on S and with values in an overfield L of K, we will call a *relation associated to v* the $<$ defined by:

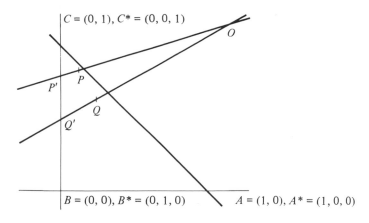

$C = (0, 1), C^* = (0, 0, 1)$

O

P' P

Q

Q'

$B = (0, 0), B^* = (0, 1, 0)$ $A = (1, 0), A^* = (1, 0, 0)$

Figure 1

$a < b$ if and only if $v(a) < v(b)$ (then, defining $\equiv$ in the usual way, $a \equiv b$ if and only if $v(a) = v(b)$).

Obviously this relation doesn't always give rise to a moral in the above sense; instead, it is in every case a near order.

The next lemma will be useful in the sequel.

Lemma 6. *However a function v from the domain S with values in L and two elements $u, k \in L$ ($k > 0$) is taken, the given function and the functions v', v'', defined by $v'(s) = v(s) + u$ and $v''(s) = kv(s)$, have the same associated relation.*

We will skip the obvious demonstration.

Now we can provide an example of a coherent non-Pascalian moral.

Example 2. Let $n = 3$ and, using a system of Cartesian coordinates Bxz, represent P_3 in the plane, with the points of the triangle of vertices $B = (0, 0)$, $A = (1, 0)$, $C = (0, 1)$ precisely associating the point of coordinates (p, r) at the triple (p, q, r) (see Figure 1).

Take a point $O = (a, c)$ with $a > 1$, $c > 0$ or $a < 0$, $c < 0$, and define a near order on the points of the triangle:

$P < Q$ if and only if OP and OQ are distinct and the intersection P' of OP with BC precedes the intersection Q' of OQ with BC in the order from B to C.

It is easy to verify that $<$ is a near order.

267

If two triples (p, q, r) and (p', q', r') differ only in the first two places $(r = r')$ with (say) $p < p'$ and $q > q'$, then it is clear that $(p', q', r') < (p, q, r)$; likewise for the cases $p = p'$ and $q = q'$. In this way the chosen moral is coherent, and -- writing M^* for the action associated with the point M -- we have $A^* < B^* < C^*$.

Now take (ab absurdo) u, v, w as values to be ordinately attributed to the three situations concerned, so that $<$ turns out to be Pascalian with these values; $v = 1 < w$.

The points to which are associated action of the same value as action $(0, 1 - k, k)$ (associated to $(0, k)$) form the line of equation

$$x + (1 - w)z = k(1 - w);$$

for proper O, it is impossible that O is on all the lines obtained by varying k. Thus, for proper O, we have a non-Pascalian moral. Note that for *improper* O of the form $(a, c, 0)$ in homogeneous coordinates (with $ac > 0$), we have all the Pascalian morals with $A^* < B^* < C^*$.

Now let a, b, c, d be actions with $a < b$ and $c \leq d$ (we write "$e \leq f$" for "$e \equiv f$ or $e < f$"). What can we say about an action h that with probability $\frac{1}{3}$ produces the same results as a and with probability $\frac{2}{3}$ the same results as c, when compared with an action k that with probability $\frac{1}{3}$ produces the same results as b and with probability $\frac{2}{3}$ the same results as d? Much depends on what we mean by "probability," but to say that in such a case $h < k$ seems to conform to the majority of uses of the concept, and to the majority of preferences in morals. We are led, that is, to require the following condition of $<$.

Condition of linearity. If $0 \leq r \leq 1$, $0 \leq s \leq 1$, r is different from 0, $r + s = 1$, $a < b$, and $c \leq d$, then $ra + sc < rb + sd$. (Here, if $a = (p_1, p_2, \ldots, p_n)$ then for $m \in K$ we set $ma = (mp_1, mp_2, \ldots, mp_n)$; if $b = (q_1, q_2, \ldots, q_n)$ then $a + b = (p_1 + q_1, p_2 + q_2, \ldots, p_n + q_n)$.)

It is easy to see that the following corollary obtains.

Corollary 1. *For every linear near order on P_n, if we let some b_i, c_i $(i = 1, 2, \ldots, k)$ be actions and r_i be nonnegative elements of K with $\sum r_i = 1$, and if we suppose that $b_i \leq c_i$ and (in particular) that $b_i < c_i$ and $r_i > 0$ for some i then $\sum r_i b_i < \sum r_i c_i$.*

The proof is easy.

Corollary 2. *For every linear near order on P_n, if we let some b_i, c_i, r_i be as in Corollary 1, and if $b_i \equiv c_i$ for all i, then $\sum r_i b_i \equiv \sum r_i c_i$.*

The proof is easy.

Corollary 3. *Every linear near order on P_n is coherent.*

Proof. Let $a_r < a_s$ and suppose that $a > c$, $b < d$, and $a + b = c + d$. Consider two actions d, e that have the same components p_i for i different from r, s, with (respectively) a, c as rth components and b, c as sth components. For $i > 1$, set $b_i = c_i = a_i$ and $b_1 = d$, $b_2 = e$. Now, for a convenient choice of the r_i, we have from the previous corollary that $d < e$. □

The following lemma is now immediate.

Lemma 7. *If $<$ is Pascalian then it is linear.*

We are now in a position to invert our results, proving the following.

Theorem. *$<$ is Pascalian if and only if it is coherent and linear (i.e., if and only if it is a linear near order).*

Proof. We have already seen that every Pascalian $<$ is coherent and linear. Let $<$ be coherent and linear. For convenience, let's say that a_i stands for the action that has 1 in the ith place and hence 0 in the other places. Excepting changes in order, we can suppose that $a_1 \le a_2 \le a_3 \le \cdots \le a_n$ (where, as usual, "$x \le y$" means "$x < y$ or $x \equiv y$"). It is easy to see that, for linearity, if some a_i are connected by $\equiv$ then everything is reduced to one case with lower n. We can therefore suppose that $a_1 < a_2 < \cdots < a_n$.

Let's work on n using induction. If $n = 1$ or $n = 2$ then the result is obvious. Let $n > 2$, and suppose that the result has been established for any number of situations less than n. If $a \in P_{n-1}$ and $a = (p_1, p_2, \ldots, p_{n-1})$, write $(a; 0)$ or a^* for $(p_1, p_2, \ldots, p_{n-1}, 0)$, which is in P_n. Consider the $<'$ defined on P_{n-1} by

$$a <' b \text{ if and only if } a^* < b^* \quad (a, b \in P_{n-1}).$$

Of course, $<'$ is coherent and linear; therefore an overfield L of K and certain $v_1, v_2, \ldots, v_{n-1} \in L$ exist such that, putting $v'(p_1, p_2, \ldots, p_{n-1}) = \Sigma\{p_i v_i : 1 \le i \le n-1\}$ and $v(a^*) = v'(a)$ for $a \in P_{n-1}$, we obtain, whatever $a, b \in P_n$ may be with the nth component equal to zero:

$$a < b \text{ if and only if } v(a) < v(b)$$

and hence

$$a \equiv b \text{ if and only if } v(a) = v(b).$$

269

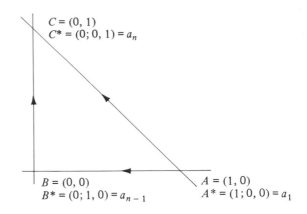

Figure 2

Note that, because of the condition of linearity, the same thing happens if a and b have the same nth component and we set $v(r_1, r_2, ..., r_n) = v'(r_1, r_2, ..., r_{n-1})$.

Now, for clarity, we make a digression – the material within square brackets.

[Consider actions in which all the components different from the first, the $(n-1)$th, and the nth are 0; for convenience we will write $(p; q, r)$ for $(p, 0, ..., 0, q, r)$. We will also refer to Figure 2, where $(p; q, r)$ is represented by the point (p, r). (If $M = (p, r)$ we put $M^* = (p; 1 - p - r, r)$.)

Because of the condition imposed by Lemma 5, the actions represented by the points on the sides are arranged by $<$ in the direction of the arrows. On the side AC (i.e., among the actions with $p_{n-1} = 0$) there are actions preferable to a_{n-1} (for example a_n) and actions to which a_{n-1} is preferable (for example a_1). There may also be actions equivalent to a_{n-1}.

We will denote by M' the set of m' of K for which $(1 - m'; 0, m') < a_{n-1}$, and by M'' the set of m'' of K for which $(1 - m''; 0, m'') > a_{n-1}$. It is easy to see that:

(i) every element of M' is less than every element of M'';
(ii) however an element $d > 0$ in K is taken, there exist certain $m' \in M'$ and $m'' \in M''$ with $m'' - m' < d$.

(That is, M' and M'' form in K a pair of contiguous classes. The non-mathematical reader should recall high-school notions about the introduction of real numbers.)

At most there exists an $m \in K$ with $(1-m; 0, m) \equiv a_{n-1}$. If so then it is easy to see that, taking $1/m$ as v_n, the given moral result is Pascalian. If this is not the case then it may also be that there are no separating elements in K or in L (though of course there are separating elements in a suitable overfield of L). If $n = 3$ then we have just proved the theorem, but if $n > 3$ then it is convenient to work as follows.]

We can suppose that $v_1 = 0$ and $v_{n-1} = 1$. For every action a in P_n, write $a^{(i)}$ for the component i of a so that

$$a = (a^{(1)}, a^{(2)}, \ldots, a^{(n)}).$$

Let us define a function w from $P_n \times L$ to L, setting

$$w(a, k) = a^{(1)}v_1 + a^{(2)}v_2 + \cdots + a^{(n-1)}v_{n-1} + a^{(n)}k.$$

We can define three subsets of L: W, W', W'', putting:

(7.1) $k \in W$ if and only if there are $a, b \in P_n$ with

$$a^{(n)} > b^{(n)}, \quad a > b, \quad w(a, k) \le w(b, k);$$

(7.2) $k \in W'$ if and only if there are $a, b \in P_n$ with

$$a^{(n)} \ne b^{(n)}, \quad a \equiv b, \quad w(a, k) = w(b, k);$$

(7.3) $k \in W''$ if and only if there are $a, b \in P_n$ with

$$a^{(n)} > b^{(n)}, \quad a < b, \quad w(a, k) \ge w(b, k).$$

Now $1 \in W$, because:

$$(0; 0, 1) > (0; 1, 0) \quad \text{and} \quad w((0; 0, 1), 1) = 1 = w((0; 1, 0), 1).$$

It is easy to see that:

(7.4) If k is in W and $h \le k$ then h is in W; if k is in W'' and $h \ge k$ then h is in W''.

It is possible that $W' = W'' = \emptyset$. For example, take as K the real field and an overfield L of K in which there is an m with $m > x$ for every real number x.

Put $u_1 = 0$, $u_2 = 1$, $u_3 = m$, and for every action a put

$$u(a) = a^{(1)}u_1 + a^{(2)}u_2 + a^{(3)}u_3.$$

Define a relation $<$ in $P_3(K)$, setting

$$a < b \text{ if and only if } u(a) = a^{(2)} + a^{(3)}m < b^{(2)} + b^{(3)}m = u(b).$$

The defined $<$ is Pascalian, and of course if $c, d \in P_3(K)$ and $c^{(3)} > d^{(3)}$ then $c > d$. We can also show:

271

(7.5) If $h \in W$ and $k \in W' \cup W''$, or $h \in W'$ and $k \in W''$, then $h < k$. If $h, k \in W'$ then $h = k$.

Proof. Let h be in W and k in W'. Then there are a, b, c, d in $P_n(K)$ such that $a \equiv b$, $c > d$, $a^{(n)} > b^{(n)}$, $c^{(n)} > d^{(n)}$, $w(a, k) = w(b, k)$, and $w(c, h) \le w(d, h)$. Put:

$$t = a^{(n)} - b^{(n)} + c^{(n)} - d^{(n)};$$
$$r = (c^{(n)} - d^{(n)})/t;$$
$$s = (a^{(n)} - b^{(n)})/t.$$

Of course $r, s > 0$ and $r + s = 1$, so applying linearity to the pairs (a, b), (d, c) we obtain

$$ra + sd < rb + sc.$$

It is easy to verify that these actions have the same nth component, whence

$$(a^{(n)}c^{(n)} - b^{(n)}d^{(n)})/t.$$

Now, from the inductive hypothesis,

(7.5.1)
$$\Sigma\{(ra + sd)^{(i)}v_i : i = 1, 2, \ldots, (n-1)\}$$
$$< \Sigma\{(rb + sc)^{(i)}v_i : i = 1, 2, \ldots, (n-1)\}.$$

From $w(a, k) = w(b, k)$ we have

(7.5.2) $\quad k = (\Sigma\{(b^{(i)} - a^{(i)})v_i : i = 1, 2, \ldots, (n-1)\})/(a^{(n)} - b^{(n)})$,

and, from $w(c, h) \le w(d, h)$,

(7.5.3) $\quad h \le (\Sigma\{(d^{(i)} - c^{(i)})v_i : i = 1, 2, \ldots, (n-1)\})/(c^{(n)} - d^{(n)})$.

Now, from (7.5.1)–(7.5.3), we obtain $h < k$.

In a similar way we can demonstrate the other propositions of 7.5. It is now clear that, setting

$$v_n = \begin{cases} \text{the unic element of } W' \text{ if } W' \ne \emptyset; \\ \text{an element } u \text{ of a suitable overfield } L^* \text{ of } L \text{ such that } h < u \text{ for} \\ \text{every } h \in W \text{ and } u < k \text{ for any } k \in W'' \textit{ otherwise}. \end{cases}$$

For every action a in P_n,

$$v(a) = \Sigma\{a^{(i)}v_i : i = 1, 2, \ldots, n\},$$

we have

$$a < b \text{ if and only if } v(a) < v(b), \quad (a, b \in P_n),$$

and hence $<$ is Pascalian (we use of course the field L^*) and the theorem follows. $\qquad \square$

272

It is now appropriate to explain why we have turned to the use of possibly non-Archimedean ordered fields. First, we show with a simple example that the field of real numbers is inadequate to satisfy the requirements. (It is interesting to note that it is usually not necessary to use the entire field of real numbers; the field of rational numbers may be sufficient with some appropriate non-Archimedean widening.)

Example 3. Taking K as the field of real numbers, suppose that in P_3, $(1, 0, 0) < (0, 1, 0) < (0, 0, 1)$. Complete the definition of an order in P_3 by assuming that (p_1, p_2, p_3) is preferable to (q_1, q_2, q_3) in the following cases (and only in these cases):

(i) if $q_3 < p_3$;
(ii) if $p_3 = q_3$ and $q_2 < p_2$.

In particular, for every positive real number less than 1, we obtain

$$(0, 1, 0) < (1 - r, 0, r).$$

If the moral in question were Pascalian with values in K, we would get as appropriate v_1, v_2, v_3:

$$rv_3 + (1 - r)v_1 > v_2,$$

of which $v_3 > (v_2 - v_1)/r + v_1$ whatever r is, which is absurd, because the expression at the second member is unlimited at its upper end (to pass to a non-Archimedean field means taking an actual infinite for v_3).

Reference to a concrete situation may help clarify this example. For a certain person, let s_1 be an unpleasant situation (say, a toothache), s_2 an indifferent situation, and s_3 a situation in which he saves a human life. The hypothesis made in the example, which may have seemed bizarre, becomes understandable and acceptable. That is, confronted with a choice between two actions – one that will certainly produce a situation to which our person is indifferent and the other that, together with a small probability of saving a human life, will almost certainly cause him a toothache – our individual will choose the second.

We should not, however, let ourselves be taken in by imagining that morals and behavior of this type are frequent. Few people, suffering from a toothache and owning a car, would hesitate to drive to a dentist in their car, even though the possibility of a road accident, perhaps even with victims, is marginally increased thereby. In this case, at least at first glance, the positive value of human life, however great, is not taken as being infinite (common sense would say priceless) with respect to the negative value of a toothache. (On second thought, a deeper analysis may lead to the reflection that a prolonged toothache could cause (say) inattention and thereby increase the probability of fatal accidents of other sorts.)

On the basis of the results demonstrated here, we can conclude that – much to the relief of the student of metamorals – every moral that satisfies certain minimum requirements is of the Pascalian type. (Of course, we do not conclude that Pascalian processes of calculation occur in the subject's mind, but only that any given moral that satisfies the aforesaid requirements can appropriately be considered Pascalian.)

NOTES

1. All things considered, this is not a happy choice. A series of imposed conditioning and behavior are united in the "superego," and in the best of cases it has the same function as a stroller for children learning to walk. The "free" man's "moral conscience" (without giving an absolute value to this term) mediates, balances, and synthesizes the most diverse among impulses and longings, including what remains acceptable to the superego, fashioning them into a form that is relatively coherent and lacking in destructive conflicts. Hence man exists in whom reason (in the widest sense of the word) – or the "ego," in a certain sense – has reached maximum development.
2. As is obvious, for example, in the case of moralists with sex phobias.
3. It is well known that, in general, what scholars call "utility" is not – even in extremely simplified situations and in the absence of other elements of value not concerned with money – a *linear* function of money possessed.
4. Indeed, a morality that does not foresee the search for and establishment of new values as its own ulterior value would appear narrow and limited, which means that my (but not only my) moral impulses would force me to reject it. A hypothetical Robinson who believed himself to be the only living being, and who also lacked the capacity to explicitly imagine the possible existence of other living beings, obviously could not set up the lives or happiness of others as explicit values. His morality could be said to be lacking if it didn't propel him toward a continual search that – without his being able to predict it – could lead to the discovery of other life and therefore the establishment of new values.
5. Savage, Leonard J., 1954, *The Foundations of Statistics*. New York: Wiley.
6. The main representative of this approach is Bruno de Finetti; in particular, see *Teoria delle Probabilità* (Theory of Probability) (Turin: Einaudi, 1970).
7. Of course we also identify different moves in this way, but we can suppose that the considered situations are the entire history of the game; thus different moves yield different elements of P_n.

274

16

The logic of Ulam's game with lies

Daniele Mundici

Someone thinks of a number between one and one million
(which is just less than 2^{20}).
Another person is allowed to ask up to twenty questions,
to each of which the first person is supposed to answer only yes or no.
Obviously the number can be guessed by asking first:
Is the number in the first half million?
then again reduce the reservoir of numbers in the next question
  by one-half, and so on.
Finally the number is obtained in less than $\log_2(1000000)$.
Now suppose one were allowed to lie once or twice,
then how many questions would one need to get the right answer?

S. M. Ulam (1976), *Adventures of a Mathematician*
(New York: Scribner, p. 281)

1. Playing Ulam's game

The questions and answers exchanged between Questioner and Responder in Ulam's game with k lies are *propositions*. In this chapter we show that the Lukasiewicz $(k+2)$-valued sentential calculus ([14], [15]) provides a natural logic for these propositions. Throughout this chapter, the Responder is identified with Pinocchio. Author and Reader will often impersonate the Questioner. Unless otherwise stated, *in this section we consider Ulam's game with at most one lie*.

Initially, Pinocchio and the Questioner agree to fix a *search space* $S = \{0, 1, \ldots, 2^n - 1\}$. Writing numbers in binary notation, S is more conveniently represented by the n-cube $\{0, 1\}^n$. Pinocchio arbitrarily chooses a number $x \in S$, writes it on a sheet of paper, and puts the paper in his pocket. We shall henceforth call x the *written number*. To guess the written number, the Questioner chooses his first *question Q*. In Ulam's game, Q is (uniquely determined by) a subset of S. To fix ideas, let us suppose Q is the set of even numbers in S. Thus, the Questioner asks Pinocchio,

"Is the number x even?"

The *opposite question* $\bar{Q}$ is given by $S - Q$, the complement of Q in S. Pinocchio's answer to Q can be either "yes" or "no." Since a negative answer to Q is the same as a positive answer to $\bar{Q}$, let us consider without loss of generality the effect (on our state of knowledge of x) of a positive answer to Q: The original search space S is partitioned into two components A and B, where $A = Q$ and $B = \bar{Q} =$ the odd numbers in S. We cannot exclude the possibility that the written number is odd. Indeed, this can be the case if and only if (iff) Pinocchio's positive answer to Q is a lie. Now we ask our second question G, say,

"Is $x \le 6$?"

Again, G is identified with the set $G = \{y \in S \,|\, y \le 6\}$, and $\bar{G} = \{y \in S \,|\, y > 6\}$. If Pinocchio's answer to G is *positive* then, in our new state of knowledge, S is partitioned into three components C, D, E, where:

$C = A \cap G$ are the numbers *satisfying* both answers, namely the even numbers ≤ 6;

$E = B \cap \bar{G}$ are the numbers *falsifying* both answers. These can be safely excluded from our search, for if the written number were an odd number > 6 then Pinocchio would have lied in both answers, which is impossible;

$D = S - (E \cup C) = (A \cap \bar{G}) \cup (B \cap G)$ are the numbers falsifying exactly one of Pinocchio's answers, namely the even numbers > 6, together with the odd numbers ≤ 6. Although their record is not so clean as for the numbers in C, they cannot be excluded from consideration as we did for the numbers in E. In fact, the written number is equal to a number in D iff exactly one of the two answers is a lie.

If Pinocchio's answer to G is *negative,* then C, D, and E are similarly obtained from A and B by interchanging the roles of G and $\bar{G}$.

Proceeding by induction on the number of questions, suppose Pinocchio gives answers $a_1, \ldots, a_t$ to questions $Q_1 = Q$, $Q_2 = G$, $Q_3, \ldots, Q_t$, respectively. Let our current state of knowledge about x be represented by a triple of pairwise disjoint sets (C_t, D_t, E_t), where

C_t is the set of numbers in S satisfying all answers $a_1, \ldots, a_t$;
E_t is the set of numbers of S falsifying at least two answers; and
D_t is the set of numbers falsifying exactly one answer.

If Pinocchio's answer a_{t+1} to the next question Q_{t+1} happens to be *positive,* then our state of knowledge is transformed into a new triple $(C_{t+1}, D_{t+1}, E_{t+1})$, where

$C_{t+1} = C_t \cap Q_{t+1}$, the numbers satisfying all answers $a_1, \ldots, a_t, a_{t+1}$;
$E_{t+1} = E_t \cup (D_t \cap \bar{Q}_{t+1})$, the numbers falsifying at least two answers;

$D_{t+1} = (C_t \cap \bar{Q}_{t+1}) \cup (D_t \cap Q_{t+1})$, the numbers falsifying exactly one answer.

If the answer to Q_{t+1} is *negative,* one similarly obtains $(C_{t+1}, D_{t+1}, E_{t+1})$ from (C_t, D_t, E_t) by interchanging the roles of Q_{t+1} and $\bar{Q}_{t+1}$. It is easy to see that C_{t+1}, D_{t+1}, and E_{t+1} are pairwise disjoint. The *initial* state of knowledge (C_0, D_0, E_0) – that is, the state before the first answer – is of course given by the triplet $(S, \emptyset, \emptyset)$ with no falsified answers.

A state of knowledge (C_u, D_u, E_u) is called *final* iff all numbers of S except one, say w, are in E_u. Then we can confidently say that the written number equals w. Indeed, we have the additional information that if w belongs to C_u then Pinocchio has never lied; if w belongs to D_u then Pinocchio has lied exactly once.

Already from this simple example, we see that Pinocchio's answers do not behave as propositions in classical logic in the following respects.

1. Two opposite answers a_i and a_j to the same (repeated) question, such as "yes, the written number is even" and "no, the written number is odd," do not lead to the inconsistent state of knowledge $(\emptyset, \emptyset, S)$ – to the effect that no number can be equal to the written number. On the contrary, from a_i and a_j we obtain the information that Pinocchio's reservoir of lies is reduced by one. We conclude that the connectives in the logic of Ulam's game are so arranged that the *conjunction of two opposite answers need not express an unsatisfiable property* of numbers in the search space.

2. The conjunction of two answers, each saying "the written number is odd," is generally more informative than a single answer "the written number is odd"; for instance, under the present stipulation that Pinocchio can lie at most once, two answers of "the written number is odd" suffice to establish that the written number is in fact odd, whereas a single answer need not suffice. Specialists would say that the logic of Ulam's game with lies does not obey the "contraction" rule; we prefer to say that this logic obeys the *repetita juvant* (repetitions are helpful) principle.

3. Although – for people having direct access to the written number – each answer of Pinocchio is either true or false, this *absolute* truth value is of little significance to the Questioner's strategy or to his current state of knowledge. The Questioner tries to make the best use of all answers $a_1, ..., a_t$, since in general he cannot discriminate between informative and misleading answers. Besides, if required to minimize the length of the game (as in Ulam's problem), the Questioner will carefully balance the questions in such a way that either answer "yes" or "no" is equally informative, in a sense that can be made precise ([3], [4]). The Questioner's current state of knowledge (C_t, D_t, E_t) assigns, to each point $y \in S$,

a "falsity value" given by the quantity $q(y)$ of answers that are falsified by y. Here, $q(y)$ ranges in the set {zero, one, "too many"}, and $C_t = q^{-1}(0)$, $D_t = q^{-1}(1)$, $E_t = q^{-1}(\text{"too many"})$. We conclude that the logic of Ulam's game is *many-valued*.

Similarly, if Pinocchio is allowed to lie up to k times, then his answers determine a function $q: S \to \{0, 1, 2, ..., k, \text{"too many"}\}$ assigning to each point y the quantity $q(y)$ of answers falsified by y. Given a question $Q \subseteq S$, a positive answer to Q naturally transforms q into a new assignment q' as follows:

(1)   if y satisfies the answer (i.e., if $y \in Q$) or if $q(y) = \text{"too many,"}$ then $q'(y) = q(y)$; otherwise $q'(y) = q(y) + 1$, where $k + 1 = \text{"too many."}$

The effect on q of a negative answer to Q is similarly defined with reference to the opposite question $\bar{Q}$. To record the Questioner's current state of knowledge, instead of using the function q we will find it convenient to use the *relative distance* function $d: S \to \{0, 1/(k+1), 2/(k+1), ..., k/(k+1), 1\}$ given by

(2)
$$d(y) = \frac{k + 1 - q(y)}{k + 1},$$

where $k + 1 - \text{"too many"} = 0$. Thus, $d(y) = 0$ iff y falsifies more than k answers (i.e., iff y is an *excluded* number); $d(y) = 1/(k+1)$ iff y falsifies exactly k answers; ...; $d(y) = k/(k+1)$ iff y falsifies exactly one answer; $d(y) = 1$ iff y satisfies all answers. Intuitively, the rational number $d(y)$ is the distance (relative to $k+1$) of y from the set of excluded numbers. We shall henceforth identify each state of knowledge with its corresponding relative distance function.

2. States of knowledge are Lukasiewicz conjunctions of Post functions

For any two numbers x and y in the real unit interval $[0, 1]$, the *Lukasiewicz conjunction* $x \bullet y$ is the amount by which the sum $x + y$ exceeds 1 (this amount being 0 in case $x + y \leq 1$). In symbols,

(3)
$$x \bullet y = \max(0, x + y - 1).$$

The Lukasiewicz *disjunction* $x \oplus y$ is the truncated sum of x and y,

(4)
$$x \oplus y = \min(1, x + y) = (\text{"}x^* \bullet y^*\text{"})^*,$$
$$\text{where } x^* = 1 - x \text{ is the } negation \text{ function.}$$

If x and y are only allowed to range over the 2-element set $\{0, 1\}$, then (3) and (4) take (respectively) the more familiar form $x \bullet y = \min(x, y)$ and

278

$x \oplus y = \max(x, y)$, yielding the truth table of Boolean conjunction and disjunction; of course, x^* then gives Boolean negation. For each $m = 2, 3, 4, \ldots$, we shall denote by I_m the m-element *Lukasiewicz chain*

$$\{0, 1/(m-1), 2/(m-1), \ldots, (m-2)/(m-1), 1\},$$

equipped with the operations $*$, $\bullet$, $\oplus$.

In the light of [6, Thms. 16–18], any function $f: S \to I_m$ is called a *Post function* (on S) *of order m*. By (2), every relative distance function d in Ulam's game with k lies is a Post function of order $k+2$. Because states of knowledge are (uniquely determined by) conjunctions of Pinocchio's answers, the latter, too, must be representable as Post functions. According to (1) and (2), the positive answer to a question Q penalizes each $y \notin Q$, decreasing by $1/(k+1)$ its relative distance from the set of excluded numbers unless this distance already equals 0; this motivates the following definition.

Definition 1. Given a question Q in Ulam's game with k lies, the *positive answer* to Q is the function $f_Q: S \to \{k/(k+1), 1\}$ given by

(5)
$$f_Q(y) = \begin{cases} 1 & \text{if } y \in Q, \\ k/(k+1) & \text{if } y \notin Q. \end{cases}$$

The *negative answer* to Q is the positive answer $f_{\bar{Q}}$ to $\bar{Q} = S - Q$. The *Post function* f of Pinocchio's answer to Q is given by $f = f_Q$ or $f = f_{\bar{Q}}$ when Pinocchio's answer to Q is (respectively) positive or negative.

Remark. The inconsistency-tolerance property mentioned in the previous section follows from our definition of "answer." As a matter of fact, except in the 0-lie game, the negative answer $f_{\bar{Q}}$ to a question Q is different from the negation $1 - f_Q$ of the positive answer to Q.

Proposition 2. *Let d and d' be the relative distance functions immediately before and after Pinocchio answers a question Q. Let f be the Post function of Pinocchio's answer to Q. Then $d' = d \bullet f$.*

Proof. It is sufficient to analyze the effect of Pinocchio's positive answer f_Q. Let q and q' be the functions corresponding to d and d', as given by (2). For each $y \in S$, we have two possible cases.

Case 1: $d(y) = d'(y)$; that is, y is not penalized by the positive answer. Then, by (1) and (2), either $d(y) = 0$ (i.e., y is already in the set of excluded numbers) or $f_Q(y) = 1$ (i.e., y satisfies Pinocchio's answer). Recalling (3) and (5), in the first subcase we have $d(y) \bullet f_Q(y) = 0 \bullet f_Q(y) = 0 = d(y) = d'(y)$. In the second subcase we have $d(y) \bullet f_Q(y) = d(y) \bullet 1 = d(y) = d'(y)$.

279

Case 2: y is penalized by the positive answer. Then, by (1) and (2), y falsifies f_Q without being an excluded number. In particular, we then have that $d(y) \geq 1/(k+1)$, $q'(y) = q(y) + 1$, and $d'(y) = d(y) - 1/(k+1)$. By (3) and (5), we obtain

$$d(y) \cdot f_Q(y) = \max(0, d(y) + f_Q(y) - 1) = \max(0, d(y) - 1/(k+1)) = d'(y).$$

In any case, we have $d'(y) = d(y) \cdot f_Q(y)$, whence $d' = d \cdot f_Q$ as required. $\square$

The *initial* state of knowledge is the Post function on S that is constantly equal to 1. A Post function $g: S \to I_m$ is a *final* state of knowledge iff we have $g(x) \neq 0$ for exactly one $x \in S$. The Questioner's states of knowledge are given by Pinocchio's answers, as follows.

Corollary 3. *The relative distance function d after Pinocchio's answers to questions $Q_1, \ldots, Q_t$ is given by the Lukasiewicz conjunction of the Post functions of the answers.*

3. Formalization of Ulam's game in Lukasiewicz logic

For each $m = 2, 3, 4, \ldots$, the set of *variable* and *connective* symbols in the m-valued sentential calculus of Lukasiewicz ([14], [15]), as well as the set of *formulas,* are exactly the same as for the 2-valued (Boolean) calculus. Just as in the 2-valued calculus every formula represents a Boolean function, in the m-valued calculus each formula p with variables $X_1, \ldots, X_n$ represents the function $f_p: (I_m)^n \to I_m$, according to the following inductive definition: Each variable X_i represents the projection onto the ith axis; if we know f_q and f_r, then $f_{\text{not } q} = 1 - f_q$, $f_{q \text{ and } r} = f_q \cdot f_r$, and $f_{q \text{ or } r} = f_q \oplus f_r$. Two formulas p and q are *logically equivalent* iff $f_p = f_q$. A *tautology* is a formula $p = p(X_1, \ldots, X_n)$ such that f_p is the constant function 1 on $(I_m)^n$. For example, "(not X) or X" is a tautology in each m-valued calculus, whereas "(not (X or X)) or X" is a tautology only in the 2-valued calculus. In fact, for each $m \geq 3$, the formula "X or X" is not logically equivalent to the formula "X" in the m-valued calculus. This is just a reformulation of the *repetita juvant* principle.

We shall formalize Ulam's game with at most k lies in the $(k+2)$-valued sentential calculus of Lukasiewicz. By Corollary 3, we must code Post functions of Pinocchio's answers only by means of formulas of the Lukasiewicz calculus. Our formalization is then a natural generalization of the familiar representation of Boolean functions by formulas in the 2-valued calculus. For the importance of the formalization procedure (already in the 2-valued case), see [2, p. 5].

We cannot identify the search space $\{0, 1\}^n$ with the set of extreme points of the cube $(I_{k+2})^n$. As a matter of fact, McNaughton's fundamental representation theorem [8] states that the set of functions $\{f_p \mid p$ is a formula in the $(k+2)$-valued calculus$\}$ is the set of restrictions of McNaughton functions to the n-dimensional cube $(I_{k+2})^n$. By definition, a *McNaughton function* $f: [0, 1]^n \to [0, 1]$ is a piecewise linear (continuous) function all of whose pieces have integral coefficients. Let $x = (x_1, \ldots, x_n) \in (I_{k+2})^n$. Write each rational coordinate x_i as a fraction a_i/b_i, where a_i, b_i are integers and $\gcd(a_i, b_i) = 1$ ($a_i \geq 0$, $b_i > 0$). Let v be the least common multiple of the denominators b_i, and let V_x be the set of possible values of McNaughton functions at x. Then a straightforward computation shows that

(6) $$V_x = \{t/v \mid t = 0, 1, \ldots, v\}.$$

Thus, in particular, a McNaughton function can only take the values 0 and 1 on the Cantor cube $\{0, 1\}^n$, and hence its restriction to $\{0, 1\}^n$ cannot be a Post function of order > 2. To find a substitute for the original search space $\{0, 1\}^n$, let us consider the cube $C(n, k) = \{1/(k+1), k/(k+1)\}^n$. Note that $C(n, k)$ is indeed a cube for each $k \geq 0$ except $k = 1$. In order to avoid notational complications, *in the rest of this chapter we shall assume $k \neq 1$.*

Proposition 4.
(i) *For each function $f: C(n, k) \to I_{k+2}$ there is a formula p such that f equals the restriction of f_p to $C(n, k)$.*
(ii) *For each formula r, the restriction of f_r to $C(n, k)$ is a Post function of order $k+2$ on $C(n, k)$.*

Proof. (i) For each $x \in C(n, k)$, let $g_x: [0, 1]^n \to [0, 1]$ be any McNaughton function such that $g_x(x) = f(x)$. The existence of g_x is ensured by (6). Let U_x be an open neighborhood of x. We can safely assume $U_x \cap U_y = \emptyset$ whenever $x \neq y$. By [9, 4.17], for each x there is a McNaughton function h_x such that $h_x(x) = 1$ and such that $h_x = 0$ outside U_x. Therefore, the McNaughton function $h = \sup\{g_x \wedge h_x \mid x \in C(n, k)\}$ equals f on $C(n, k)$. By McNaughton's theorem, there is a formula p such that $f_p = h$.
(ii) Trivial, using (6). $\square$

We are now ready to formalize Ulam's game with k lies in the $(k+2)$-valued calculus. Let the map $\mu: \{0, 1\} \to \{1/(k+1), k/(k+1)\}$ be defined by $\mu(0) = 1/(k+1)$ and $\mu(1) = k/(k+1)$. Then μ induces a one–one correspondence $x = (x_1, \ldots, x_n) \to x^\mu = (\mu(x_1), \ldots, \mu(x_n))$ from the original

281

search space $S = \{0, 1\}^n$ onto $C(n, k)$; in symbols, $\mu: \{0, 1\}^n \cong C(n, k)$. By Proposition 4, we have a natural one–one correspondence

(7) $\qquad f \to f^\mu, \quad$ where $f(x) = f^\mu(x^\mu)$ for all $x \in \{0, 1\}^n$

between Post functions on S of order $k+2$, and restrictions to $C(n, k)$ of McNaughton functions. Under this correspondence, the *initial* state of knowledge corresponds to the function constantly equal to 1 on $C(n, k)$. The latter is in turn represented by any formula p such that f_p equals 1 on $C(n, k)$ (e.g., any tautology will do). Similarly, a *final* state of knowledge is represented by any formula r that is *uniquely satisfiable* in $C(n, k)$, in the sense that there is precisely one $x \in C(n, k)$ such that $f_r(x) \neq 0$.

Let $Q \subseteq S$ be a *question*. Then the correspondence $\mu: \{0, 1\}^n \cong C(n, k)$ canonically transforms Q into a subset Q^μ of $C(n, k)$, by the stipulation $x \in Q$ iff $x^\mu \in Q^\mu$. In light of Definition 1, together with the correspondence (7), Pinocchio's *positive answer* f_Q is then represented by any formula a such that $f_a(x) = 1$ for all $x \in Q^\mu$ and $f_a(x) = k/(k+1)$ for all $x \in C(n, k) - Q^\mu$. Formulas representing *negative answers* are similarly defined, with $\bar{Q}$ in place of Q. Formulas representing the actual *answer* to Q are now defined according as the answer is "yes" or "no." Suppose d is the Questioner's state of knowledge after Pinocchio's answers to questions $Q_1, \ldots, Q_t$, and suppose that formula a_i represents the answer to Q_i for each $i = 1, \ldots, t$. Then, by Corollary 3, the (Lukasiewicz) conjunction $s = (a_1$ and a_2 and $\ldots$ and $a_t)$ represents d, in the sense that the Post function d^μ coincides with the restriction of f_s to $C(n, k)$.

Summing up our analysis, we have the following table.

Ulam game	Lukasiewicz logic
maximum number k of lies	$k+2$ truth values $0, 1/(k+1), \ldots, k/(k+1), 1$
search space $S = \{0, 1\}^n$	$C(n, k) = S^\mu = \{1/(k+1), k/(k+1)\}^n$
number y in S	point y^μ in $C(n, k)$
initial state of knowledge	tautology
final state of knowledge	formula uniquely satisfiable in $C(n, k)$
current state of knowledge d	formula s such that $d^\mu = f_s$ in $C(n, k)$
question Q	corresponding subset Q^μ of $C(n, k)$
opposite question $\bar{Q}$	complementary subset of $\bar{Q}$ in $C(n, k)$
positive answer f_Q to Q	formula p with $(f_Q)^\mu = f_p$ in $C(n, k)$
negative answer $f_{\bar{Q}}$ to Q	formula p with $(f_{\bar{Q}})^\mu = f_p$ in $C(n, k)$
state d after answers to $Q_1, \ldots, Q_t$	conjunction s of corresponding formulas
set of excluded numbers, $d^{-1}(0)$	points of $C(n, k)$ falsifying s

3. Concluding remarks

1. A long-standing problem in m-valued logic – one to which Lukasiewicz himself devoted considerable attention – is to give natural interpretations

to truth values when $m \geq 3$ [15, p. 275]. In our interpretation, truth values are distances (measured in units of $m-1$) from the set of excluded numbers, in Ulam's game with $m-2$ lies. The contraction rule fails because the *repetita juvant* principle holds – just as in everyday life. Noncontradictory coexistence of opposite answers, which is inevitable in the presence of lies, can be handled using more than two truth values.

2. In the 0-lie case, an optimal strategy [1, pp. 6, 62] to guess a number $x \in \{0,1\}^n$ is by definition a sequence of questions (*alias* Boolean functions $f_1, \ldots, f_n$ of n variables) such that, for each choice of the parameters $\epsilon_1, \ldots, \epsilon_n \in \{\text{yes}, \text{no}\}$, the function $f_1^{\epsilon_1} \wedge \cdots \wedge f_n^{\epsilon_n}$ is nonzero in exactly one point. Here, $f^{\text{yes}} = f$ and $f^{\text{no}} = 1 - f$. Equivalently, $f_1, \ldots, f_n$ is an *independent* set of n elements in the free Boolean algebra with n generators [12, p. 39]. Observe that the binary notation system is a by-product of the particular optimal strategy $b_1, \ldots, b_n$, where each Boolean function b_i asks, "Is the ith digit of x equal to 1?"

Considering now the case where at most k lies are allowed, in the (nonadaptive, static, predetermined [1, p. 9]) case when all questions precede all answers, optimal searching strategies immediately yield optimal k-error correcting codes. However (see [7]), very few optimal codes are so known when $k \geq 2$. The situation is much better in the (adaptive, dynamical, sequential) case of Ulam's game, where questions may depend on Pinocchio's answers. For small values of k, optimal strategies are known ([11], [3], [4], [10]). These strategies can also be used in the equivalent version of Ulam's game where Pinocchio need not know when he lied. We can, for instance, assume that an honest Pinocchio is sending us his answers from a distant place, using a low-power transmitter; distortion can affect Pinocchio's transmission. On the other hand, using powerful transmitters, we can send our questions to Pinocchio without any distortion. We must correct the distortions (lies?) of Pinocchio's answers by asking the minimum number of questions. In this way, Ulam's game naturally fits in the theory of communication with feedback ([5], [13]). The representation of searching strategies in terms of formulas is a starting point to measure the complexity of the underlying coding and decoding procedures.

3. Since the many-valued sentential calculus of Lukasiewicz is deeply related to AF C^*-algebras [9], and since AF C^*-algebras are useful in the description of quantum spin systems, our interpretation of Ulam's game may be of help in the analysis of the logical aspects of such systems.

REFERENCES

[1] Aigner, M. (1988), *Combinatorial Search*. New York: Wiley and Stuttgart: Teubner.
[2] Birkhoff, G. (1971), "The Role of Modern Algebra in Computing." *SIAM-AMS Proceedings* 4: 1–47.

[3]   Czyzowicz, J., Mundici, D., and Pelc, A. (1988), "Solution of Ulam's Problem on Binary Search with Two Lies." *Journal of Combinatorial Theory, Series A* 49: 384–8.

[4]   Czyzowicz, J., Mundici, D., and Pelc, A. (1989), "Ulam's Searching Game with Lies." *Journal of Combinatorial Theory, Series A* 52: 62–76.

[5]   Dobrushin, R. L. (1958), "Information Transmission in a Channel with Feedback." *Theory of Probability and Applications* 34: 367–83.

[6]   Epstein, G. (1960), "The Lattice Theory of Post Algebras." *Transactions of the American Mathematical Society* 95: 300–17.

[7]   MacWilliams, F. J., and Sloane, N. J. A. (1978), *The Theory of Error-correcting Codes*. Amsterdam: North Holland.

[8]   McNaughton, R. (1951), "A Theorem about Infinite-valued Sentential Logic." *Journal of Symbolic Logic* 16: 1–13.

[9]   Mundici, D. (1986), "Interpretation of AF C^*-algebras in Lukasiewicz Sentential Calculus." *Journal of Functional Analysis* 65: 15–63.

[10] Negro, A., and Sereno, M. (to appear), "Solution of Ulam's Problem on Binary Search with Three Lies." *Journal of Combinatorial Theory, Series A*.

[11] Pelc, A. (1987), "Solution of Ulam's Problem on Searching with a Lie." *Journal of Combinatorial Theory, Series A* 44: 129–40.

[12] Sikorski, R. (1960), *Boolean Algebras*. Berlin: Springer.

[13] Slepian, D. (ed.) (1974), *Key Papers in the Development of Information Theory*. New York: IEEE Press (contains [5]).

[14] Tarski, A., and Lukasiewicz, J. (1956), "Investigations into the Sentential Calculus." In J. Corcoran (ed.), *Logic, Semantics, Metamathematics,* 2nd ed. (1983). Indianapolis: Hackett.

[15] Wojcicki, R. (1988), *Theory of Logical Calculi*. Dordrecht: Kluwer.

17

The acquisition of common knowledge

MICHAEL BACHARACH

INTRODUCTION

I shall say that there is *mutual knowledge of degree 1* between two people of a fact if they both know it, *mutual knowledge of degree 2* if they both know that both know it, and so on, for every finite string of knowings of knowings of this sort. There is *common knowledge* between them of it if there is mutual knowledge (MK) of every degree between them of it. In the multiperson theories of rational behavior – game theory, economic theory, speech-act theory – states of MK nowadays play a central role. Often, actions are defended as rational by appeal to MK between the agents of something: in philosophy of language, acts of Gricean meaning are so defended; in game theory, persuasive solutions such as "nastiness" (confession at the first round) in the repeated Prisoner's Dilemma. These claims place two obligations on the theories that advance them: first, to provide an etiology – how do knowers *get* MK? – and second, to cope with a charge of psychological impossibility. The required degree of MK may be high (sometimes even infinitely high); and if it is then the ability to be in the required epistemic state seems to be beyond mere humans.

Three-quarters of this chapter is about the conditions of possibility of mutual knowledge. I consider first the etiological problem, then the problem of human limitations. For the MK-involving theories of rational action, my conclusions are cheering on the first score, depressing on the second. In Section 1 I show, by means of a formal theory (T_I) set in epistemic logic, that possible knowers acquire MK of arbitrarily high degree. T_I is inspired by Lewis's and Schiffer's informal accounts of the acquisition of MK by two knowers of the occurrence of a conspicuous event that unfolds between them.

The knowers of T_I conjure their limitless body of knowledge from an initial stock that is finite and modest. But in the deductive powers these knowers are paragons. In Section 2 I consider how mere humans fare, by amending T_1; I bound the complexity of knowers' inferences in a natural (though particular) way once proposed by Hintikka. I show that we too can acquire MK; but so far and no further, for all that may be shown. Common knowledge, it seems, is not for us. For it is a theorem (the No-Such-Thing Theorem) that, for high enough n, it is *not* a theorem of my formal theory that humans acquire MK of degree n of the conspicuous event.

In Section 3 I consider an argument (the Bringing-To-See Argument) that seems to get 'round the No-Such-Thing Thoerem. It seeks to trade on the fact that there is no limit to how far humans can, by diligence or by suitable instruction, penetrate along the reasoning path of the paragons. But there is, I show, no profit in this argument.

In the last section, I turn from the theory of mutual knowledge to the client theory of rational action. I consider an argument (the Mimetic Argument) that seeks to trade on the fact that humans understand arguments which show that perfectly rational agents reach the decisions given in the classical solutions. Thus, we understand the Backward Induction Argument and thereby know that perfectly rational players are nasty. Does it matter for game theory that we can't have MK of high degree, so long as we know what we would do if, like these epistemic paragons, we did? I show that it does, and that the human players depicted in my formal theory of MK have, for all we know, no reason to be nasty.

1. How common knowledge is acquired

1.1. Current views

It is widely believed that there are facts of which we humans have common knowledge – we know them, that each other knows them, that each other knows that each other knows them, and so on endlessly; we have mutual knowledge of them of every degree. They include, it is held, the tautologies; but those to have aroused most interest in theories of rational multiperson behavior are *public occurrences*. If there is a candle on the table between us two normal, open-eyed people, or if an announcement is made as we three sit – normal, open-eared and -eyed – in a quiet, well-lighted room, there rises up between us a mounting tower of knowings: of the presence of the candle, or the fact of the announcement. I *know* that you know that I know that there's a candle; and you know that I do; and I know that . . . . Yes, but *how?*[1]

The recent literature in informational economics and game theory, though brimming with assumptions of common knowledge, has given this question scant attention. This is not surprising, for the models of

286

MK used in this literature easily suggest that there is no problem. These models – stemming from Aumann (1976) – are set-theoretical; in effect if not in design, they are semantical models of a deductive system that is itself not presented. An elegant recent model of this kind is that of Monderer and Samet (1989). Let Ω be a set (of *states of the world*), and let K_i ($i = 1, 2$) be an operator on the subsets of Ω (the *events*). For each event E, the event $K_i E$ is interpreted as "the event that i knows that E." The latter event is introduced in such a way that, for all $E \subseteq \Omega$, $K_i E \subseteq E$; that is, K_i is "factive" – knowledge of E implies its occurrence. For each person i there are some events E such that $E \subseteq K_i E$; if E occurs, i knows so. These events are called *evident for i*. If E is evident for both 1 and 2, it follows easily that if E occurs then there is common knowledge of this between 1 and 2; that is, $E \subseteq K_1 E$, $E \subseteq K_2 E$, $E \subseteq K_1 K_2 E$, $E \subseteq K_2 K_1 E$, $E \subseteq K_1 K_2 K_1 E$, and so on. For, from the factivity of K_i and evidentness, first $K_1 E = E$ and $K_2 E = E$, whence $K_2 K_1 E = K_2 E = E$ and $K_1 K_2 E = K_1 E = E$, whence $K_1 K_2 K_1 E = K_1 E = E$, and so forth.

Such models appear to show that there are facts that are common knowledge (namely, the occurrence of events evident for all concerned); and common knowledge is apparently established in these models on very weak assumptions: the existence of such events together with the innocuous principle that the known is true. But these appearances are deceptive. Two things warn of rocks beneath the smooth surface of the argument.

First, there are events evident for two persons of which there clearly is no MK even of the second degree between the persons – for example, those events unfolding on the street in front of the adjacent windows out of which, unknown to one another, both are gazing. Second, the proof consists of substitutions of varying expressions referring to a certain set, these expressions being operands of operators interpreted as "*i* knows that." Such substitutions are valid steps in the set theory in which these models are cast. But their use means that these models build a measure of suspect *extensionality* into the knowledge systems of the creatures modeled.

It may be that there is a special kind of evident event whose evidence to others is known to each – and that might deal with the first difficulty. It may be that the substitutions on which the proof rests can be interpreted as valid inferences that *rational* knowers would make – and that might deal with the second. But whether these repairs can be carried out is certainly not obvious. To find out, we need a treatment that is explicit about these issues. Set-theoretic models of the above kind are of heuristic value, pointing enquiry in promising directions; but they cannot, by themselves, answer the question of how mutual knowledge is acquired.

1.2. Schiffer's account

Lewis (1969) and Schiffer (1972) give versions of an argument – I shall call it the Transparency Argument – that is more explicit both about evidentness and what knowers know of it, and about knowers' inferences. I briefly sketch Schiffer's version of the argument, from which the theory of how mutual knowledge is acquired that I shall propose directly descends. At the center of Schiffer's theory is the property of persons of being "normal" – that is, having normal sense faculties, intelligence, and experience. Schiffer makes two epistemic claims concerning normality, which I shall call the Recognition Postulate and the Double Sufficiency Postulate.

Postulate 1 (Recognition). Being normal and present suffices for knowing that everyone normal and present is normal and present.[2]

Postulate 2 (Double Sufficiency). If being normal and present suffices for knowing that p, then it suffices for knowing that it suffices for knowing that p.

Someone is "present" in the sense intended if she is one of a certain pair who are transmitting characteristic signs of their normality to each other in a certain situation – for example, they are sitting at a table opposite each other with sense organs operating and behaving normally. Let p be something that any normal and present person knows – say, that there is a candle in the middle of the table, or that a waiter has entered, called for silence, and announced the death of the dictator.[3] Schiffer argues, informally, that the two postulates imply that between two present and normal persons there is MK of every degree that p. The gist of Schiffer's proof is as follows.

Suppose that

(1)   being normal and present suffices for knowing that p.

Call the two persons 1 and 2. Since

(2)   1 and 2 are normal and present,

we have at once from (1) and (2) that

(3)   1 knows that p.

By (2) and Postulate 1,

(4)   1 knows that 2 is normal and present.

By Postulate 2 and (2), 1 knows (1). Hence, putting two and two together,

1 knows that 2 knows that p.

Now 1's normality and presence is all that was needed to establish that she knows (1). As much would be so of anyone; that is,

(5)   being normal and present suffices for knowing that (1).

By Postulate 2 and (2), 1 knows (5). But then (4) and a simple deduction by 1 yield

(6)   1 knows that 2 knows (1).

Also,

(7)   1 knows that 2 knows that 1 is normal and present,

for according to Postulate 1, being normal and present suffices for recognizing someone normal and present as being so. Hence by Postulate 2 and (2), 1 knows of this sufficiency. Since she also knows that 2 is normal and present and that she herself is (by Postulate 1), her putting two and two together assures (7). Finally, (6) and (7) mean that, with a further bit of deduction by 1 (this time concerning a deduction of 2's),

1 knows that 2 knows that 1 knows that p.

We can clearly go on like this indefinitely.

The argument of the last two paragraphs is neither complete nor convincing in all its details. We shall see that there is, however, a fully rigorous argument to common knowledge whose essential postulational basis is a pair of postulates of Schifferian stamp. Like that of the above argument, its broad strategy is to show that two persons may acquire MK by valid inferences. Like Schiffer's argument, its method for showing this is the narrative one of *tracking* the steps that the knowers make.

Two features in Schiffer's treatment of the Transparency Argument are obscure. What is the force of "suffices for" in the statements of Recognition and Double Sufficiency? And what deductive inferences must it be supposed that the knowers make? The formal theory presented in the next section clarifies both these matters, which turn out to be linked.

1.3. The theory T_1

I explicate Schiffer's argument by combining elements of it with elements of a third recent account of mutual learning, which may be called the *epistemic logical* account. Here, explicit rules govern what the knowers come to know inferentially. Suppose that 1 and 2 are two persons who satisfy one of the standard epistemic logics – say that of Hintikka (1962). Then it is easy to show that they have MK, of every degree, of every *theorem* of the logic (see, e.g., Bacharach 1985) – in particular, they have common knowledge of all ordinary logical truths. This result is largely due to the presence in standard epistemic logics of the inference rule known as the Rule of Epistemization, which allows one to infer, from any theorem, knowledge of it.

It seems, then, that a promising strategy for formalizing Schiffer's account is to embed the Recognition and Double Sufficiency Postulates in an epistemic logic, and to do this in such a way that what normal and present people know by virtue of being so is given in theorems. This is what the theory T_I does.

T_I is a formal axiomatic theory developed in a variant of the epistemic predicate logic T. T_I is defined by three components: the axioms and inference rules of a quantifier-free predicate calculus (PC);[4] the (non-PC) axioms of T; and non-T axioms and an inference rule that is a weakened Rule of Epistemization. Consider two knowers 1 and 2. For $i = 1, 2$, "κ_i" is an operator on formulas with the interpretation "i knows that." The second component, TA, consists of two axioms:[5]

(A1) $$\kappa_i P \to P,$$

(A2) $$\kappa_i P \wedge \kappa_i (P \to Q) \to \kappa_i Q,$$

where "P", "Q" range over formulas and "$\wedge$" binds more strongly than "$\to$". (A1) says that what is known is the case; (A2) that i's knowledge is closed under Modus Ponens.[6]

The third component of T_I gives a "theory of normality": It describes the epistemic situation and resources of two knowers who are normal in Schiffer's sense. It consists of three nonlogical axioms or postulates and one inference rule. Some extra syntax is needed. "1" and "2," which already appear as indexes on knowledge operators, will also serve as individual constants, intended to name 1 and 2; "N" and "Z" are one-place predicates with the intended interpretations "is normal" and "is present," respectively; things are neater if we also have the predicate "$\hat{N}$" ("is both normal and present"):

(D1) $$\hat{N}i \stackrel{\text{def}}{=} Ni \wedge Zi.$$

"A" is a sentence letter of T_I, which expresses something in the vein of "there is a candle on the table between 1 and 2" or "a waiter asks for everyone's attention and announces that the dictator is dead." The postulates of T_I are three ($i, j = 1, 2$):

(P1) $$Zi;$$

(P2) $$\hat{N}i \rightarrow \kappa_i A;$$

(P3) $$\hat{N}i \rightarrow (\hat{N}1 \wedge \hat{N}2 \rightarrow \kappa_i(\hat{N}1 \wedge \hat{N}2)).$$

The inference rule is

(R_I) if $\vdash \hat{N}1 \wedge \hat{N}2 \rightarrow \kappa_j P$ then $\vdash \hat{N}i \rightarrow \kappa_i(\hat{N}1 \wedge \hat{N}2 \rightarrow \kappa_j P)$,

where "$\vdash$" stands for "it is a theorem of T_I that."

This completes the description of T_I. Under its intended interpretation, its third component bears a strong resemblance to Schiffer's model, but there are some important differences. (P3) is a formulation of the Recognition Postulate, in which Schiffer's "suffices for" is cashed out as a material implication of axiomatic status; in semantic terms this means that, in every theoretically possible world in which someone is normal and present, she knows of herself and of anyone else who is normal and present that they both are.

The Double Sufficiency Postulate appears as the inference rule (R_I). The first two of the Schifferian "suffices for"s are cashed out as theorematic material implications (the third, embedded, one is material). (R_I) says that necessarily any normal-and-present knower knows truths of a certain form, provided that these truths are *derivable*. (R_I) thus also offers a possible explanation of *how* the knower comes by her knowledge of these truths: by seeing how they follow from basic truths.[7] On this interpretation, (R_I) ascribes certain *inferential* abilities to normal knowers – namely, the ability to track derivations in T_I. I shall return to this point in the next section.

Notice that the hypothesis in (R_I) is a little weaker than the antecedent of Postulate 2 (and so my version of Double Sufficiency is stronger than Schiffer's): It posits only that the normality-and-presence of *both* knowers is sufficient for P's being known to one of them. This strengthening is needed for the desired results to follow.

Postulate (P2) corresponds to the hypothesis (1) in my sketch of Schiffer's proof, and expresses the evidentness of A for normal knowers in a certain situation (such as being at the table, attentive). A fortiori, (P2) holds for any A knowledge of which requires *only* normality, including simple tautologies and pieces of elementary contingent knowledge such as that money doesn't grow on trees. (P1) corresponds to one-half of the hypothesis (2) in Schiffer's proof, the presence of 1 and 2. My interest in this chapter is in the role of normality rather than presence in producing MK, and this role is thrown into sharper relief by treating only normality (and not presence) as an antecedent in my theorems; so I make presence an axiom.[8]

Before we proceed, let's check the (syntactic) consistency of the theory T_I. That T_I is consistent is immediately clear when we note that (R_I) is a restricted version of the Rule of Epistemization "if $\vdash Q$ then $\vdash \kappa_i Q$," in which Q has a special form and the conclusion licensed is weaker in being conditional on $\hat{N}i$. Thus T_I is a subtheory of the theory – T_I^+, say – obtained by subjoining (P1)–(P3) to T. But T_I^+ is consistent since it certainly has a model; for example, any model of T in every world of which $Z1$ and $Z2$ are true and $\hat{N}1$ and $\hat{N}2$ are false.

Henceforth, for concreteness I shall take A to express the fact that a certain announcement has been made. The first main result is the following theorem.

Theorem 1 (Common Knowledge Theorem). *It is a theorem of* T_I *(in its intended interpretation) that if 1 and 2 are normal then there is common knowledge between them that the announcement has been made.*

Proof. Write κ_i^n for the n-long operator string $\kappa_i \kappa_j \kappa_i \kappa_j \cdots$ $(j \neq i)$. It suffices to show that, for any positive integer n and for $i = 1, 2$:

(8)
$$\vdash N1 \wedge N2 \to \kappa_i^n A,$$

which we shall do by induction on n.

(i) *Inductive basis:* Assume $N1 \wedge N2$. For each i, (P1) and (D1) yield $\hat{N}i$, by *PC*. Then $\kappa_i^1 A$ by (P2) and *PC*.[9]

(ii) *Inductive step:* Consider any $n \geq 1$. Suppose that (8) is true for $i = 1, 2$. Assume (H): $N1 \wedge N2$. It suffices to derive $\kappa_j^{n+1} A$ for $j \neq i$ within T_I, as follows.

Inductive hypothesis (8):

$$N1 \wedge N2 \to \kappa_i^n A.$$

Hence, from (D1) by *PC*,

$$\hat{N}1 \wedge \hat{N}2 \to \kappa_i^n A.$$

By (R_I),

(9)
$$\hat{N}j \to \kappa_j(\hat{N}1 \wedge \hat{N}2 \to \kappa_i^n A) \quad (j \neq i).$$

From (H), (P1), and (D1),

(10)
$$\hat{N}1 \wedge \hat{N}2;$$

from (9) and (10),

(11)
$$\kappa_j(\hat{N}1 \wedge \hat{N}2 \to \kappa_i^n A);$$

from (10) and (P3),

(12)
$$\kappa_j(\hat{N}1 \wedge \hat{N}2).$$

From (11), (12), and (A2),

$$\kappa_j^{n+1} A. \qquad \square$$

I note two points about the proof. (i) It uses all but one of the non-*PC* elements of T_I. The unused one is the factivity axiom (A1); thus there is a parallel theorem for common belief.[10] (ii) The weaker inference rule "if $\vdash \hat{N}i \rightarrow \kappa_i P$ then $\vdash \hat{N}j \rightarrow \kappa_j(\hat{N}i \rightarrow \kappa_i P)$" (which corresponds more closely to Double Sufficiency à la Schiffer) would not do in the above proof. It would do if for all n it were a theorem in the weakened system that $\hat{N}i \rightarrow \kappa_i^n A$ – for instance, a theorem that $\hat{N}i \rightarrow \kappa_i \kappa_j A$ – but it is clear that to derive this latter theorem we would need the premise $\hat{N}j$ ($j \neq i$).

At the core of Theorem 1 and at the root of common knowledge are two ideas. They are expressed in (R1) and (P3), and they descend from the two Schifferian postulates. (R1) makes certain inferential powers part of normality; (P3) makes normal beings placed together be aware of each other's normality. This engenders awareness of a constitutive feature of that normality – these inferential powers – and this in turn engenders awareness of the knowledge each other obtains by dint of these powers. But this latter knowledge is itself obtained by nothing more than the use of the powers, so it too becomes the object of each other's awareness. A tower of MK rises up.

The theorem shows that between knowers of a certain kind there is mutual knowledge of every degree of certain propositions: any for whose knowing it is enough to be normal and present. The fact of this infinite knowledge is yielded by a finitely expressed axiom basis. In particular, the basis includes no assumption of any piece of knowledge of degree higher than 2, so that T_I provides a non-question-begging answer to the query: "How can we explain the possession of higher MK?" The finite basis feature is reassuring for those who maintain that people have common knowledge, for it keeps alive the possibility of reconciling this claim with the finiteness of our cognitive machinery. This possibility would seem foreclosed if for the epistemically infinitary *explanandum* we could find only epistemically infinitary *explanantia*.

2. HOW RATIONAL HUMANS ACQUIRE SOME MUTUAL KNOWLEDGE

2.1. Difficult inferences

There is a threat of another kind to the explanation of mutual knowledge offered by T_I. We discern it if we turn our attention to the *inferences* of the knowers of T_I. We are struck by the complexity of the inferences that T_I portrays them as making, not least by the complexity of the items they are portrayed as manipulating when making these inferences.[11]

The proof of Theorem 1 is broadly narrative; it tracks the knowers' reasoning. Nevertheless, it gives little idea of the "nitty-gritty" of the inferences that 1 and 2 make in arriving at MK of degree n, of what it's like on the ground. Both its inductive form and my use of the abbreviation "κ_i^n" diminish the graphicness of the narrative. Consider one step in such an inference more concretely. Let us say henceforth that *i has MK of degree n with j that P* ($j \neq i$) if i knows that j knows that i knows . . . P, with n occurrences of "knows that" (thus there is MK of degree n between i and j if and only if each has MK of degree n with the other). I shall also abbreviate "MK of degree n" by "MK^n."

Suppose that 1 and 2 are you and me. It's a theorem that if we are normal and present then there is fourth-degree MK between us that the waiter has made the announcement. The story then goes: On the one hand, you recognize both of us to be normal and you see we're both present. On the other, being normal, you have grasped the truth of the said theorem: You have seen that given that we both are both normal and present, it must be that I know that you know that I know that you know that the waiter made the announcement. Putting two and two together, therefore, you also know that I know that you know that I know that you know that he did. In this way you achieve fifth-degree MK.

Such inferences, simple in the form, are formidable in the execution. One locus of cognitive strain is the handling of the formulas containing stacked epistemic operators. Though we easily achieve theoretical mastery of the *numerical* notion of fourth-degree MK, we have great difficulty in "getting the feel" of the situation of my knowing that you know that I know that you know something; nor can we read expressions like the one I have just used (of more than a certain degree) without resorting to counting. As the degree reaches six, seven, or eight, the strain becomes overwhelming. What chance have humans of achieving MK of really high degree, as T_I would have it?

This tension between T_I and what we know of our own information-processing abilities is a special case of a general difficulty for standard epistemic logics. The presence in them of certain schematic axioms and of the Rule of Epistemization or its equivalents makes knowers out to have limitless deductive competences, to be "logically omniscient," in certain tasks. More than one approach has been proposed for modifying these logical systems to make them suitable models of the rational inferential knowledge of humans; they are surveyed by Vardi (1986). In this chapter I shall follow the approach of Hintikka (1970)[12] and attempt to characterize the class of inferences that are "obvious" for rational humans, then restrict the systems so that they have knowers making only obvious inferences.[13] Hintikka's discussion is in terms not of the Rule of Epistemization but rather of the equivalent Rule of Conservation of Proved Implication: "if $\vdash P \to Q$ then $\vdash \kappa_i P \to \kappa_i Q$." His proposal is to restrict this rule to cases in which the proof of "$P \to Q$" is obvious; it amounts to replacing the Rule of Conservation by the rule:

(13)   if $\vdash_* P \to Q$ then $\vdash \kappa_i P \to \kappa_i Q$,

where "$\vdash_* R$" indicates that there is a proof of R within the logic that is obvious according to a defined standard, and "$\vdash R$" that there is a proof of R within the logic.

The problem of characterizing obviousness or "visibility" of inferences in a general way is a challenging one. For the problem in hand, fortunately, a general characterization is not needed; we need only assess for relative obviousness a specific range of inferences – those which the Transparency Argument has it that people make in acquiring higher MK. I submit that the obviousness of an inference tends to diminish, ceteris paribus, with increase in the *epistemic degree* of its contained formulas.[14] Two things argue for this: first, the introspective evidence that we have noticed; second, the general conjecture that the difficulty of an inference increases with the "arithmetical depth" of the expressions occurring in it. The arithmetical depth of a function is the maximum "nesting" of basic operations in it. Von Neumann (1958) argues that the language used by the central nervous system is marked by low arithmetical depth, and Hintikka (1970), noting that quantifiers are equivalent to certain (Skolem) functions, proposes "quantificational depth" as a standard of visibility of proofs in quantificational languages. Epistemic degree is arithmetical depth with respect to the operations κ_i. In Section 2.2, accordingly, in order to model the acquisition of mutual knowledge by rational humans, I modify T_I by restricting its rule (R_I) in the manner of (13), and use epistemic degree to define a standard of visibility.

It may be objected that there is an alternative (and much simpler) approach to modeling the way human depth limits rule out common knowledge. Instead of putting restrictions on knowers' ability to make inferences of high degree, why not just "cap" the attainable degree of MK, restricting the domains of the operations κ_i themselves? The first reason for preferring an inference-based model is that it yields a nontrivial impossibility result, in which the impossibility of high-degree MK is derived from weaker premises (and so, if there *are* bounds on the degree of MK, it yields a nontrivial explanation of these). For it is beyond question that mutual knowledge is inferential knowledge and, this being so, showing that the only plausible inferential path is blocked is enough to show that high-degree

295

MK is impossible; while inability to make high-degree inferences falls short of logically entailing the absence of such MK.

The second reason is that there is a substantial counterargument to the inference-based account (namely, the Bringing-to-See Argument). If, as it seems, the boundedness of inferential abilities provides the best case for the thesis that high-degee MK and common knowledge are impossible, then this thesis turns crucially on the strength of the counterargument, which we must therefore examine. And to do this, we must first set out a version of the inference-based account of impossibility against which the counterargument is ranged.

2.2. The theory T_H

T_I is so labeled because it is a theory of *ideal* knowers with unlimited powers of inference. In this subsection I present a theory, which I label T_H, of *human* knowers with limited ones. As a formal system, T_H differs little from T_I. Discard the predicate "N"; let its place be taken by the one-place predicate "H" (interpretation: "is a normal human"); discard "$\hat{N}$" and introduce "$\hat{H}$" ("is both a normal human and present") by:

(D2) $$\hat{H}i \overset{\text{def}}{=} Hi \wedge Zi.$$

Substitute "$\hat{H}$" for "$\hat{N}$" in the postulates (P2) and (P3), so that these are replaced by:

(P4) $$\hat{H}i \to \kappa_i A,$$

(P5) $$\hat{H}i \to (\hat{H}1 \wedge \hat{H}2 \to \kappa_i(\hat{H}1 \wedge \hat{H}2)).$$

Crucially, replace the Double Sufficiency rule (R_I) for the idealized knowers of T_I by the Double Sufficiency rule for humans:

(R_H)   If $\vdash_H \hat{H}1 \wedge \hat{H}2 \to \kappa_j P$ then $\vdash \hat{H}i \to \kappa_i(\hat{H}1 \wedge \hat{H}2 \to \kappa_j P)$.

Finally, $\vdash_H$ must be given an effective syntactic definition. I shall adopt:

(D3) $\vdash_H P \overset{\text{def}}{=}$ there is a proof of P in T_H of epistemic degree $\leq h$,

where the epistemic degree of a proof is defined to be that of the set of formulas occurring in it. I shall call h the *visibility bound* of the rule (R_H). Notice that the visibility bound is the degree of the deepest theorem that a human knower sees, not of the formula expressing the knowledge state that she thereby enters. If $h = 4$ and P is a theorem of degree 4, then (R_H) allows us to infer the formula $\kappa_i P$, and this is of degree 5. To summarize, the theory T_H consists of: *PC*; *TA*; (P1), (P4), (P5); and (R_H). The consistency of T_H is shown in the same way as that of T_I.

Comments on (R_H). (i) "*h*" is to be thought of as standing for a particular though unspecified nonnegative integer. If $h = 0$, (R_H) never has application. For concreteness one may think of h as 4 or 5 or 6.

(ii) It might be objected to (D3) that whether a proof is easy varies with the person, with her situation, and with her effort. However, what determines the aptness of a restriction on the Double Sufficiency rule is the answer to this question: When are we fully justified in holding, of a certain theorem, that j knows it *solely in virtue of* being normally human and present? That is, what counts is the low-water mark of normal humans' inferential capacities as one ranges over persons, their situations, and their efforts.

(iii) One can imagine a more refined theory which on the one hand recognizes more than two degrees of difficulty, and which on the other allows one to infer not j's knowledge of the theorem but rather his probable knowledge of it, or even his probable partial degree of belief in its truth. But simple cases first.

(iv) I note that substituting (R_H) for (R_I) does not make T_H exclude that j knows the difficult theorems; at most it makes it unprovable in T_H that he does so.

There are two main results concerning T_H. The first is positive: Mere normal-human knowers do acquire MK of announcements they attend.

Theorem 2. *For each $n \leq h+1$, it is a theorem of T_H (in its intended interpretation) that if 1 and 2 are human and normal then they have mutual knowledge of degree n that the announcement has been made.*

Proof. It suffices to show that in T_H, for any positive integer $n \leq h+1$ and for $i = 1, 2$,

$$\vdash H1 \wedge H2 \rightarrow \kappa_i^n A.$$

This is so if the proof of Theorem 1, with the substitutions H/N and $\hat{H}/\hat{N}$ and the inductive step restricted to $n \leq h$, is sound. The inductive basis clearly is; in the inductive step, each line but the third is justified as before or by (P5) in place of (P3). The third line is sound provided that, for each $n \leq h$,

(14) $$\vdash_H H1 \wedge H2 \rightarrow \kappa_i^n A.$$

This too may be shown by induction on n. Formula (14) holds for $n = 1$ because the inductive basis of the proof of Theorem 1 is of degree 1. For arbitrary $n \leq h-1$ assume (14), that is, that there is a proof of $H1 \wedge H2 \rightarrow \kappa_i^n A$ of degree $\leq h$. Inspection of the inductive step of the proof of Theorem 1

297

shows that the deepest of its constituent formulas has degree $n+1$, and so does not exceed h. Since the degree of a proof is preserved under concatenation, there is a proof of $H1 \wedge H2 \to \kappa_i^{n+1}A$ of degree $\leq h$; that is, (14) holds for $n+1$. $\qquad \square$

At the core of Theorem 2 and at the root of human mutual knowledge are two ideas. They are expressed in (R_H) and (P5), and descend from the two Schifferian postulates. (R_H) makes certain inferential powers *constitutive* of being normal-human; (P5) makes normal human beings placed together be aware of each other's normal humanity. This engenders awareness of the constitutive feature of that normality, these inferential powers, and this in turn engenders awareness of the knowledge each other obtains by dint of the powers. But this latter knowledge is itself obtained by nothing more than the use of these powers, so it too becomes the object of each other's awareness. A tower of MK rises up.

The second main result about T_H is negative: It tells us that the tower may have a top floor. It is *not* a truth of T_H that normal humans present at a conspicuous event have MK^n of it, if n exceeds by more than 1 the visibility bound h. There is, in T_H, no common knowledge.

Theorem 3 is circumspect; it speaks only of what can and cannot be shown in the theory T_H. But the theory T_H is incomplete; that a piece of MK^n is not provable in T_H does not mean that its negation is,[15] and so the possibility is not closed that such MK might be come by otherwise than by virtue of the epistemic resources that T_H ascribes. If, however, the Transparency Argument is, as I believe, the only serious argument for the possibility of MK, and if theories such as T_I and T_H are adequate vehicles for Transparency Arguments, then Theorem 3 entitles us to the bald assertion that, for human beings, *there is no such thing as common knowledge.*

Theorem 3 (No-Such-Thing Theorem). *For each $n > h+1$, it is unprovable in T_H (in its intended interpretation) that if 1 and 2 are human and normal then there is mutual knowledge of degree n between them that the announcement has been made.*

We must show that the formulas $H1 \wedge H2 \to \kappa_i^n A$ for $n > h+1$ are not theorems of T_H. One way of doing this is to set up an appropriate semantics[16] and display counterexamples in the shape of worlds in which these formulas are false. Here, to avoid the setup costs, I shall give a metasyntactic proof of Theorem 3 that revolves around properties of proofs within T_H. As a preliminary, I make some general points about proofs in T_H.

A *proof* in T_H of a formula P is a sequence $\langle P_1, \ldots, P_m \rangle$ of formulas of T_H with $P_m = P$ and containing a subsequence Λ (the *lemmas* of the proof) such that:

(i) if $P_t \in \Lambda$ then there is a subset $\mathbf{M}_{t-1}$ of $\{P_1, \ldots, P_{t-1}\}$ (the *premises* of P_t) from which P_t is directly deducible by ρ_t, where ρ_t (the *license* for P_t) is one of the two inference rules of T_H, Modus Ponens and (R_H); and

(ii) if $P_t \notin \Lambda$ then P_t is an axiom of T_H.

The proof of Theorem 3 plays on the epistemic degree of the formulas it concerns. The formulas that are to be shown unprovable in T_H are of high degree. The strategy of the proof is to show first (Lemma 1) that the derivation in T_H of formulas of more than a certain degree can be achieved only by calling repeatedly on the inference rule (R_H); then that this process is blocked at a certain level by the restriction written into (R_H) on the degree of its premise.

Write $\delta(P)$ for the epistemic degree of a formula P, and $\delta(\mathbf{P})$ for the *epistemic degree* of a set $\mathbf{P}$ of formulas, defined as $\delta(\mathbf{P}) = \max\{\delta(P) \mid P \in \mathbf{P}\}$. Write $\mathbf{P}_t = \{P_1, \ldots, P_t\}$ for the first t formulas, and $\mathbf{L}_t = \mathbf{P}_t \cap \Lambda$ for the subset of these that are lemmas. Define ι_t, the increment at t in the degree of the lemma stock, by

$$\iota_t \stackrel{\text{def}}{=} \delta(\mathbf{L}_t) - \delta(\mathbf{L}_{t-1}) \quad (t = 2, 3, \ldots),$$

where we set $\delta(\mathbf{L}_{t-1}) = 0$ if $\mathbf{L}_{t-1} = \emptyset$.

Lemma 1. *If* $\langle P_1, \ldots, P_m \rangle$ *is a proof in* T_H *then, for* $t = 1, \ldots, m$:
(i) $\iota_t \leq 1$; *and*
(ii) *if* $\iota_t = 1$ *then either* $\delta(\mathbf{L}_t) = 1$ *or the license for* P_t *is* (R_H).

Proof. Let P_t be a lemma in $\mathbf{P}_m$. Suppose that

(15) $\qquad\qquad\qquad \rho_t$ is Modus Ponens.

Then

(16) $\qquad\qquad\qquad \delta(P_t) \leq \delta(\mathbf{M}_{t-1}),$

for there is a formula P such that $\mathbf{M}_{t-1} = \{P, P \to P_t\}$; whence, by the definition of δ, $\delta(P_t) \leq \delta(P \to P_t) \leq \delta\{P, P \to P_t\} = \delta(\mathbf{M}_{t-1})$. Next,

(17) If $\mathbf{M}_{t-1}$ contains a *T*-axiom then it contains a lemma L with $\delta(P_t) \leq \delta(L)$,

for the *T*-axioms are (A1) and (A2). Suppose one premise of P_t is an instance $M = (\kappa_i P \wedge \kappa_i (P \to Q)) \to \kappa_i Q$ of (A2); then the other – L, say – is either $\kappa_i P \wedge \kappa_i (P \to Q)$ or of the form $[(\kappa_i P \wedge \kappa_i (P \to Q)) \to \kappa_i Q] \to R$. By

299

inspection, in neither case is L an axiom or postulate of T_H. Further, in each case $\delta(M) \leq \delta(L)$, whence by (16), $\delta(P_t) \leq \delta(L)$. The case in which one premise of P_t is an instance of (A1) is dealt with similarly.

Exactly one of the following four cases holds:

(a) (15) is true and M_{t-1} contains a T-axiom;
(b) (15) is true and M_{t-1} contains no T-axiom and $\delta(L_{t-1}) \geq 1$;
(c) (15) is true and M_{t-1} contains no T-axiom and $\delta(L_{t-1}) = 0$; or
(d) (15) is false.

In case (a), $\iota_t = 0$ by (17). In cases (b) and (c), M_{t-1} contains only PC-axioms, postulates, and lemmas. Noting that the set consisting of PC-axioms and of postulates is of degree 1, $\delta(P_t) \leq \delta(M_{t-1}) \leq \max\{1, \delta(L_{t-1})\}$ by (16). In case (b), then, $\delta(P_t) \leq \delta(L_{t-1})$ and $\iota_t = 0$. In case (c), $\delta(P_t) \leq 1$ and $\iota_t \leq 1$. In case (d), $\iota_t = 1$ by the definition of (R_H). □

Proof of Theorem 3. Let $\langle P_1, \ldots, P_m \rangle$ be a proof in T_H of $H1 \wedge H2 \rightarrow \kappa_i^n A$. Because P_1 is not a lemma, $\delta(L_1) = 0$. Because P_m is (by inspection) not an axiom of T_H, it is a lemma; since P_m is of degree n, $\delta(L_m) = n$.

If $n = 1$ then $n \leq h+1$, since h is nonnegative. If, on the other hand, $n \geq 2$, then $\delta(L_m) - \delta(L_1) \geq 2$. Thus, by (i) of Lemma 1, there are integers s, t such that $1 < s < t \leq m$, with $\delta(L_s) \geq 1$, $\delta(L_{t-1}) = n-1$, and $\delta(L_t) = n$. Since $\delta(L_t) > 1$, the license for P_t is (R_H), by (ii) of Lemma 1. Since $\delta(L_t) = n$, the premise of P_t is of order $n-1$, by the definition of (R_H). Hence, by the definition of (R_H), in this case too $n \leq h+1$. □

That completes my discussion of the formal aspects of T_H. On the one hand, it has enabled us to establish rigorously that people have mutual knowledge, of at least some modest degree, of events that conspicuously unfold between them. On the other hand, it creates a presumption that, where inferential capacities are bounded, so is MK – for beyond that modest degree the rigorous basis does not take us. But this can be no more than a presumption; we can not yet rule out other bases, productive of further layers of MK. Such bases would have to endow knowers with richer epistemic resources, more favorable external circumstances, or both. In Section 3 I shall consider one such enrichment as well as the claim that it has such an effect.

The theories T_I and T_H that I have described in this section deal only with two special cases: that of ideally rational knowers, and that of humanly rational knowers characterized by inferential capacities limited in a very specific way. It is possible to imagine more inclusive theories. To begin with, T_I and T_H may simply be conjoined; the resulting theory is consistent. Going further, we may construct a general mixed theory T with many epistemic types τ, each with its type-specific inferential abilities captured in a rule (R_τ), and the members of each (by their nature) knowing cospecifics when they meet them. Such a theory would allow us to analyze the epistemic consequences of mixed-type encounters. What happens if a human and an ideal knower are present at the announcement; more generally, if knowers of distinct types τ and τ' are? To get further, we must specify more: which types recognize which? What becomes of Double Sufficiency in mixed-type cases? I shall not pursue these questions in detail in this chapter, but a natural line of development is as follows.

Let the set of types have index set $\mathfrak{J} = \{0, 1, 2, \ldots; \infty\}$, where knowers of type-index τ satisfy a restricted rule of inference (R_τ) whose visibility bound is τ. Suppose all types recognize each other. Then we will obtain that, if 1 and 2 are respectively a normal type τ and a normal type τ', where $\tau < \tau'$, and if both are present at the announcement, then there is MK of it between them of degree $\tau + 1$; in addition, 2 has MK with 1 of the announcement of degree $\tau + 2$ – that is, $\kappa_2^{\tau+1}A$. Finally, there is – for all that can be shown in T – no further iterated knowledge of it on either side. This gives 2 the advantage of 1 in certain games.[17]

3. THE BRINGING-TO-SEE ARGUMENT

The theory of acquisition of mutual knowledge described in Section 2 implies that the bounds on humans' powers of inference preclude mutual knowledge of high degree between them. Schiffer seeks to deny this consequence by arguing that a certain legitimate enrichment of the basis of MK in T_H produces indefinitely many further layers. I shall show in this section that the enrichment, legitimate or not, does not have this effect. Schiffer acknowledges that the deductions of the knowers recounted in the Transparency Argument might not be executed by ordinary mortals. He argues, however, that this would not mean that the latter lack the MK in question. Schiffer's argument is complex; it depends both on denying a distinction between knowledge and occurrent knowledge, and on the following claim, which I shall call the Demonstrability Claim: *Humanly rational agents can be "brought to see" each step in the string of inferences executed by the knowers in the argument. They can be brought to see it by having it pointed out and being persuaded of it as the need may be.*[18] Schiffer argues that, since knowledge need not be occurrent, there *is* MK^n if such knowledge would occur were the knowers brought to see enough steps; thus the Demonstrability Claim implies the MK^n. I call this the Bringing-to-See Argument.

My target here is not the Demonstrability Claim, but this is not because it is plausible. There must be a demonstrator, someone who does the

301

bringing-to-see; who is to fill the role? Even the cleverest instructor will herself eventually buckle under the cognitive strain as the degree of the difficult inferences mounts. If so, the Demonstrability Claim is false and the most we can hope for from bringing-to-see is to jack up the degree of attainable MK from h to something a bit higher – from five or so to ten or twenty or thirty; at most by an order of magnitude.

These considerations weaken the Bringing-to-See Argument but are not fatal to it. Something else is: Even if the Demonstrability Claim is true, it gets us nowhere. The nub of why is this: For two knowers to achieve high-degree MK, what is needed is not just that each be brought to see the steps but that *there be high-degree MK between them so that each has been brought to see the steps.* The question of how actual knowers can ever have MK exceeding a certain degree is therefore begged.

For consider: We are both at the announcement. Because *I am present at the announcement and normally human,* it follows that I know it was made. So you are in a position to know that I do (for you know I am there and am normally human). But suppose (to keep the example as brief as can be) that humans can't be relied on to see even first-degree arguments; that is, suppose the visibility bound h is 0. Then, although you are in a position to know that I do, it is not assured that you know it. Having a qualified demonstrator go through the argument with you would no doubt bring you to see. Let us be magnanimous: Suppose that it *follows* that you would see. But for me to know that you see *I must know that you had instruction.* Suppose that somehow I do know this (I might in all kinds of ways). Then I am in a position to know that you know that I know that the announcement was made. But seeing that you do means seeing a second-degree argument; and I am only human, so it is not assured that I know that you know that I know the announcement was made. Suppose I too have instruction; then I certainly see the second-degree argument and acquire the third-degree knowledge. But for you to know that I have seen, you must know that I had instruction; only then will you know that I know that you know that I know that the announcement was made. Moreover, if you are to have this fourth-degree knowledge, you must also know something else: that I know the truth of the premises of the argument in which I had instruction. One of these is that you know that I know the announcement was made. But my knowledge of this, we have seen, depends on my knowing that you were instructed. So for you to know that I know that you know that I know the announcement was made, *you must know that I know that you were instructed.* And so it goes.

This rough sketch suggests that, in the case $h = 0$, the most we may say concerning bringing-to-see is this.[19]

Proposition 1. *For each $n > 1$, if 1 and 2 are normally human and present at the announcement, that they have mutual knowledge of it of degree n follows if and only if they have mutual knowledge of degree $n - 2$ that both have instruction.*

Let us see how Proposition 1 might be formulated in the style of the last two sections. I shall limit myself to one such formulation in a variant of T_H, to be labeled T_H^*, which contains the new primitive predicate D ("has a demonstrator"). We would like D to have this interpretation: If a normal human satisfies D then she sees any truth of the theory of the sort that appear in the Transparency Argument. This interpretation is made appropriate by subjoining to the axioms of T_H the new inference rule

(R_H^*) $\quad$ if $\vdash \hat{H}^*1 \wedge \hat{H}^*2 \to \kappa_i P$ then $\vdash \hat{H}^*j \to \kappa_j(\hat{H}^*1 \wedge \hat{H}^*2 \to \kappa_i P)$,

where $\hat{H}^*$ is defined by

(D4) $\qquad\qquad\qquad\qquad \hat{H}^*i \overset{\text{def}}{=} \hat{H}i \wedge Di.$

That is (broadly speaking): If joint instructed human-normality[20] implies that a person knows that P, then instructed human-normality is sufficient for knowing this implication. Why should the truths of the sort that appear in the Transparency Argument include ones about what joint *instructed* human-normality implies a person knows? Otherwise, a Transparency Argument could gather no momentum. The initial phase of such an argument does not depend on our including such truths, because perceptual knowledge needs no instruction; later steps do, because inferred knowledge does.[21]

Remark now that by adding a Recognition Postulate for $\hat{H}^*$ we obtain an extension of the theory T_H^*, which is just T_I with $Hi \wedge Di$ substituted uniformly for Ni; for (R_H^*) contains no visibility restriction. This shows that if being normally human and instructed shares the basic properties of being normally human – Double Sufficiency *and* Recognition – then indeed bringing-to-see does the trick. And this fact throws new light on Proposition 1, since if being normally human and instructed shares these properties then MK of *it* begins to be explicable. But as a tactic for showing the efficacy of bringing-to-see in producing MK, a Recognition Postulate for instructedness is a nonstarter. For there is nothing in the notion of receiving instruction, of being brought to see, which makes it necessary – or even common – that two people who receive instruction know that each other receive it.

The elements of the theory T_H^* are: *PC*; the *T*-axioms; (P1), (P4), (P5); and (R_H^*) – that is, T_H^* is T_H with (R_H^*) replacing (R_H). Like (R_H), (R_H^*) is a restricted version of the Rule of Epistemization: instead of limiting the complexity of the theorems that it guarantees knowers know, it issues a conditional guarantee that they know them. Proposition 1 can now be formulated as a fact about T_H^*, as follows.

Proposition 2.[22] *The formula* $N1 \wedge N2 \wedge \kappa_i^m(D1 \wedge D2) \to \kappa_i^n A$ *is a theorem of* T_H^* *if and only if* $m \geq n - 2$.

Proposition 2 means that the obstacle to mutual knowledge between humans posed by their inferential limits is not disposed of by bringing-to-see but only shunted sideways: from mutual learning of the announcement to mutual learning of the bringing-to-see. How could the latter come about? Suppose we could concoct a setup in which the occurrence of $D1 \wedge D2$ is observed in the manner of an announcement, that is, in which tutorials in the hard steps conspicuously unfold in the joint presence of 1 and 2. This would achieve MK of $D1 \wedge D2$, but of degree at most $h + 1$. Let $h = 4$, and suppose it is important that 1 and 2 have MK of degree 20 of the announcement. They cannot achieve MK in the untutored case, as $20 > 5$; they would achieve it if they could achieve MK of degree 18 that $D1 \wedge D2$. But this is no help, for 18 is still greater than 5. Conversely: In the simple announcement case, there is a low finite bound of 5 on the degree of MK of the announcement they may achieve. In the concocted setup, there is still a low finite bound on the degree of MK they may achieve of the announcement – namely, 7. Mutual knowledge of modest finite degree remains the most that 1 and 2 can aspire to.[23]

The reasons for the failure of the Bringing-to-See Argument are the clearest proof that the enterprise of mutual knowledge is radically collective. Instruction sessions are, we have seen, useless by themselves; they must be mutually known-of if MK is to be assured. So, then, is trying harder useless by itself: There must be MK of the efforts, and of their success. When an individual's epistemic goal is a state of higher MK, instruction and effort, the trusted paths to epistemic achievement, lead only back to the starting point. Likewise, individual gifts are of no avail: A human paragon at an announcement gets no further than a human mediocrity. Nor do two! What is needed transcends all means by which individuals single-mindedly advance their knowledge, it is something that *we cannot do by ourselves*. We must advance in those epistemic capacities that are constitutive of our humanness. We must collectively become more like paragons. We must become creatures who can do certain things we cannot now do and who, by virtue of being the creatures they are, are endowed with the knowledge that each other can do them.

4. Mimetic rationality

4.1. Mutual knowledge and manageable games

In this last section I consider certain implications of the preceding arguments for rational decision making. Theorem 2 tells us that if n does *not* exceed $h+1$ then the occurrence of an announcement made where attentive, normal, rational humans are present *does* become mutual knowledge of degree n between them. Let knowledge of the fact of the announcement go with knowledge of the truth of the announcement.[24] Then T_H gives a formal demonstration of the disposition of rational persons to acquire MK of modest degree of what is said in their joint presence. This is good news for game theory: There are games that possess clear solutions if, but only if, the specification of the game is MK of some modest degree n between the players; we shall see, for example, that any repeated Prisoner's Dilemma of n rounds or fewer is such a game. Call these *manageable* games.[25] In the past, accounts of how manageable games are rationally played have simply assumed the modest MK. The theory T_H of acquisition of mutual knowledge allows us to complete the formal solution of manageable games, beginning from the point of the umpire's announcement. T_H "endogenizes" the acquisition of the modest MK.

But not all the news for game theory is good. We have found that truths about human cognitive limits unwarrant the assumption of higher-degree MK among humans. For according to Theorem 3, humans cannot acquire – by virtue of the epistemic resources T_H credits them with – such MK; the failure of the Bringing-to-See Argument and affine arguments[26] means there is no extant plausible proposal for securing them that knowledge by enriching their resources. And so Theorem 3 subverts all accounts of human behavior that depend on the assumption of such MK, let alone common knowledge. It subverts the standard accounts of rational action in long repeated Prisoner's Dilemmas and other nonmanageable games, as well as Schiffer's account of acts of S-meaning. In short, Theorem 3 seems to show that we don't know how to rationalize behavior that we thought we could. In particular, it seems to show that, for all we know, it is, after all, not rational to adopt the "nasty" strategy of confessing from the outset in long sequences of Prisoner's Dilemmas.

So it seems. But there is a plausible defense of the rationality of this and other traditional solutions. The Bringing-to-See Argument sought to show that assumptions of common knowledge are tenable, notwithstanding cognitive frailty. Had it succeeded, it would have provided an *epistemic* argument for the traditional solutions. By contrast, the "mimetic" defense, to which I now turn, argues for the traditional solutions by appeal to a thesis about *practical reason*. It seeks not to reinstate higher MK, but to do without it; it seeks to show that even for inferentially puny humans who can't manage higher MK, the actions picked out by the traditional classical solutions are rational. I shall conclude, however, that like the Bringing-to-See Argument it fails, and that there is (for all that we now know) no warrant for the classical acts, no reason to be nasty.

4.2. The Mimetic Argument

The Mimetic Argument is quite general, but for vividness I shall present it in terms of a particular decision problem, perhaps the most familiar of all those in which MK is essentially adduced in identifying rational actions: the finitely repeated Prisoner's Dilemma. I shall mean by an *N-round Prisoner's Dilemma* (PD_N) a finitely repeated Prisoner's Dilemma of the standard kind; that is, a repeated game with N "stage games" each with payoff matrix **M**, where **M** is a payoff matrix of a Prisoner's Dilemma, where a player's payoff in the game is the sum of her stage-game payoffs, and where at each round all earlier choices are known by both players. By *being nasty* I mean adopting, in a PD_N, any strategy that includes confessing at the first round. It is widely held that, if there is common knowledge of the specification of a PD_N between players of it, then if these players are rational they will both be nasty. The claim rests on the so-called Backward Induction Argument (BIA).[27]

It is perhaps natural to think, and it is often thought, that *the BIA gives an actual person good reason to be nasty in a PD_N*.[28] This supposition is one reason why the BIA has been found disturbing, and why so much effort has been spent looking for "escapes from the N-round Dilemma" – by appeals to uncertainty about the game specification or the players' rationality, and by otherwise varying the original problem.[29] Call it the Natural Supposition; I shall argue in this section that it is mistaken.

The results of Sections 2 and 3 go against the Natural Supposition. As we shall see, the BIA for a PD_N adduces mutual knowledge of degree N of the game. Grant that if such MK is acquired then it is acquired by the players' joint presence at an umpire's announcement of the rules. But Theorem 3 and the failure of the Bringing-to-See Argument create a presumption that human players do not acquire MK of degree greater than h by attending such announcements. So it appears that the BIA lacks application to human game-players in all PD_N's in which $N > h$.

The results of Sections 2 and 3 do not, however, quite settle the matter. One reason for caution is respect for our own intuitions. For many of us, the conviction that the BIA rationalizes human nastiness cannot be

306

suppressed: we recognize that high-degree MK is psychologically impossible for humans, yet the intuition survives. Its resilience is most lively when the BIA is fresh in our minds – when we have just read it, grasped it, and felt its force. Before we dismiss the Natural Supposition, we should explain and answer this testimony of our intuitions on its behalf.

A second reason for caution is that there are more ways than one in which reading the BIA might give you reason to be nasty. There is a straightforward way: Check that you fit the description of some player i in the BIA; if so, conclude that you should do what the BIA shows i should do. This is the way blocked for human players by Theorem 3: You do not fit the bit that says that i has MK of degree $N+1$. But it is not the only possible way. There might be a property R of agents such that: (i) the description of i in the BIA implies that she is R; and (ii) you should do whatever you have reason to believe that an agent who is R would do in your situation. And then, since the BIA shows you that i should be nasty, you may reasonably conclude that you should be nasty.

The two reasons for caution are, I suggest, linked: the intuitive pull of the Natural Supposition comes from the belief that there is a certain property with features (i) and (ii). That property is: *being perfectly rational*. It has feature (i) right enough; it has feature (ii) if the following principle holds:

> *It is rational to do whatever you have good reason to believe a perfectly rational agent in the same situation as you would do.*

Call this principle the Mimetic Maxim (MM).

The Mimetic Maxim endorses as rational parasitism on the rationality of others. It contains no proviso that the agent must be aware of the perfectly rational agent's reasons – nor, therefore, that she act *for* these reasons. This explains how it is that the maxim can endorse as rational for a person an action (being nasty) the reasons for which include items of knowledge (high-degree MK) that the person is incapable of having.

The Mimetic Maxim makes a good deal of sense. A more sweeping version (with "highly" for "perfectly" and "competent" for "rational") provides what seems to be a good guide in a range of practical cases – trusting a boffin; cribbing; following an experienced mountaineer on a difficult descent. Parasitic as it is, the maxim cannot provide a sufficient basis for an account of rational action. It stands in need of supplementation by other principles that identify first-order reasons for actions. But this essential incompleteness does not disqualify it as a perfectly good component of a more comprehensive theory. Let us accept MM, and pass to formulating the argument according to which it justifies the Natural Supposition.

The argument for MM (call it the Mimetic Argument) rests on five premises:

(i) the BIA is sound;
(ii) the human agent (g, say) understands the BIA;
(iii) the BIA shows that a perfectly rational agent in a PD_N is nasty;
(iv) g has good reason to believe that a perfectly rational agent in a PD_N is in the same situation as g in a PD_N; and
(v) the Mimetic Maxim.

By (i) and (ii), the BIA gives g good reason to believe the conclusion of the BIA. Thus, by (iii), the BIA gives g good reason to believe that a perfectly rational agent in a PD_N is nasty. Hence, by (iv), the BIA gives g good reason to believe that a perfectly rational agent in the same situation as hers in a PD_N is nasty. And so, by (v), the BIA gives g good reason to be nasty in a PD_N.

To see whether the Mimetic Argument works, a crucial unclarity in MM must be removed. When is one agent "in the same situation" as another? Too lax criteria of sameness and MM lacks plausibility; too tight criteria of sameness and it lacks application. Is it enough to adopt the (type-)identity criteria of traditional game theory – same extensive form, same initial knowledge about the extensive form – so that any two PD_N's, say, count as the same "situation"? Not if we need what is not among those criteria: sameness of the inferential capacities of the agent. But adding this would be on the tight side: You in a PD_N and a perfectly rational agent in a PD_N would necessarily count as agents in different situations, and so MM would *never* help. Are there nongame criteria of identity, other than inferential powers, that should be adopted? I shall say that there are; I shall take the identity criteria of the "situation of an agent" to include, at least, both the traditional *differentiae* of games, and one more.

Let us say that, for an agent in a game G, an action is *expedient* if, of those available to her G, this action maximizes her payoff in G. Someone's being rational does not imply that she chooses an expedient action. It might, if her rationality guaranteed that she recognized expedient acts for what they were; but we cannot say this in general. An expedient act is one that it is *in fact* best to do. But determining which acts are expedient is often the crux of the problem faced by a deliberating agent; and then, which of these acts *are* constitutes the (burden of the) solution of her decision problem. It is clear that the validity of the advice to imitate an expert problem solver – the advice expressed in MM – is confined to cases in which the expert's problem has the same solution as yours. With all this in mind, I adopt for the situations of MM the following necessary

308

condition for (type-)identity: *Two agents are in the same situation only if they face the same game and have the same expedient actions.*

How is expediency of actions related to players' inferential powers? It is not hard to see that whether an action is expedient for a player depends on the inferential powers of her opponent.[30] It is much less obvious whether its expediency depends on her own powers. It may. Call a game *inferentially essential* if the set of acts expedient for a player varies with *that* player's inferential powers. If G is such a game then there are agents who are in different situations when they play the same game G, even if they play against identical opponents. I shall argue that for all we know the PD_N is such a game. Specifically, for all we know, if $N > h + 3$ then a human agent g in a PD_N and a perfectly rational agent i in a PD_N, both pitted against the same opponent, have different expedient actions. But then they are in different situations. Thus premise (iv) of the BIA lacks foundation, and the Mimetic Argument fails.

Write $PD_N(\tau, \infty)$ for the PD_N between 1 and 2 in which 1 and 2 have type-indexes $\tau \in \mathfrak{I}$ and ∞ respectively, and types recognize each other. Then we have the following proposition.

Proposition 3. *It is necessary and sufficient for showing that being nasty is expedient for 1 in a $PD_N(\tau, \infty)$ that $\tau \geq N - 2$.*

The argument for Proposition 3 falls into two parts. The first is the Backward Induction Argument,[31] an argument by mathematical induction. The inductive basis is just the claim that if a player knows she faces a simple Prisoner's Dilemma, she confesses. The inductive hypothesis is:

(IH) If $k_j^r E$ then j confesses from round $N - r + 1$ on
      $(j = 1, 2; r = 1, \ldots, N)$.

Here, $k_j^r E$ abbreviates "j has MK of degree r with i $(i = 1, 2; i \neq j)$ of the extensive form of the PD_N."[32] Putting $r = N$ and $j = 1$ in (IH), we have the conclusion of the BIA as it applies to $PD_N(\tau, \infty)$: *1 is nasty in $PD_N(\tau, \infty)$ if she has MK of degree N with 2 of the extensive form.*

It now follows (by the results of Section 3) that player 1 is nasty if $\tau \geq N - 1$. In particular, if 1 is an ideally rational player then, whatever the length N of the PD_N, 1 is nasty. We have confirmed premise (iii) of the Mimetic Argument if we interpret PD_N in (iii) as $PD_N(\tau, \infty)$; the BIA does show that a perfectly rational agent is nasty in a PD_N.

What does the BIA show about the *expediency* of nastiness for 1? It has this immediate corollary: *If $k_2^r E$ then confession at $N - r$ is expedient for 1.* For (IH) implies that if $k_2^r E$ then 2 confesses from $N - r + 1$ on, and does so independently of what 1 does at $N - r$. In particular, then, through this corollary the BIA shows that if $k_2^{N-1}E$ then being nasty is expedient for 1. Now I take it that we can show the expediency of nastiness for 1 *only* by appeal to this corollary of the BIA, and thus only by showing that $k_2^{N-1}E$. And this we can do, we have seen, only if $\tau \geq N-2$. This establishes the "only if" clause of Proposition 3. The "if" clause follows easily from the hypothesis that 2 has type-index ∞.[33]

Proposition 3 shows that PD_N's whose length exceeds the visibility bound of normal human players by more than 2 are inferentially essential (for all that we can show). Specifically, they are so in such a way that humans and paragons are in different situations when playing them, by our criteria of sameness. The Mimetic Maxim therefore *has no application* for humans in long PD_N's. However perfectly a human being may understand why a perfectly rational agent would act as he would in the very same game, this understanding is useless to her in deciding how to act. The paragon is no practical guide, but a distant star in the firmament of reason.[34]

The phenomenon of inferential essentiality is odd, but readily understandable in terms of our theory of acquisition of mutual knowledge. Being nasty is expedient for 1 only if the degree of 2's mutual knowledge with her of something is high enough. But it takes two to do this tango: The degree of 2's MK can be so high only if that of 1's MK is (almost) as high, and that is a matter of 1's inferential powers. More exactly, it is a matter of her epistemic type τ. For in her personal powers she may diverge from others of her type; she can do better than her kin – grasp proofs of degree τ plus one, or two, or even ten – if she takes lessons, or tries, or changes notation, or excels. But that would change nothing, for it would not change her type; it would be only a private quirk, devoid of public significance.

NOTES

1. The public occurrence is not the only scenario that has been suggested for the production of common knowledge. A closely related one is found in Roth and Murnighan's (1982) experimental setup. Each knower is given essentially the following self-referential message: "*p*, and I am giving the other subject this whole message." This scenario is a way of implementing directly Harman's (1977) self-referential definition of common knowledge. A rather different scenario posits an indefinite to-and-fro of messages, each telling of the delivery of the last, as in the Coordinated Attack Problem (see, e.g., Rubinstein 1989). But this setup does not present one aspect of the puzzle of common knowledge: how infinitary knowledge is apparently generated from finitary knowledge.
2. That is, knowing that A is and that B is and ..., where A, B, ... are the persons who are. We might have put "anyone" here rather than "everyone." The latter formulation is stronger, because a conjunction of knowings does not quite entail knowing the conjunction of the things known. But the extra strength

is negligible in a two-person analysis like ours, and it saves some irrelevant complications.

3. There is a close connection between p's satisfying this condition and its being jointly evident. Say 1 and 2 are present just if $p \wedge q$; for instance, p might be the proposition that there's a candle on the table, and q that 1 and 2 are sitting at that table. Then the condition is equivalent to: If q is true, then p is evident for both 1 and 2.

4. Since there are no quantifiers, this theory has only the axioms and inference rules of propositional calculus. The reason for using predicate and individual constants is mnemonic: We can write "$N2$" instead of an unstructured letter for the formula that will be interpreted as "2 is normal," and so forth.

5. I shall henceforth use the term "axiom" to include schemata.

6. T is described (in its alethic form) in Hughes and Cresswell (1968). Hintikka's (1962) system for knowledge consists of first-order predicate calculus plus the modal component of $S4$, which is just that of T with the additional axiom schema $\kappa_i P \rightarrow \kappa_i \kappa_i P$.

7. Similarly, the natural rationale for the unrestricted Rule of Epistemization is that being rational enables a knower to see epistemic-logical theorems generally.

8. Making Pi an axiom means that my theorems apply only to knowers who satisfy Z under its intended interpretation. I remark, however, that to carry through a Transparency Argument we don't need on-the-spot presence but only joint membership of any community small enough for people to know each other. This reinterpretation of Z gives (P1) a wide application. It also has the further effect of changing the class of A's for which (P2) is plausible: occasional events drop out, and less ephemeral facts that are significant for the community come in.

9. I shall stop citing PC when its role is obvious.

10. The belief in question is rational belief of a variety a little weaker than Hintikka's "defensible belief"; the latter is characterized by the epistemic logic $S4$ minus the factivity axiom.

11. In spite of the difficulties we have in carrying out the inferences described in the proof of Theorem 1, with a little mathematical training we have no trouble in understanding its proof. How is it that – though we are unable to emulate the inferences of the knowers described in it – we can grasp an argument that takes us through their steps? The first answer is that it redescribes those steps in terms of number notions, which are easier to grasp than raw strings of operators (furthermore, it "concertinas" them by adopting the quantificational device of mathematical induction). The second answer is that one can understand a proof, in a sense that suffices for knowing that it is valid, without grasping the meanings of its formulas; one need only see that its steps are substitution instances of valid rules, and this is a syntactic exercise. Similar points apply to the Backward Induction Argument which I consider in Section 4.

12. Rescher and Parks (1974) make an alternative suggestion about implementing Hintikka's proposal.

13. Eberle (1974) criticizes Hintikka's metalogical approach on the grounds that no fixed standard of obviousness can succeed in capturing the customary notion of knowledge, and instead introduces within the logic a binary operator with the interpretation "i infers • from •" axiomatized by appeal only to

the "very meaning" of "inferring." The approach of Kaneko and Nagashima (1988) has some features in common: the "investigator's" formal theory distinguishes i's inferred knowledge from her knowledge *tout court,* and credits the former only with Modus Ponens closure, withholding a Rule of Epistemization.

14. The epistemic degree of a formula, like modal degree of any kind, may be defined recursively: a formula is of degree 0 if it contains no operators, and of degree n if the greatest degree of the arguments of contained operators is $n-1$. Thus both $\kappa_1^3 p$ and $\kappa_1(p \to \kappa_2 \neg \kappa_1 q)$ are of degree 3. See for example Hughes and Cresswell (1968) for further details. I shall henceforth say that there is MK of degree 0 that P if P is true.

15. Theorem 3 will show that P is not a theorem of T_H. T_H is a subtheory of T_H^+, the consistent theory got by subjoining the postulates of T_H to T. Hence a theorem of T_H is verified by every Kripke T_H^+-model. Let $n > h+1$. It is easy to see that every Kripke T_H^+-model verifies the formula $H1 \wedge H2 \to \kappa_i^n A = P$, say (it suffices to note that P is derivable in T from the postulates of T_H). Hence $\neg$P is not a theorem of T_H either. The source of the incompleteness is that certain inferences, sound with respect to the Kripke semantics, are blocked by the restrictive clause in (R_H).

16. We need a semantics with respect to which T_H^+ is sound, and in which the formulas in question are not valid: If the former condition is met, then by finding counterworlds we show that the formulas are not derivable; but only if the latter is met is there hope of finding counterworlds. The Kripke semantics for T will not do because it fails the latter condition. An appropriate semantics is the impossible-world semantics of Rantala (1982), with respect to (appropriate versions of) which restricted inference modal systems like T_H^+ are both sound and complete.

17. For example, the finitely repeated Prisoner's Dilemma. The deeper player will be able to carry the standard backward induction argument back one round further than the shallower, and will thus gain the maximum payoff for one round. What this and other effects of differences in type τ imply about the best τ to *choose* (if one is about to play this game) is analyzed in Bacharach, Shin, and Williams (1991).

18. Schiffer emphasizes the first, but also indicates the second, of these tasks for the bringer-to-see. There is only the first task when the inference in question is easy and has simply not occurred to the knower; if it is difficult, there is also the second. My account of the Transparency Argument recognizes difficulty but not oversight, assuming in effect that the knower leaves no inferential stone unturned, even if sometimes fazed by what she finds. It might be possible to treat oversight formally by restricting the Rule of Epistemization to premises whose derivations are both easy and conspicuous; but providing a standard of conspicuousness would be a daunting task. In the present context the issue evaporates: To show that bringing-to-see is *not* adequate to cope with two difficulties, it's enough to show it inadequate to cope with one.

19. Theorem 2 assures us that bringing-to-see is not needed for there to be MK of degree up to $h+1$. A general statement of Proposition 1 requires n to exceed $h+1$.

20. Here, to avoid clutter I omit mention of being present.

21. An alternative way of introducing D is the simpler inference rule "if $\vdash$P then $\vdash Di \to \kappa_i P$." But by making the bringing-to-see facility a *deus ex machina* of unlimited epistemic bounty, this formulation cheapens the Bringing-to-

See Argument. And to no avail: For nth degree knowledge of A, $(n-2)$th knowledge of instruction remains necessary just the same. (R_H) is also preferable because it allows an illuminating comparison of T_H^* with T_I, as we shall see in a moment.

22. I shall not prove Proposition 2 here. The "if" part is straightforward; the "only if" part can be proved by the semantic method described previously.

23. Matters have improved very slightly – the attainable degree of MK has been raised by 2. Could it be further raised by repeating the maneuver? We may imagine publicly unfolding second-order tutorials in which 1 and 2 are instructed in the hard steps of the acquisition of MK of the first-order tutorials in the hard steps of the original argument; and so on, indefinitely. But this cannot be taken seriously. Very early in such a sequence, the evidentness postulate (P2) applied to rth order tutorials loses all plausibility.

24. The equivalence requires understanding of the announcer's utterance and justified reliance on its truth. These matters, the concern of the theories of meaning and of testimony, I must here leave untouched but for one remark. Mutual knowledge of the truth of P, the content of the announcement, will require MK not only of occurrence but also of understanding and reliance (or of the equivalence). This may be shown by the methods of Section 2 if it is assumed in our formal theory that normal knowers know that the announcer's utterance means that P, and that what the announcer says is to be believed; more peremptorily, we might assume that normal knowers know that $A \rightarrow P$.

25. The convincing solutions in question consist of actions that are dominant in a generalized sense (recursively dominant). There are much wider classes of solutions – Nash equilibria variably refined – that have been defended by appeal to common or higher mutual knowledge or expectation; in these defenses the object of the mutual expectation (or whatever) includes the *actions*. See for example Lewis (1969) and Tan and Werlang (1988). Bacharach (1992) discusses the general role of common belief assumptions in theories of solutions of games.

26. These include the Change-of-Notation Argument, which depends on supposing that knowers replace "chunks" of operator strings by new operators (e.g., replacing $K_1 K_2 K_1 K_2 K_1$ by "1 has 5-knowledge with 2 that"). This argument is directly analogous to the Bringing-to-See Argument. Another related argument is that of Clark and Marshall (1981), which is based on the knowers' merely *appreciating* that they could be brought to see; this is discussed in Bacharach (1989).

27. Luce and Raiffa (1957) apply this argument to the repeated Dilemma; Selten (1978) both to this and to the Chain Store game; another well-known application yields the Surprise Test paradox. Its key steps will be described presently.

28. On the other hand, many rational people who are convinced by the BIA would nevertheless opt for "nice." Selten calls the analogous clash of convictions for the Chain Store game the Chain Store Paradox. The arguments of this section may be regarded as one way of resolving the "Repeated Dilemma Paradox."

29. See the well-known paper of Kreps, Milgrom, Roberts, and Wilson (1982).

30. It was noted long ago that what is to be gained from a "maximin" strategy depends on the intelligence of the opponent; recently we have seen how, in dynamic games, best replies are sensitive to the possibility that the opponent is a preprogrammed automaton. See respectively von Neumann and Morgenstern (1944), and Kreps et al. (1982).

31. A fuller version of this argument is given in Bacharach, Shin, and Williams (1991). The argument may be given formally in a theory that unites T with a formal theory of rational choice in games (see Bacharach 1987). Here I give only a sketch of a formal argument of this sort.

32. The inductive step goes like this. Assume (IH) to be proved for r; then, by epistemic logic, 1 knows it to be true for r and for $j = 2$. Assume the antecedent of (IH) for $r + 1$ and $j = 1$ – that is that $k_1^{r+1}E$; then, by epistemic logic, 1 knows that 2 confesses from round $N - r + 1$ on, and does so independently of what she does; thus she knows confession to be expedient for her and (by her practical rationality) confesses at $N - r$. By epistemic logic, $k_1^{r+1}E$ also implies that $k_1^q E$ for $q = r, r - 1, ..., 2$, whence (similarly) that 1 confesses at $N - r + 1, N - r + 2, ..., N$.

I remark that, in a full version of the BIA, (IH) is shown conditionally on what amounts to iterated knowledge of the players' "rationality."

33. This follows because, in this case, 1 has MK of degree $\tau + 1$ with 2 of the extensive form.

34. However, the Mimetic Argument does accomplish something. Just as the Bringing-to-See Argument, while incapable of demolishing the barrier to MK created by humans' inferential limits, did dent it, so the Mimetic Argument dents the barrier to rational action created by these limits. It raises by 2 the length of repeated Dilemmas to which the BIA on its own provides clear rational solutions.

REFERENCES

Aumann, R. (1976), "Agreeing to Disagree." *Annals of Statistics* 4: 1236–9.
Bacharach, M. (1985), "Some Extensions of a Claim of Aumann in an Axiomatic Model of Knowledge." *Journal of Economic Theory* 37: 167–90.
Bacharach, M. (1987), "A Theory of Rational Decision in Games." *Erkenntnis* 27: 17–55.
Bacharach, M. (1989), "Mutual Knowledge and Human Reason." Presented at the Workshop on "Knowledge, Belief, and Strategic Interaction" (June 1989), Castiglioncello, Italy.
Bacharach, M. (1992), "Backward Induction and Beliefs about Oneself." *Synthese* 90.
Bacharach, M., Shin, H. S., and Williams, M. E. (1991), "Cooperation and the Depth of Players' Knowledge." Mimeo, Institute of Economics and Statistics, Oxford.
Clark, H., and Marshall, C. (1981), "Definite Reference and Mutual Knowledge." In A. K. Joshi, B. L. Webber, and I. A. Jay (eds.), *Elements of Discourse Understanding*. Cambridge: Cambridge University Press.
Eberle, R. A. (1974), "A Logic of Believing, Knowing and Inferring." *Synthese* 26: 356–82.
Harman, G. (1977), "Review of *Linguistic Behavior* by Jonathan Bennett." *Lanaguage* 53: 417–24.
Hintikka, J. (1962), *Knowledge and Belief*. Ithaca, NY: Cornell University Press.
Hintikka, J. (1970), "Knowledge, Belief, and Logical Consequence." *Ajatus* 32: 32–47.
Hughes, G. E., and Cresswell, M. J. (1968), *An Introduction to Modal Logic*. London: Methuen.
Kaneko, M., and Nagashima, T. (1988), "Players' Deductions and Deductive Knowledge, and Common Knowledge on Theorems." Mimeo, Department of Economics, Virginia Polytechnic Institute.
Kreps, D., Milgrom, P., Roberts, J., and Wilson, R. (1982), "Rational Cooperation in the Finitely Repeated Prisoner's Dilemma." *Journal of Economic Theory* 27: 244–52.
Lewis, D. K. (1969), *Convention: A Philosophical Study*. Cambridge, MA: Harvard University Press.

Luce, D., and Raiffa, H. (1957), *Games and Decisions*. New York: Wiley.

Monderer, D., and Samet, D. (1989), "Approximating Common Knowledge with Common Beliefs." *Games and Economic Behavior* 1: 170–90.

Rantala, V. (1982), "Impossible Worlds Semantics and Logical Omniscience." *Acta Philosophica Fennica* 35: 106–15.

Rescher, N., and Parks, Z. (1974), "Restricted Inference and Inferential Myopia in Epistemic Logic." In N. Rescher et al. (eds.), *Studies in Modality*, American Philosophical Quarterly Monograph No. 4. London: Basil Blackwell.

Roth, A., and Murnighan, J. K. (1982), "The Role of Information in Bargaining: An Experimental Study." *Econometrica* 50: 1123–42.

Rubinstein, A. (1989), "The Electronic Mail Game: Strategic Behaviour under 'Almost Common Knowledge'." *American Economic Review* 79: 385–91.

Schiffer, S. (1972), *Meaning*, Oxford: Oxford University Press.

Selten, R. (1978), "The Chain Store Paradox." *Theory and Decision* 9: 127–59.

Tan, T., and Werlang, S. (1988), "The Bayesian Foundations of Solution Concepts of Games." *Journal of Economic Theory* 45: 370–91.

Vardi, M. (1986), "On Epistemic Logic and Logical Omniscience." In J. Y. Halpern (ed.), *Theoretical Aspects of Reasoning about Knowledge: Proceedings of the 1986 Conference*. San Mateo, CA: Morgan Kaufman.

Von Neumann, J. (1958), *The Computer and the Brain*. New Haven, CT: Yale University Press.

Von Neumann, J., and Morgenstern, O. (1944), *Theory of Games and Economic Behavior*. Princeton, NJ: Princeton University Press.

315

18

The electronic mail game: Strategic behavior under "almost common knowledge"

ARIEL RUBINSTEIN

A very basic assumption in all studies of game theory is that the game is "common knowledge." Following John Harsanyi (1967), situations without common knowledge are analyzed by a game with incomplete information. A player's information is characterized by his "type." Each player "knows" his own type and the prior distribution of the types is common knowledge. Jean-Francois Mertens and Samuel Zamir (1985) have shown that under quite general conditions one can find type spaces large enough to carry out Harsanyi's program and to transform a situation without common knowledge into a game with incomplete information in which the different types may have different states of knowledge. Harsanyi's method became the cornerstone of all modern analyses of strategic economic behavior in situations with asymmetric information (i.e., most of the theoretical Industrial Organization literature).

What does it mean that the game G is "common knowledge"? Following David Lewis (1969), Stephen Schiffer (1972), and Robert Aumann (1976), this concept has been studied thoroughly by relating it to concepts of "knowledge" and "probability" (for a recent presentation of this literature see Ken Binmore and Adam Brandenberger, 1987). Intuitively speaking, it is common knowledge between two players 1 and 2 that the played game is G, if both know that the game is G, 1 knows that 2 knows that the game is G and 2 knows that 1 knows that the game is G, 1 knows that 2 knows that 1 knows that the game is G, and 2 knows that 1 knows that 2 knows that the game is G, and so on and so on.

Department of Economics, Tel Aviv University, Tel Aviv, Israel. My thanks to Ken Binmore, Edi Dekel, John Geanakopolos, Avner Shaked, Chuck Wilson, Asher Wolinsky, and a referee of *The American Economic Review* for the very useful comments. Reprinted from Ariel Rubinstein, "The electronic mail game: Strategic behavior under 'almost common knowledge'," *The American Economic Review*, June 1989, vol. 79, no. 3, pp. 385–91, by permission of the American Economic Association.

One of the main difficulties with this intuitive definition (and with the formal definitions which capture this perception) is that even "simple" sentences like "I do not know that you do not know that I know that you do not know that I know" are very difficult to visualize, thus making an assessment of their validity problematic. Therefore it would be interesting to understand whether a game-theoretic informational structure, referred to as "almost common knowledge," in which only a finite (but large) number of propositions of the type "1 knows that 2 knows that 1 knows . . . that the game is G" are true, is very different from the situation where the game G is common knowledge. In this short paper I will present a simple example of a situation with "almost common knowledge" of the game. The situation is analyzed using, as a tool, the idea of a game with incomplete information. It is shown that the game-theoretic "prediction" for the "almost common knowledge" situation is very different from the situation with common knowledge.

The example is similar to the "coordinated attack problem" which is well known in the distributed systems literature.[1] A description of the problem and a comparison with this paper analyzed appears in Section IV.

I. Coordination through Electronic Mail

Two players, 1 and 2, are involved in a coordination problem. Each has to choose between two actions A and B. There are two possible states of nature, a and b. Each of the states is associated with a payoff matrix as follows:

The game G_a

	A	B
A	M, M	$0, -L$
B	$-L, 0$	$0, 0$

state a
probability $1 - p$

The game G_b

	A	B
A	$0, 0$	$0, -L$
B	$-L, 0$	M, M

state b
probability p

In the state of nature a (b) the players get a positive payoff, M, if both choose the action A (B). If they choose the same action but it is the

318

"wrong" one they get 0. If they fail to coordinate, then the player who played B gets $-L$, where $L > M$. Thus, it is dangerous for a player to play B unless he is confident enough that his partner is going to play B as well. The state a is the more likely event; b appears with a priori probability of $p < \frac{1}{2}$.

The information about the state of nature is known initially only to player 1. Without transferring the information, the players cannot achieve an expected payoff higher than $(1-p)M$. If the information could become common knowledge they would be able to achieve the payoff M. However, imagine that the two players are located at two different sites and they communicate only by electronic mail signals. Due to "technical difficulties" there is a "small" probability, $\epsilon > 0$, that the message does not arrive at its destination. At the risk of creating discord, the electronic mail network is set up to send a confirmation *automatically* if any message is received, including not only the confirmation of the initial message but a confirmation of the confirmation; and so on. To be more precise, it is assumed that, when player 1 gets the information that the state of nature is b, his computer automatically sends a message (a blip) to player 2 and then player 2's computer confirms the message and then player 1's computer confirms the confirmation and so on. If a message does not arrive, then the communication stops. No message is sent if the state of nature is a. At the end of the communication phase the screen displays to the player the number of messages his machine has sent. Let T_i be a variable for the number of messages i's computer sent (the number on i's screen).

Notice that sending the messages is not a strategic decision by the players. It is an automatic device carried out by the computers. The designer of the system sets up the communication network between the players and they can choose between A and B only after the communication phase has ended.

If the two machines exchange an infinite number of messages, then we may say that the two players have common knowledge that the game is G_b. However, since only a finite number of messages are transferred, the players never have common knowledge that the game they play is G_b.

In choosing between A and B after the end of the communication phase, player 1 (and similarly player 2) faces uncertainty: Given that he sent T_1 messages he does not know whether player 2 did not get the T_1th message, or whether player 2 got the T_1th message, but the T_1th confirmation has been lost. Any number on the screen corresponds to a state of knowledge not only about the state of nature but also about the other player's knowledge. For example if player 1's computer sent two messages it means that:

319

$K_1(b) -$ 1 knows that b
$K_1 K_2(b) -$ 1 knows that 2 knows that b

(by the fact that he has received confirmation of his first message). However, it is not true that $K_1 K_2 K_1 K_2(b)$ – 1 does not know that 2 knows that 1 knows that 2 knows that b. Player 1 assigns probability $z = \epsilon/[\epsilon + (1-\epsilon)\epsilon]$ to $T_2 = 1$ and $(1-z)$ to $T_2 = 2$. Therefore player 1 believes that:

with probability $1 - z$ $K_2 K_1 K_2(b)$ and
with probability z that
     2 believes that
with probability $1 - z$ $K_1 K_2(b)$ and
with probability z that
     1 believes that

with probability z 2 believes that with probability $(1-p)/(1-p\epsilon)$, a, and with probability $(1-z)$, 2 knows that b.

     The statements of higher order are even more complicated. Notice that, under the model's assumption that player 1 gets accurate information about the state of nature, "x" and "$K_1(x)$" are two equivalent statements.

     Similarly, any number on a player's screen at the end of the communication stage corresponds to a sequence of propositions describing the player's knowledge about the state of nature, about his opponent's belief about the state of nature, about his opponents's belief about his belief about the opponent's belief about the state of nature, and so on. The larger is T_1, the more statements of the type $K_1 K_2 K_1 \ldots K_1 K_2(b)$ are true, and the closer we are to the common knowledge situation.

     How could we analyze the situation when the two players have the numbers T_1 and T_2 on their screens? To calculate his best action when $T_1 = 2$, for example, player 1 may have to form beliefs about player 2's actions when T_2 is 1 or 2. The optimality of these would have to be checked given player 1's behavior when $T_1 = 1$, 2, or 3, and so on. Harsanyi's method suggests that we analyze a situation given any pair of numbers on the screens, as part of a game of incomplete information which I will refer to as "the electronic mail game" (to distinguish from the coordination games). The set of types in the electronic mail game is the set of natural numbers and the distribution of the pairs of types is deduced from the electronic mail technology (namely, the probability of (T_1, T_2) being respectively $(0, 0)$, $(n+1, n)$, and $(n+1, n+1)$ are $1-p$, $p\epsilon(1-\epsilon)^{2n}$, and $p\epsilon(1-\epsilon)^{2n+1}$, respectively). Define player i's strategy in the electronic mail game, S_i, to be a function from the set of natural numbers $0, 1, 2, \ldots$ into the action space $\{A, B\}$. Then $S_i(t)$ is interpreted as i's action if his machine sent t messages.

II. The Analysis of the Electronic Mail Game

Proposition 1. *There is only one Nash equilibrium in which player 1 plays A in the state of nature a. In this equilibrium the players play A independently of the number of messages sent.*

Proof. Let (S_1, S_2) be a Nash equilibrium such that $S_1(0) = A$. We will prove by induction that $S_1(t) = S_2(t) = A$ for all t. If $T_2 = 0$ then player 2 did not get a message. He knows that it might be because player 1 did not send him a message (this could occur with probability $1 - p$) or because a message was sent but did not arrive (this happens with probability $p\epsilon$). In the first case, player 1 plays A ($S_1(0) = A$). If player 2 plays A, then, whatever $S_1(1)$ is, player 2's expected payoff is at least $[(1-p)M + p\epsilon 0]/[(1-p)+p\epsilon]$ and if he plays B he gets at most $[-L(1-p) + p\epsilon M]/[(1-p)+p\epsilon]$. Therefore it is strictly optimal for 2 to play A, that is $S_2(0) = A$.

Assume now that we have shown that, for all $T_i < t$, players 1 and 2 play A in equilibrium. Assume $T_1 = t$. Player 1 is uncertain whether $T_2 = t$ (in the case where player 2 received the tth message but 2's tth message was lost) or $T_2 = t - 1$ (in the case where 2 did not receive the tth message). Given that he did not receive confirmation of his tth message, his conditional probability that $T_2 = t - 1$ is $z = \epsilon/[\epsilon + (1-\epsilon)\epsilon] > \frac{1}{2}$. Thus it is more likely that player 1's last message did not arrive than that player 2 got the message. (This fact is the key to our argument.) By the inductive assumption, player 1 assesses that, if $T_2 = t - 1$, player 2 will play A. If player 1 chooses B, player 1's expected payoff is at most $z(-L) + (1-z)M$. If he chooses A, then his utility is 0. Given that $L > M$ and since $z > \frac{1}{2}$, his only best action must be A. Thus $S_1(t) = A$. Similarly we show that $S_2(t) = A$. $\square$

Thus even if both players know that the actual played coordination game is G_b and even if the noise in the network (the probability ϵ) is arbitrarily small, the players ignore the information and play A. The best expected payoff the players can obtain in any equilibrium is still $(1-p)M$, just as if no electronic mail system existed!

Remark 1. Consider the mechanism described above but with the addition that, after a commonly known fixed finite number of messages, T, the system stops, if it has not stopped before. If $\epsilon(-L) + (1-\epsilon)M > 0$ then there is an equilibrium in which each player plays B if he receives confirmations of all his messages. The expected payoffs of this equilibrium, conditional on the state b are: $(1-\epsilon)^T M$ to the last player who is supposed to get a message and $(1-\epsilon)^{T-1}[\epsilon(-L) + (1-\epsilon)M]$ to the other player.

321

Notice that these two numbers are decreasing in T and therefore the only "efficient" schemes might be those with $T = 1$ and $T = 2$. The mechanism with $T = 1$ is a better scheme for player 2 and $T = 2$ is a better scheme for player 1. If the communication channel is so noisy that $\epsilon(-L) + (1 - \epsilon)M < 0$ then the efficient equilibrium is the one where the messages are ignored (the argument is similar to the proof of the proposition).

III. The coordinated attack problem

As was mentioned in the introduction the electronic mail game is strongly related to the coordinated attack problem known in the distributed systems folklore. The problem as described in Joseph Halpern (1986, p. 10) is the following:

Two divisions of an army are camped on two hilltops overlooking a common valley. In the valley awaits the enemy. It is clear that if both divisions attack the enemy simultaneously they will win a battle, whereas if only one division attacks it will be defeated. The divisions do not initially have plans for launching an attack on the enemy, and the commanding general of the first division wishes to coordinate a simultaneous attack (at some time the next day). Neither general will decide to attack unless he is sure that the other will attack with him. The generals can only communicate by means of a messenger. Normally, it takes the messenger one hour to get from one encampment to the other. However, it is possible that he will get lost in the dark or, worst yet, be captured by the enemy. Fortunately, on this particular night, everything goes smoothly. How long will it take them to coordinate an attack?

Suppose the messenger sent by general 1 makes it to general 2 with a message saying "Let's attack at dawn." Will general 2 attack? Of course not, since general 1 does not know he got the message, and thus may not attack. So general 2 sends the messenger back with an acknowledgment. Suppose the messenger makes it. Will general 1 attack? No, because now general 2 does not know he got the message, so he thinks general 1 may think that he (general 2) didn't get the original message, and thus not attack. So general 1 sends the messenger back with an acknowledgment. But of course, this is not enough either. I will leave it to the reader to convince himself that no amount of acknowledgments sent back and forth ever guarantee agreement. Note that this is true if the messenger succeeds in delivering the message every time.

The question asked in the quoted paragraph is whether there is a common knowledge of the attack plan at the end of the information transmission stage. The above "communication protocol" cannot result in the players' having common knowledge about the time of the attack. However, the fact that the generals could not achieve common knowledge does not exclude the possibility that with positive probability they will both attack at dawn. This sounds plausible especially if the probability of a messenger failure is very small.

For this reason it is interesting to analyze the problem in the explicit form of a game. This is the minor contribution of this paper. In order to address the problem as a game, we need to add more structure to the problem and, in particular, we have to specify the probability conditions under which general 1 decides to initiate an attack at dawn. In terms of Section II, state b can be interpreted as the conditions which make an attack at dawn likely to succeed, while state a is the "status quo" state. Action B is "attack at dawn" and action A is the default action. The payoffs in Section I represent an assumption that, in case of an uncoordinated attack, only the general who attacks loses. If, alternatively, we assume that both generals' utilities are $-L$ if an uncoordinated attack is launched, then there is an equilibrium in which general 2 attacks as soon as he gets at least one message, provided that ϵ is small enough (less than $M/(M+L)$). This last fact emphasizes the importance of addressing the problem within a game-theoretic framework.

IV. Final comments

A. Is "almost common knowledge" close to "common knowledge"?

It should be emphasized that the game about which knowledge is being hypothesized in the above is the coordination game and not the electronic mail game. One is concerned with what the two players do or do not know about the payoffs in the coordination game and with what the players do or do not know about the knowledge of their opponent. The story of the interchange of messages by electronic mail is intended only to provide a precise, albeit rather special, model of how knowledge on those questions may come to be shared by the players.

The main message of this paper is that players' strategic behavior under "almost common knowledge" may be very different from that under common knowledge. To emphasize, by "almost common knowledge" I refer to the case when the numbers on the screens are "very large." Then a "very large" number of statements of the type "player i knows that player j knows that . . . the coordination game is G_b" are correct. Still, the players will not coordinate on the action B whereas they are able to coordinate on the action B if it is common knowledge that the coordination game is G_b.

B. The electronic mail game as a perturbed game

Selten's perfection definitions and the Kreps–Milgrom–Roberts–Wilson (1982) approach used small perturbations in a game in order to select an equilibrium in a game with multiplicity of equilibria and to create new equilibria in the absence of a reasonable equilibrium. If we think of ϵ as being small then the noisy electronic mail game is a perturbation of a nonnoisy electronic mail game (the electronic mail game $\epsilon = 0$). The non-noisy game has several equilibria (since it is just a coordination problem); however the perturbation unfortunately excludes the more reasonable equilibria. Notice that the difference between a game and a perturbed version of the game has already been demonstrated many times in the past and I feel less paradoxical about this as compared to the paradoxical features of the present example.

C. The paradoxical aspect of the example

What would *you* do if the number on your screen is 17? It is hard to imagine that when L is slightly above M and ϵ is small a player will not play B. The sharp contrast between our intuition and the game-theoretic analysis is what makes this example paradoxical.

The example joins a long list of games such as the finitely repeated Prisoner's Dilemma, the chain store paradox, and Rosenthal's game, in which it seems that the source of the discrepancy is rooted in the fact that in our formal analysis we use mathematical induction while human beings do not use mathematical induction when reasoning. Systematic explanation of our intuition that we will play B when the number on our screen is 17 (ignoring the inductive consideration contained within Proposition 1's proof) is definitely a most intriguing question.

D. Games with incomplete information

As mentioned earlier the situation without common knowledge is analyzed, à la Harsanyi, as a game with incomplete information. Notice that almost all the nonabstract literature uses the distinction between types to reflect differences in knowledge about payoff-relevant items. The current example is exceptional in that it demonstrates a family of natural game-theoretic scenarios in which the main difference between the types is in their knowledge about other players' knowledge.

E. A formal presentation of the type spaces and the information partitions[2]

Those readers who are familiar with Aumann (1976) may find it helpful to have a formal statement of the type spaces and the information partitions in the electronic mail game. The type spaces of the two players are

324

the sets which include $(a, 0, 0)$ and the triples (b, t, t') where $t > 0$ and t' is either t or $t - 1$. Array the set in the following order:

$$(a, 0, 0)(b, 1, 0)(b, 1, 1)(b, 2, 1)(b, 2, 2)(b, 3, 2)(b, 3, 3) \dots .$$

Player 1's information partition is:

$$\{(a, 0, 0)\}\{(b, 1, 0), (b, 1, 1)\}\{(b, 2, 1), (b, 2, 2)\}\{(b, 3, 2), (b, 3, 3)\} \dots$$

and player 2's information partition is:

$$\{(a, 0, 0), (b, 1, 0)\}\{(b, 1, 1), (b, 2, 1)\}\{(b, 2, 2), (b, 3, 2)\}\{(b, 3, 3) \dots .$$

The meet of the two partitions is the trivial partition which contains only the entire type space. Thus the event "b" consists of the entire type space with the exception of $(a, 0, 0)$ and is never common knowledge. Notice that when $\epsilon = 0$, the feasible states are just $(a, 0, 0)$ and (b, ∞, ∞).

F. Topology

Two of the readers of the first version of this paper, both experts in the literature on common knowledge, raised objections to the way I use the term "almost common knowledge." They based their objection on the fact that when $\epsilon \to 0$ the information partitions of the players do not converge to the information partitions when $\epsilon = 0$ (see this section, Part E). A referee suggested several topologies in which alternative concepts of "almost common knowledge" make sense.

Before reacting to this criticism let me emphasize again that I use the term "almost common knowledge" not for stating that the electronic mail game with ϵ close to 0 is almost the game with $\epsilon = 0$. What I am saying is that the situation with a high T_1 is close to the common knowledge situation. However, I would like to use this objection to spell out my opinion on the role that topology (in common with most other fields of "fancy mathematics") should play in economic theory. Topology should be used in one of two ways: (1) as technical tool for phrasing a meta-claim about a family of models, or (2) as a substantial tool to formalize natural intuitions about "closeness." I envisage the high T_1 situation as being close to the common knowledge situation in the sense of (2). This may be unhelpful from a technical point of view and a conclusion from the example is indeed that the Nash equilibrium is not upper hemicontinuous in this convergence. However, lack of technical usefulness is not an argument against the perception that a situation with high T_1 is close to a situation with common knowledge. Obviously other definitions of convergence may be useful not only as technical methods but also for expressing other intuitions of closeness.

1. I should like to thank John Geanakopolos for referring me to the "coordinated attack problem."
2. In this section I am closely following a referee's suggestion.

REFERENCES

Aumann, Robert J. (1976), "Agreeing to Disagree." *Annals of Statistics* 4: 1236–9.

Binmore, Kenneth, and Brandenberger, Adam (1987), "Common Knowledge and Game Theory." Discussion Paper No. TE/88/167, STICERD, London School of Economics.

Halpern, Joseph Y. (1986), "Reasoning about Knowledge: An Overview." In J. Y. Halpern (ed.), *Reasoning about Knowledge*. San Mateo, CA: Morgan Kaufmann, pp. 1–18.

Harsanyi, J. C. (1967), "Games with Incomplete Information Played by Bayesean Players," Parts I, II, III. *Management Science* 14: 159–82, 320–34, 486–502.

Kreps, D., Milgrom, P., Roberts, J., and Wilson, R. (1982), "Rational Cooperation in the Finitely Repeated Prisoner's Dilemma." *Journal of Economic Theory* 27: 245–52.

Lewis, David (1969), *Convention, A Philosophical Study*. Cambridge, MA: Harvard University Press.

Mertens, Jean-Francois, and Zamir, Samuel (1985), "Foundation of Bayesian Analysis for Games with Incomplete Information." *International Journal of Game Theory* 14: 1–29.

Schiffer, Stephen R. (1972), *Meaning*. Oxford: Oxford University Press.

19

Knowledge-dependent games: Backward induction

CRISTINA BICCHIERI

1. INFORMATION AND META-INFORMATION

Although notions of rationality have been extensively discussed in game theory, the epistemic conditions under which a game is played – though implicitly presumed – have seldom been explicitly analyzed and formalized. These conditions include the players' reasoning processes and capabilities, as well as their knowledge of the game situation. Some aspects of information about chance moves and other players' moves are represented by information partitions in extensive-form games. But a player's knowledge of the structure of information partitions themselves is different from his information about chance moves and other players' moves. The informational aspects captured by the extensive form have nothing to do with a player's knowledge of the structure of the game.

A common epistemic presumption is that the structure of the game is *common knowledge* among the players. By "common knowledge of *p*" is meant that *p* is not just known by all the players in a game, but is also known to be known, known to be known to be known, . . . ad infinitum. The very idea of a Nash equilibrium is grounded on the assumptions that players have common knowledge of the structure of the game and of their respective priors. These assumptions, however, are always made outside the theory of the game, in that the formal description of the game does not include them.[1]

Statements about players' rationality, the specification of the structure of the game, and players' knowledge of all of them should be part of the theory of the game. Recent attempts to formalize players' knowledge as part of a theory of the game include Bacharach (1985), Brandenburger and Dekel (1985a,b), Mertens and Zamir (1985), Gilboa (1986), Bacharach (1987), Kaneko (1987), and Samet (1987). In these works, a common-knowledge axiom is explicitly introduced, stating that the axioms of logic, the axioms of game theory, the behavioral axioms, and the structure of the game are all common knowledge among the players.

Common knowledge of the theory of the game is not always necessary for a solution to be inferred. Different solution concepts may require different amounts of knowledge on the part of the players to have predictive validity at all. For example, while common knowledge is necessary to attain an equilibrium in a large class of normal form games, it may lead to inconsistencies in finite, extensive-form games of perfect information (Reny 1987, Bicchieri 1989).[2]

In this chapter, only this latter class of games is considered. These games are solved by working backward from the end, and this procedure yields a unique solution. It is commonly assumed that backward induction is supported by common knowledge of rationality (and of the structure of the game). In Section 2 it is proved instead that much less than common knowledge need be assumed. In fact, only a finite number of levels of knowledge of the theory of the game (hence, of players' rationality) is needed to infer the backward induction solution. That limited knowledge is sufficient to infer a solution for this class of games is not a striking result. More interesting is the fact that it is also a necessary condition. In Section 3, the concepts of knowledge-dependent games and knowledge-consistent play are introduced, and it is proved that knowledge must be limited for a solution to obtain. More specifically, it is proved that backward induction equilibria are knowledge-consistent plays of knowledge-dependent games. Conversely, every knowledge-consistent play of a knowledge-dependent game is a backward induction equilibrium. For the class of games considered, there exist knowledge-dependent games that have no knowledge-consistent play. For example, a player might be unable – given what she knows – to "explain away" a deviation from equilibrium on the part of another player, in that reaching her information set is inconsistent with what she knows. In this case, predictability is lost.

If the theory of the game were to include the assumption that every information set has a small probability of being reached (because a player can always make a mistake), then no inconsistency would arise. In this case, the solution concept is that of "perfect equilibrium" (Selten 1975), which requires an equilibrium to be stable with respect to "small" deviations. The idea of perfect equilibrium (like other refinements of Nash equilibrium) has the defect of being ad hoc, as well as of assuming less than perfect rationality.

This chapter has a different goal. It seeks to establish the epistemic conditions under which a unique prediction about the outcome of the game can be derived from a rationality axiom. Because the players (as well as the game theorist) must reason to an equilibrium, the theory of the game must contain a number of meta-axioms stating that the axioms of the theory are known to the players. The interesting result is that the theory of the game T can contain a meta-axiom A_n stating that the set of game-theoretic axioms A_1–A_{n-1} is k-level group knowledge among the players, but not a meta-axiom A_{n+1} saying that A_n is group knowledge among the players. If A_{n+1} is added to T, it become group knowledge that the theory is inconsistent at some information set. In this case, the backward induction solution cannot be inferred.

2. BACKWARD INDUCTION EQUILIBRIUM

In this section noncooperative, extensive-form games of perfect information are defined, and it is proved that the number of levels of knowledge needed to infer the backward induction equilibrium is finite.

Definition 2.1. A noncooperative game is a game in which no precommitments or binding agreements are possible.

Definition 2.2. A finite n-person game G of perfect information in extensive form consists of the following elements:
  (i) A set $N = \{1, 2, \ldots, n\}$ of players.
 (ii) A finite tree (a connected graph with no cycles) T, called the game tree.
(iii) A node of the tree (the root), called the first move. A node of degree 1 and different from the root is called a terminal node. The set of all terminal nodes is denoted W.
 (iv) A partition $P^1, \ldots, P^n$ of the set of nonterminal nodes of the tree, called the player partition. The nodes in P^i are the moves of player i. The union of $P^1, \ldots, P^n$ is the set of moves for the game.
  (v) For each $i \in N$, a partition $I^{i1}, \ldots, I^{ik}$ of P^i (I^{ij} denotes the jth information set ($j \geq 1$) of player i) such that, for each $j \in \{1, \ldots, k\}$:
      (a) each path from the root to a terminal node can cross I^{ij} at most once; and
      (b) since there is perfect information, I^{ij} is a singleton set for every i and j.
 (vi) For each terminal node t, an n-dimensional vector of real numbers, $f^1(t), \ldots, f^n(t)$, called the payoff vector for t.
Every player in G knows (i)–(vi).

Definition 2.3. A *pure strategy* s^i for player i is a k-tuple that specifies, for each information set of player i, a choice at that information set. The set of i's pure strategies is denoted by $S^i = \{s^i\}$. Let $\mathbf{S} = S^1 \times \cdots \times S^n$.

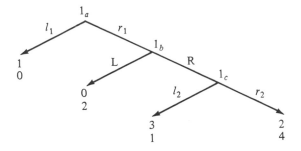

Figure 1. Game G^1.

Definition 2.4. The function $\pi^i\colon S^1 \times \cdots \times S^n \to \Re$ is called the *payoff function* of player i. For an n-tuple of pure strategies, $s = (s^1, \ldots, s^n) \in \mathbf{S}$, the expected payoff to player i, $\pi^i(s)$, is defined by

$$\pi^i(s) = \sum_{t \in \Omega} p_s(t)\,\pi^i(t),$$

where $p_s(t)$ is the probability that a play of the game ends at the terminal node t when the players use strategies $s^1, \ldots, s^n$.

Definition 2.5. A pure strategy n-tuple $s = (s^1, \ldots, s^n) \in \mathbf{S}$ is an *equilibrium point* for Γ if

$$\pi^i(s \,|\, y^i) \leq \pi^i(s) \quad \text{for all } y^i \in S^i,$$

where $s \,|\, y^i = (s^1, \ldots, s^{i-1}, y^i, s^{i+1}, \ldots, s^n)$.

We also say that $s^i \in S^i$ is a *best reply* of player i against s if

$$\pi^i(s \,|\, s^i) = \max_{y^i \in S^i} \pi^i(s \,|\, y^i).$$

Definition 2.6. A subgame $\Gamma_j \in \Gamma$ is a collection of branches of the game that start from the same node, where the branches and node together form a game tree by itself.

Theorem 2.1 (Kuhn 1953). *A game G of perfect information has an equilibrium point in pure strategies.*

The equilibrium can be found by working backward from the terminal nodes to the root. At each information set, a player chooses the branch that leads to the subtree yielding him the highest equilibrium payoff. To illustrate this method, consider the following two-person, extensive-form game of perfect information with finite termination (see Figure 1). In

330

game G^1, $N = \{1, 2\}$. The game starts with player 1 moving first at a. The union $P^1 \cup P^2$ is the set of moves $\{a, b, c\}$. $P^1 = \{a, c\}$, $P^2 = \{b\}$, $I^{11} = \{a\}$, $I^{21} = \{b\}$, and $I^{12} = \{c\}$. Each player has two pure strategies: either to play left – thus ending the game – or to play right, in which case it is the other player's turn to choose. $S^1 = \{l_1 l_2, l_1 r_2, r_1 r_2, r_1 l_2\}$; $S^2 = \{L, R\}$. The payoffs to the players are represented at the endpoints of the tree, the upper number being the payoff of player 1, and each player is assumed to be rational (i.e., to wish to maximize his expected payoff).

The equilibrium is obtained by backward induction as follows: At node I^{12}, player 1 (if rational) will play l_2, which grants him a maximum payoff of 3. Note that player 1 need not assume 2's rationality in order to make his choice, since what happened before the last node is irrelevant to his decision. Thus, node I^{12} can be replaced by the payoff pair $(3, 1)$. At I^{21}, player 2 (if rational) will only need to know that 1 is rational in order to choose L. That is, player 2 need consider only what she expects to happen at subsequent nodes (i.e., the last node) as, again, the preceding part of the tree is now strategically irrelevant. The penultimate node can thus be replaced by the payoff pair $(0, 2)$. At node I^{11}, rational player 1 – in order to choose l_1 – will have to know that 2 is rational *and* that 2 knows that 1 is rational (otherwise, he would not be sure that at I^{21} player 2 will play L). From right to left, nonoptimal actions are successively deleted, and the conclusion is that player 1 should play l_1 at his first node. Thus $\bar{s}^1(I^{11}) = l_1$, $\bar{s}^2(I^{21}) = L$, $\bar{s}^1(I^{12}) = l_2$, and $(p^1(\bar{s}), p^2(\bar{s})) = (1, 0)$.

In the classical account of this game, $(l_1 L l_2)$ represents the only possible pattern of play by rational players because the game is one of *complete information;* that is, the players know each other's rationality, strategies, and payoffs. Player 1, at his first node, has two possible choices: l_1 or r_1. What he chooses depends on what he expects player 2 to do afterward. If he expects player 2 to play L at the second node, then it is rational for him to play l_1 at the first node; otherwise he may play r_1. His conjecture about player 2's choice at the second node is based on what he thinks player 2 believes would happen if she played R. Player 2, in turn, must conjecture what player 1 would do at the third node, given that she played R. Indeed, both players must conjecture each other's conjectures and choices at each possible node, until the end of the game.

In our example, complete information translates into the conjectures $p(l_1) = 1$, $p(R) = 0$, and $p(r_2) = 0$. The notion of complete information does not specify any particular number of levels of knowledge on the part of the players, but it is customarily assumed that the structure of the game and players' rationality are common knowledge among them.

Note, again, that specification of the solution requires a description of what both agents expect to happen at each node, were it to be reached,

331

even though in equilibrium play no node after the first is ever reached. The central idea is that if a player's strategy is to be part of a rational solution, then it must prescribe a rational choice of action in all conceivable circumstances – even those that are ruled out by some putative equilibrium. An equilibrium is thus endogenously determined by considering the implications of deviating from the specified behavior. The backward induction requirement calls for considering equilibrium points that are in equilibrium in each of the subgames and in the game considered as a whole. This means that it only matters where you are, not how you arrived there, as history of past play has no influence on what individuals do.

Because a strategy specifies what a player should choose in every possible contingency (i.e., at all information sets at which he may find himself), and because a player's contingency plan ought to be rational in the contingency for which it was designed, it is necessary to give meaning to the idea of a choice that is conditional upon a given information set having being reached. What counts as rational behavior at information sets not reached by the equilibrium path depends on how a player explains the fact that a given information set is reached, since different explanations elicit different choices. For example, it has been argued that at I^{21} it is not evident that player 2 will only consider what comes next in the game (Binmore 1987, Reny 1987). Reaching I^{21} may not be compatible with backward induction, since I^{21} can be reached only if 1 deviates from his equilibrium strategy, and this deviation stands in need of explanation. When player 1 considers what player 2 would choose at I^{21}, he must have an opinion as to what sort of explanation 2 is likely to give for being called to play, since 2's subsequent action depends on it. Binmore's criticism rightly points out that a solution must be stable with respect to forward induction also. In other words, if equilibrium behavior is determined by behavior off the equilibrium path, then a solution concept must allow the players to "explain away" deviations.

Selten's "trembling-hand" model (1975) provides the canonical answer. Whenever a player wants to make some move a, he will have a small positive probability ϵ of instead making a different and unintended move $b \neq a$ by "mistake." If any move can be made with a positive probability then all information sets have a positive probability of being reached. The relation of Selten's theory of mistakes to backward induction is straightforward. Since the backward induction argument relies on the notion of players' rationality, one must show that rationality and mistakes are compatible. Admitting that mistakes can occur means drawing a distinction between deciding and acting, but a theory that wants to maintain a rationality assumption is bound to make mistakes entirely random and uncorrelated. Systematic mistakes would be at odds with rationality, because one would expect a rational player to learn from past actions and so modify his behavior. If a deviation tells that a player made a mistake (i.e., his hand "trembled"), but not that he is irrational, then a mistake must not be the product of a systematic bias in favor of a particular type of action, as would be the case with a defective reasoning process.

In our example, when player 2 finds she has to move, she will interpret 1's deviation as the result of an unintended, random mistake. So if 1 plays (but did not mean to play) r_1, 2 knows that the probability of r_2 being successively played remains vanishingly small: $p(r_2) = p(r_2 | r_1) = \epsilon$. This makes 2 choose strategy L, which is a best reply to player 1's strategy after allowing for the possibility of trembles. Player 1 knows that, were he to play r_1, player 2 would respond with L, and that there is only a vanishingly small probability that R is played instead. For $p(R) = \epsilon$, player 1's best reply is l_1. Thus $(l_1 L l_2)$ remains an equilibrium in the new perturbed game that differs from the original one in that any move can be made with a small positive probability.

According to Binmore (1987, 1988), this characterization of mistakes is necessary for the backward induction argument to work, in that it makes out-of-equilibrium behavior compatible with players' rationality. Otherwise, Binmore argues, a deviation would have to be interpreted as proof of a player's "irrationality." This conclusion is warranted if one assumes common knowledge of rationality; then it must be explained how a player, facing a deviation, can maintain without contradiction that the deviator is rational. In what follows, it is proved that the number of levels of knowledge needed for the backward induction solution to obtain is finite, and that this number depends on the length of the game. A consequence of weakening the common-knowledge assumption is that out-of-equilibrium behavior is made compatible with rationality, without assuming that mistakes are possible.

It is easy to verify that, for backward induction to work in game G^1, different levels of knowledge of rationality (and of the structure of the game) are needed at different stages of the game. For example, if R_1 stands for "player 1 is rational," R_2 for "player 2 is rational," and $K_2 R_1$ for "player 2 knows that player 1 is rational," then R_1 alone will be sufficient to predict 1's choice at the last node. But in order to predict 2's choice at the penultimate node, one must know that rational player 2 knows that 1 is rational (i.e., $K_2 R_1$). $K_2 R_1$, in turn, is not sufficient to predict 1's choice at the first node, since 1 will also have to know that 2 knows that he is rational; that is, $K_1 K_2 R_1$ must obtain. Moreover, while R_2 only (in combination with

$K_2 R_1$) is needed to predict L at the penultimate node, $K_1 R_2$ must be the case at I^{11}.

Theorem 2.2. *In finite extensive-form games of perfect and complete information, the backward induction solution holds if the following conditions are satisfied for any player i at any information set I^{ik}:*
(a) *player i is rational and knows it, and knows his available choices and payoffs; and*
(b) *for every information set I^{jk+1} that immediately follows I^{ik}, player i knows at I^{ik} what player j knows at information set I^{jk+1}.*

Proof. The proof is by induction on the number of moves in the game. If the game has only one move then the theorem is vacuously true, since at information set I^{i1} if player i is rational and knows it, and knows his available choices and payoffs, he will choose that branch which leads to the terminal node associated with the maximum payoff to him, and this is the backward induction solution. Suppose the theorem is true for games involving at most K moves (some $K \geq 1$). Let Γ be a game of perfect and complete information with $K+1$ moves and suppose that conditions α and β are satisfied at every node of game Γ. Let r be the root of the game tree T for Γ. At information set I^{ir}, player i knows that conditions α and β are satisfied at each of the subgames starting at the information sets that immediately follow I^{ir}. Then at I^{ir} player i knows that the outcome of play at any of those subgames would correspond to the backward induction solution for that subgame. Hence, at I^{ir}, if player i is rational then he will choose the branch going out of r that leads to the subgame whose backward induction solution is best for him, and this is the backward induction solution for game Γ. $\square$

Theorem 2.2 says that only limited knowledge of rationality and of the structure of the game need be assumed for the backward induction solution to hold. All that is needed is that a player, at any of her information sets, knows what the next player to move knows. Thus the player who moves first will know more things than the players who move immediately after, and these in turn will know more than the players who follow them in the game. Because a player may have to move at different points in the game, it seems an obvious requirement that her knowledge be the same at each of her information sets. Though this requirement has a natural interpretation in the normal-form representation of the game, it cannot be fulfilled in the extensive form, pointing to a crucial difference between normal- and extensive-form representations. Consider the normal-form equivalent of game G^1:

	L	R
l_1	1, 0	1, 0
$r_1 l_2$	0, 2	3, 1
$r_1 r_2$	0, 2	2, 4

Here, strategy $r_1 l_2$ weakly dominates $r_1 r_2$, so if player 2 knows that player 1 is rational then 2 expects 1 to eliminate $r_1 r_2$. In the extensive-form representation, this corresponds to player 2 knowing that rational player 1, at the last node, will choose l_2. In order to eliminate his weakly dominated strategy, player 1 need not know whether 2 is rational. This corresponds to the last node of the extensive-form representation, where 1 does not need to consider what happened before (as it is now strategically irrelevant). Player 1 needs to know that 2 is rational only when, having eliminated $r_1 r_2$, he has made L weakly dominant over R. Note that player 1, in order to be sure that 2 will choose L, must know that 2 is rational *and* that 2 knows that 1 is rational; otherwise, there would be no weakly dominated strategy for player 2 to delete. Having thus deleted R, 1's best reply to L is l_1; this corresponds to the first node, where player 1 must know that 2 is rational and that 2 knows that 1 is rational. Evidently, player 1 needs to know *more* than player 2, even in the normal form, since the order of iterated elimination of dominated strategies starts with player 1's strategy $r_1 r_2$.

In the extensive-form representation of the game, player 1's previous knowledge has become irrelevant when the last node is reached, but this does not mean that player 1 knows less. The sequential nature of decision making makes one ask – if any node after the first is reached – why deviations took place, since any future play will depend upon the answer provided by the players. For example, reaching node c is inconsistent with what 1 knows, if we assume that 1's knowledge at any node is the union of what 1 knows at every node. Player 1 knows that player 2 is rational and knows that 1 is rational, but then 2 should not have played R. In this simple game, l_2 remains an equilibrium strategy in the subtree starting at c, but this is only because c is the last decision point in the game. A different example is depicted in Figure 2. If a player's knowledge is interpreted as the union of the information he has at each node, then at node c player 1 – knowing as much as he knows at node a – faces an inconsistency. It is not obvious that he will choose to play l_3, because player 2's deviation is inconsistent with what 1 knows of her. This seems to imply that the equilibrium $(l_1 l_3, l_2 l_4)$ is not subgame perfect, in that l_3 would not be an equilibrium strategy in the subgame starting at c. However, if

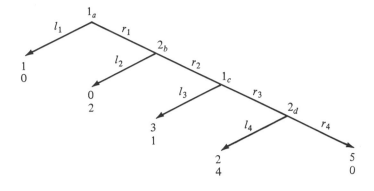

Figure 2. Game Γ^1.

we consider the subgame starting at c as a game of its own, it makes sense to interpret player 1's knowledge at c as the *intersection* of what the players that precede him in the larger game (himself included) know of him.

Interpreting a player's knowledge at any given node in this way – as the intersection of what the preceding players know of him – depicts the process of forward induction undergone by each of those players. For example, when player 1 considers his available choices at node a, what matters to him is player 2's state of knowledge at b (which includes what 2 thinks of 1's state of knowledge at c), since it is 1's knowledge of 2's state of knowledge that determines his choice and *not* what 1 knows that he would know were he to reach c. "What player 1 knows at node c," therefore, is what 1 knows at node a about what 2 knows at node b about player 1 at c, and this is precisely the intersection of what 1 and 2 know about player 1 at node c.

In this interpretation l_3 is an equilibrium strategy, as the intersection of what players 1 and 2 know of player 1 at (respectively) nodes a and b is precisely the knowledge that allows 1 to play l_3. In the normal form, on the contrary, a player can use all his available knowledge (i.e., what he knows at his first decision node) at once without incurring in a contradiction, since the game is played simultaneously.

3. KNOWLEDGE-DEPENDENT GAMES

Because the rationality of a plan of action must be tested against possible alternatives, a player's choice depends on what he expects the following players to do were they to reach their information sets. This requires a specification of what other players would expect him to do at subsequent decision nodes. However, a player's expectations are well defined only if reaching his information set is consistent with what he knows about the game. In this section it is proved that inconsistencies arise by simply increasing the number of levels of knowledge beyond what is sufficient to infer the equilibrium solution. In this case, expectations are undefined and the players are deprived of a criterion of choice between different strategies. The main result is that, for any finite extensive-form game of perfect information, the number of levels of knowledge that is sufficient to infer the backward induction solution is also necessary to infer it.

To address this issue, the players' knowledge of the game, as well as the reasoning process that leads them to choose a particular sequence of actions, must be explicitly modeled. The theory of the game must thus include a set of assumptions about what the players know about the structure of the game and the other players. More formally, if there are n players and some propositions $p_1, \ldots, p_m$, we can construct a knowledge language $\mathcal{L}$ by closing under the standard truth-functional connectives and the rule that says that if p is a formula of $\mathcal{L}$ then so is $K_i p$ ($i = 1, \ldots, n$), where $K_i p$ stands for "i knows that p." Since we are interested in modeling collective knowledge, we add the group-knowledge operator E_G, where $E_G p$ stands for "everyone in group G knows that p." If $G = \{1, 2, \ldots, n\}$ then $E_G p$ is defined as the conjunction $K_1 p \wedge K_2 p \wedge \cdots \wedge K_n p$. K-level group knowledge of p can be expressed as

$$E_G^k p \equiv \bigwedge_{i_j \in G, 1 \le j \le k} K_{i_1} K_{i_2} \cdots K_{i_k} p.$$

If p is E_G^k-knowledge for all $k \ge 1$, then we say that p is common knowledge in G; that is, $C_G p \equiv p \wedge E_G p \wedge E_G^2 p \wedge \cdots \wedge E_G^m p \wedge \cdots$. $C_G p$ implies all formulas of the form $K_{i_1} K_{i_2} \cdots K_{i_n} p$ ($i_j \in G$) for any finite n, and is equivalent to the infinite conjunction of all such formulas.

In order to reason about knowledge, we must provide a semantics for this language. Following Hintikka (1962), we use a possible-worlds semantics. The main idea is that there are a number of possible worlds at each of which the propositions p_i are stipulated to be true or false, and all the truth functions are computed at each world in the usual way. For example, if w is a possible world then $p \wedge q$ is true at w iff ("if and only if") both p and q are true at w. An individual's state of knowledge corresponds to the extent to which he can tell what world he is in, so that a world is possible relative to an individual i. In a given world one can associate with each individual a set of worlds that, given what she knows, could possibly be the real world. Two worlds w and w' are equivalent for individual i iff they create the same evidence for i. Then we can say that an individual i knows a fact p iff p is true at all worlds that i considers possible; that is, $K_i p$ is true at w iff p is true at every world w' that is equivalent to w for individual i. An individual i does not know p iff there is at least one world that i considers possible where p does not hold.

The following set of axioms and inference rules provides a complete axiomatization for the notion of knowledge we use:

A_1: all instances of tautologies;
A_2: $K_i p \Rightarrow p$;
A_3: $(K_i p \wedge K_i(p \Rightarrow q)) \Rightarrow K_i q$;
A_4: $K_i p \Rightarrow K_i K_i p$;
A_5: $\sim K_i p \Rightarrow K_i \sim K_i p$;
MP: If p and $p \Rightarrow q$, then q;
KG: If $\vdash p$ then $\vdash K_i p$.

Some remarks are in order. A_2 states that if i knows p then p is true. A_3 says that i knows all the logical consequences of his knowledge; this assumption is defensible considering that we are dealing with a very elementary (decidable) logical system. A_4 says that knowing p implies that one knows that one knows p. Intuitively, we can imagine providing an individual i with a database. Then i can look at her database and see what is in it, so that if she knows p then she knows that she knows it. A_5 is more controversial, since it says that not knowing implies that one knows that one does not know. This axiom can be interpreted as follows: Individual i can look at her database to see what she does not know, so if she doesn't know p then she knows that she does not know it. Rule KG says that if a formula p is provable in the axiom system A_1–A_5, then it is provable that $K_i p$. A formula is provable in an axiom system if it is an instance of one of the axiom schemas, or if it follows from one of the axioms by one of the inference rules MP or KG. Also, a formula p is consistent if $\sim p$ is not provable.

It is easy to verify that the rule KG makes all provable formulas in the axiom system A_1–A_5 common knowledge among the players. Suppose q is a theorem: then by KG it is a theorem that $K_i q$ ($i = 1, \ldots, n$). If $K_i q$ is a theorem then it is a theorem that $K_j K_i q$ (for all $j \neq i$), and it is also a theorem that $K_i K_j K_i q$, and so on. In the system A_1–A_5, if $\vdash p$ then $\vdash Cp$. We call the class of axioms A_1–A_5 *general axioms*.

Beside logical axioms, a theory of the game will include game-theoretic solution axioms, behavioral axioms, and axioms describing the information possessed by the players. We call this second class of axioms *special axioms*.

Let us consider as an example game Γ^1:

A_6: the players are rational (i.e., $R_1 \wedge R_2$);

A_7: at node I^{11}, $(r_1 \vee l_1) \wedge \sim (r_1 \wedge l_1)$;

A_8: at node I^{21}, $(L \vee R) \wedge \sim (L \wedge R)$;

A_9: at node I^{12}, $(r_2 \vee l_2) \wedge \sim (r_2 \wedge l_2)$;

A_{10}: $\pi^1(l_1) = 1$, $\pi^2(l_1) = 0$;

A_{11}: $\pi^1(L) = 0$, $\pi^2(L) = 2$;

A_{12}: $\pi^1(r_2) = 2$, $\pi^2(r_2) = 4$;

A_{13}: $\pi^1(l_2) = 3$, $\pi^2(l_2) = 1$;

A_{14}: at node I^{12}, $R_1 \Rightarrow l_2$;

A_{15}: at node I^{21}, $[R_2 \wedge K_2 R_1] \Rightarrow L$;

A_{16}: at node I^{11}, $[R_1 \wedge K_1(R_2 \wedge K_2 R_1)] \Rightarrow l_1$;

A_{17}: $E_G^2(A_6\text{--}A_{16})$.

A_6 is a behavioral axiom: It states that the players are rational in the sense of being expected-utility maximizers. A_7–A_9 specify the choices available to each player at each of his information sets, and say that a player can choose only one action. A_{10}–A_{13} specify players' payoffs. A_{14}–A_{16} are solution axioms, and specify what players should do at any of their information sets if they are rational and know (a) that the next player to move is rational and (b) what the next player to move knows. A_{17} says that each player knows that each player knows A_6–A_{16}. We call these axioms "special" because, even if every player knows that every player knows the axioms A_6–A_{16}, no common knowledge is assumed.

From A_1–A_{17}, the players can infer the equilibrium solution l_1. To verify that the amount of information attributed to the players is compatible with a deviation from equilibrium, consider in turn each player's reasoning. To decide which strategy to play, player 1 must predict how player 2 would respond to his playing r_1. The main stages of 1's reasoning can be described as follows.

(1) r_1 [by assumption];

(2) $K_1 K_2([R_1 \wedge K_1(R_2 \wedge K_2 R_1)] \Rightarrow l_1)$ [by axioms A_{16} and A_{17}];

(3) $K_1 K_2(\sim l_1 \Rightarrow \sim [R_1 \wedge K_1(R_2 \wedge K_2 R_1)])$ [by (1), (2), A_1, and KG];

(4) $K_1 K_2(r_1 \vee l_1) \wedge \sim (r_1 \wedge l_1)$ [by A_{17} and A_7];

(5) $K_1 K_2(\pi^1(l_1) = 1, \pi^2(l_1) = 0)$ [by A_{17} and A_{10}];

(6) $K_1(R_2 \wedge K_2 R_1)$ [by A_6 and A_{17}];

(7) $K_1 \sim K_1 K_2(K_1(R_2 \wedge K_2 R_1))$ [by A_5, (3), and A_{17}].

For all that player 1 knows, his playing r_1 can be "explained away" by player 2 as being due to $\sim K_1(R_2 \wedge K_2 R_1)$. In other words, what player 1 knows of player 2 does not conflict with his knowledge that $K_2 R_1$. Because

(8) $K_1[R_2 \wedge K_2 R_1] \Rightarrow L$ [by A_{17} and A_{15}],

player 1 knows that 2 will respond with L to r_1, hence he plays l_1.

Consider now what player 2 would think when facing a deviation on the part of player 1.

(1) r_1 [by assumption];
(2) $K_2(r_1 \vee l_1) \wedge \sim (r_1 \wedge l_1)$ [by A_{17} and A_7];
(3) $K_2(\pi^1(l_1) = 1, \pi^2(l_1) = 0)$ [by A_{17} and A_{10}];
(4) $K_2(\sim l_1 \Rightarrow \sim [R_1 \wedge K_1(R_2 \wedge K_2 R_1)])$ [by A_{17}, A_{16}, and A_1];
(5) $K_2(R_1 \wedge K_1 R_2)$ [by A_{17} and A_6];
(6) $K_2(\sim l_1 \Rightarrow \sim K_1 K_2 R_1)$ [by (4) and (5)].

Player 2 can "explain" why r_1 was played, and since this explanation does not conflict with $K_2 R_1$, she will choose strategy L.

Consider what would happen if further levels of knowledge were added to the theory in the form of the following axiom.

A_{18}: $E_G^2(A_6 - A_{17})$.

Because there is one more level of knowledge, both players now know that $K_1 K_2 R_1$ and $K_2 K_1 R_2$ obtain. This level of information implies that – were r_1 to be played – player 2 would face an inconsistency.

(1) $K_2(\sim l_1 \Rightarrow \sim [R_1 \wedge K_1(R_2 \wedge K_2 R_1)])$ [by A_1, A_{16}, and A_{18}];
(2) $K_2[R_1 \wedge K_1(R_2 \wedge K_2 R_1)]$ [by A_6 and A_{18}];
(3) $K_2 l_1$ [by (2) and A_{16}];
(4) r_1 [by assumption];
(5) $K_2(r_1 \vee l_1) \wedge \sim (r_1 \wedge l_1)$ [by A_7 and A_{18}];
(6) $K_2 \sim [R_1 \wedge K_1(R_2 \wedge K_2 R_1)]$ [by (1) and (4)];
(7) $[R_1 \wedge K_1(R_2 \wedge K_2 R_1)]$ [by A_2];
(8) $\sim [R_1 \wedge K_1(R_2 \wedge K_2 R_1)]$ [by A_2].

Because the conjunction of formulas (7) and (8) is false, and since one can deduce anything (in classical logic) from a false statement, player 2 can use this conjunction to construct a proof that "r_1." Adding axiom A_{18} makes the theory of the game *inconsistent* for player 2; thus 2 is unable to use it to predict how player 1 would respond if she were to play R. This leaves 2 uncertain as to how to play herself.

Note that the theory of the game is not inconsistent for player 1. It is easy to verify that the state of information of player 1 does not allow him to realize that – were he to play r_1 – player 2 would face an inconsistency. By A_{18}, player 1 knows $K_2 K_1 R_2$. But the number of levels of knowledge assumed in A_{18} does not let 1 know that $K_2(K_1 K_2 R_1)$. Therefore player 1 can believe that 2 will explain a deviation by assuming $\sim (K_1 K_2 R_1)$. If so, he predicts that 2's response will be L, which makes him play l_1. Hence a theory of the game that includes axiom A_{18} supports the backward induction solution.

The backward induction equilibrium can be inferred except in the case where $K_1(K_2K_1K_2R_1)$ obtains. This level of knowledge is brought forth by one additional axiom.

A_{19}: $E_G^2(A_6-A_{18})$.

In this case, player 1 would know that playing r_1 makes the theory of the game inconsistent for player 2 at I^{21}. If so, player 2 will be unable to predict what would happen were she to play R, and 1 – knowing that – will be unable to predict what would happen were he to play r_1.

Because a solution concept for the class of games examined here depends upon the number of levels of knowledge possessed by the players, we must introduce a few new definitions.

Definition 3.1. A *knowledge-dependent game* is a quadruple $G = (N, S^i, K^i, \pi^i)$, where $N = \{1, \ldots, n\}$ is the number of players; S^i is the set of strategies of player i; K^i is the knowledge possessed by player i, defined as the union of what i knows at each of his information sets:

$$K^i = \bigcup_{1 \leq j \leq k} K^i I^{ij};$$

and π^i is player i's payoff.

Definition 3.2. An n-tuple of strategies $(s^1, \ldots, s^n)$ is a *knowledge-consistent* play of a knowledge-dependent game if, for each player i, every choice s^{ij} that strategy s^i recommends at each information set $I^{ij} \in P^i$ satisfies the following conditions: (i) reaching I^{ij} is compatible with $K^i I^{ij}$; and (ii) it can be proven from $K^i I^{ij}$ that s^{ij} is a best reply for player i at I^{ij}.

Theorem 3.1. *For every finite, extensive-form game of perfect and complete information, the backward induction equilibrium is a knowledge-consistent play of some knowledge-dependent game and, conversely, every knowledge-consistent play of a knowledge-dependent game is a backward induction equilibrium.*

Proof. The first part of the proof is trivial, since Theorem 2.2 illustrates a specification of the knowledge of each player that makes the backward induction equilibrium a knowledge-consistent play. The second part of the theorem can be proven by induction on the number of moves in the game.

341

Suppose the game has only one move. In order to make a choice, the player who has to move must know his available strategies and payoffs. A rational player knows that he should choose that branch leading to a terminal node with the maximum payoff to him. Then, if the player knows his strategies and payoffs, he can infer his payoff-maximizing solution, which is the backward induction solution. Assume the theorem is true for all games involving at most M moves (some $M \geq 1$). Then it follows that the knowledge-consistent play $(s^1, \ldots, s^n)$, restricted to any of the subgames of G having no more than M moves, corresponds to the backward induction solution for that subgame. Let G be a knowledge-dependent game with $M+1$ moves, and let r be the root of the game tree T for G. At information set I^{ir} there is a recommendation of play s^{ir} for player i that can be inferred from $K^i I^{ir}$. Let $\mathbf{K} = \bigcup_{1 \leq m \leq k} K^j I^{jr+m}$ be the union of the knowledge possessed by each player j who must play at an information set that immediately follows I^{ir}. Then player i's knowledge of $\mathbf{K}$ implies the choice of the move that is the backward induction solution at I^{ir}. Therefore the union of $K^i I^{ir}$ and $\mathbf{K}$ allows one to derive both the backward induction solution for I^{ir} and the strategy s^{ir}. The two must coincide, since the union of $K^i I^{ir}$ and $\mathbf{K}$ cannot lead to an inconsistent system. $\qquad\square$

NOTES

1. Bayesian game theory has the same problem: The players' incomplete information about the structure of the game is simply described as an extensive-form game with chance moves (Harsanyi 1967/68). In this case, too, some basic assumptions of the theory are not treated as part of the theory.
2. More recently, Gilboa and Schmeidler (1988) proved that, in information-dependent games, a common-knowledge axiom is inconsistent with a rationality axiom.

REFERENCES

Aumann, R. J. (1976), "Agreeing to Disagree." *Annals of Statistics* 4: 1236–9.
Bacharach, M. (1985), "Some Extensions of a Claim of Aumann in an Axiomatic Model of Knowledge." *Journal of Economic Theory* 37: 155–67.
Bacharach, M. (1987), "A Theory of Rational Decision in Games." *Erkenntnis* 27: 17–55.
Bicchieri, C. (1989), "Self-refuting Theories of Strategic Interaction: A Paradox of Common Knowledge." *Erkenntnis* 30: 69–85.
Binmore, K. (1987), "Modeling Rational Players I." *Economics and Philosophy* 3: 179–214.
Binmore, K. (1988), "Modeling Rational Players II." *Economics and Philosophy* 4: 9–55.
Brandenburger, A., and Dekel, E. (1985a), "Common Knowledge with Probability." Research Paper no. 796R, Graduate School of Business, Stanford University.
Brandenburger, A., and Dekel, E. (1985b), "Hierarchies of Beliefs and Common Knowledge." Research Paper no. 841, Graduate School of Business, Stanford University.

Gilboa, I. (1986), "Information and Meta-information." Working Paper No. 30-86, Department of Economics, Tel-Aviv University.

Gilboa, I., and Schmeidler, D. (1988), "Information Dependent Games." *Economics Letters* 27: 215-21.

Harsanyi, J. (1967/68), "Games with Incomplete Information Played by 'Bayesian' Players," Parts I, II, III. *Management Science* 14: 159-82, 320-34, 486-502.

Hintikka, J. (1962), *Knowledge and Belief.* Ithaca, NY: Cornell University Press.

Kaneko, M. (1987), "Structural Common Knowledge and Factual Common Knowledge." RUEE Working Paper No. 87-27, Hitotsubashi University.

Kuhn, H. W. (1953), "Extensive Games and the Problem of Information." In H. W. Kuhn and A. W. Tucker (eds.), *Contributions to the Theory of Games.* Princeton, NJ: Princeton University Press.

Mertens, J.-F., and Zamir, S. (1985), "Formulation of Bayesian Analysis for Games with Incomplete Information." *International Journal of Game Theory* 14: 1-29.

Reny, P. (1987), "Rationality, Common Knowledge, and the Theory of Games." Working Paper, Department of Economics, University of Western Ontario.

Samet, D. (1987), "Ignoring Ignorance and Agreeing to Disagree." Mimeo, Department of Economics, Northwestern University.

Selten, R. (1975), "Re-examination of the Perfectness Concept for Equilibrium Points in Extensive Games." *International Journal of Game Theory* 4: 22-55.

20

Common knowledge and games with perfect information

PHILIP J. RENY

1. INTRODUCTION

It is by now rather well understood that the notion of common knowledge (first introduced by Lewis 1969 and later formalized by Aumann 1976) plays a central role in the theory of games. (An event E is common knowledge between two individuals, if each knows E, each knows the other knows E, etc. . . .) Indeed, most justifications of Nash's (1951) equilibrium concept usually include (perhaps only implicitly) the assumption that it is common knowledge among the players that both the Nash equilibrium in question will be played by all and that all players are expected utility maximizers.[1] (We shall henceforth call expected utility maximizers "rational."[2]) We hope to illustrate in an informal manner that there is in fact a large class of extensive-form games, in which each of which it is not possible for rationality to be common knowledge throughout the game.

The consequences of this for many well-known extensive-form refinements of Nash equilibrium are quite serious. Consider, for example, Selten's (1965) notion of subgame perfect Nash equilibrium. The requirements on a solution here are not only that the strategies form a Nash equilibrium of the game as a whole, but also that the strategies induce on every proper subgame a Nash equilibrium.[3] If, however, there are proper subgames beginning at (singleton) information sets at which it is not possible for rationality to be common knowledge, then Nash behavior in that subgame can no longer be justified on common-knowledge grounds.[4] At best then, significant modifications are required in our explanation of Nash behavior in such subgames. At worst, Nash behavior in such subgames should not be considered as the only possibility. In either case, a re-evaluation of the subgame perfect equilibrium notion would be called

PSA 1988, Volume 2, pp. 363–9. Copyright © 1989 by the Philosophy of Science Association. Reprinted by permission.

for. Since extensive-form refinements such as sequential equilibria (Kreps and Wilson 1982) and perfect equilibria (Selten 1975) involve even stronger restrictions upon behavior in subgames, the comments above apply as well to each of these notions.[5]

In addition, our result implies that arguments supporting the backward induction solution, based on rationality being common knowledge at every information set, begin with a false hypothesis. Hence, the elegant argument supporting backward induction advanced by Kreps et al. (1982) runs into difficulty. Also, the recent work of Bernheim (1984) and Pearce (1984) on "Rationalizability" involves heavy use of the common knowledge of rationality in both normal- and extensive-form games. If, in extensive-form games, such common knowledge is not always possible, then at the very least a reinterpretation of their extensive-form analyses is called for. These issues will be explored further at the end of the chapter.

Finally, others have also expressed certain difficulties with a variety of the above equilibrium concepts (see for instance Basu 1985, Binmore 1985, and Rosenthal 1981). All of these difficulties essentially appear to involve in one way or another a problem with the assumption that rationality is common knowledge. We now illustrate by means of the simplest sort of example that there are games containing information sets at which it is not possible for rationality to be common knowledge.

2. AN EXAMPLE

Consider the following two-player perfect information game. A referee comes equipped with n dollars and places one in front of players one and two. Player one can take the dollar thereby ending the game, or he can leave it. If he leaves it, the referee places a second dollar in front of the players. Player two now has the opportunity to take the two dollars and end the game or not, in which case the process repeats. In general, at the kth stage of the game, the referee adds one dollar to the pot bringing its total to k dollars. If k is odd (even), player one (two) may take the k dollars and end the game, or leave it. Players' payoffs are assumed strictly increasing in dollars. Finally, should the game continue until the nth stage and the player whose turn it is decides to leave the n dollars, it is then given to the other player. Call this game TOL(n) (Take it or leave it). TOL(n) for n odd is depicted in Figure 1.

In this particular game with perfect information, backward induction and the subgame perfect, sequential, and perfect equilibrium concepts each yield the same equilibrium strategies. We proceed via backward induction. At the last stage (n odd) player one's best choice is to take the

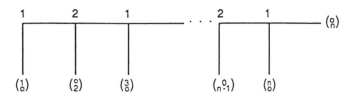

Figure 1. TOL(n).

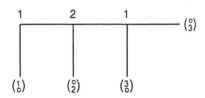

Figure 2. TOL(3).

n dollars. With this in mind, two's best response at the second last stage is to take the $n-1$ dollars and end the game. This process continues with each player choosing to take the money and end the game on his turn if he gets the chance. In the end, backward induction (and hence the subgame perfect, sequential, and perfect equilibrium concepts) yields that player one take the one dollar and end the game in the first round. This is independent of the value of n! That is, no matter how large the pot may potentially grow, the standard equilibrium notions indicate that player one will take the one dollar in the first round.[6] This sort of paradox is by no means new and is clearly reminiscent of that associated with the finitely repeated prisoner's dilemma.

We now move to the problem of common knowledge. It is enough to consider TOL(3) (see Figure 2).

We claim that if player one does not take the dollar and end the game in the first round, but instead leaves it so that player two must decide whether or not to take the two dollars, then it is no longer possible for rationality to be common knowledge (i.e., at player two's information set, it is not possible for rationality to be common knowledge). The argument is really quite straightforward and proceeds by contradiction. Suppose that rationality were common knowledge at player two's information set. Player two, believing that player one is rational (i.e., an expected utility maximizer), must believe that at stage 3, player one will take the three dollars, leaving player two with nothing. A rational player two would respond to this by taking the two dollars at the second stage of the game, leaving player one with zero. Hence, if at player two's information set it is the case that

(i) player two is rational, and
(ii) player two believes that player one is rational, then player two will take the two dollars leaving player one with zero.

But since rationality is common knowledge, it must be the case that in particular,

(i′) player one believes that player two is rational, and
(ii′) player one believes that player two believes that player one is rational.

That is, player one believes (i) and (ii) above. Finally, however, this implies that player one believes that player two will take the two dollars, leaving player one with zero and rendering player one's choice not to take the dollar in the first round (recall that player two's information set has been reached) an irrational (non–expected utility maximizing) one. Hence, player two must believe that player one is not rational, which contradicts our original assumption and completes the argument.[7] A similar argument shows that in TOL(n), as soon as player one leaves the first dollar it is not possible for rationality to be common knowledge.

This observation and the work of Kreps et al. (1982) suggest an alternative analysis of TOL(n). Recall that Kreps et al. were interested in explaining cooperation in the finitely repeated prisoner's dilemma. They showed that if from the start, there is a positive probability that one of the players is not rational, then cooperation could emerge as a sequential equilibrium of a long enough repeated prisoner's dilemma game suitably modified to take into account the incomplete information about the player's rationality.[8] It turns out that one can apply a generalized version of Kreps et al.'s analysis of the prisoner's dilemma to TOL(n) and achieve similar results. That is, if from the outset there is a positive probability that one of the players is not rational, or one of the players believes the other is not rational, or . . . , and n is large enough, then allowing the pot to build for some time can emerge as a sequential equilibrium.[9] Moreover, any such sequential equilibrium must yield both players higher expected utilities than they would get if player one took the dollar at the first stage. Hence, there is a sense in which both players are better off in TOL(n) when n is large enough and rationality is not common knowledge.

Among those elements left unexplained by Kreps et al.'s analysis and the generalized version we've applied to TOL(n) is how the positive probability that a player is not rational, or that a player believes the other is not rational, or . . . arises in the first place. What we've shown above, however, is that player one can, by not taking that first dollar, create an environment in which rationality is not common knowledge. Since for n large enough, this potentially makes player one better off, this furnishes a sound explanation of where the positive probability comes from. Hence, this exogenously introduced positive probability that one of the players is not rational or that one believes the other is not rational etc. . . . need not be exogenously introduced at all. Our observation that it is not possible for rationality to be common knowledge once player one leaves the first dollar, supplemented by Kreps et al.'s analysis once this is the environment, shows that this positive probability can arise as the result of expected utility maximizing behavior. Taken as a whole, this analysis of TOL(n) (unlike backward induction and the more traditional solution concepts) implies that two expected utility maximizing players can, acting in their own self-interests, allow the pot to grow.

It is somewhat paradoxical that we can justify in terms of rational behavior player one leaving the first dollar in TOL(n), when this very justification requires that when player one does so, player two believes that player one is not rational or that two believes that one believes that two is not rational or etc. . . . For if an explanation based on the players' rationality is possible, won't players believe one another are rational? And won't each then believe that each believes this etc. . . . ? That is, won't then rationality be common knowledge?

The answer to the last question is: "Absolutely not." We have already argued that once player one leaves the first dollar, rationality *cannot* be common knowledge. This is simply inconsistent with the structure of the game and the current position in it. The answer to each of the other questions is "not necessarily." Since rationality can not be common knowledge when player one leaves the dollar, some statement of the form: Two believes that one believes that two believes that . . . that one (or two) is *not* rational *must* be true. Hence the answer to at least one of the other questions *must* be no. That is, a formal proof can be constructed to demonstrate this. On the other hand, although such beliefs are possible, no formal proof can be constructed to demonstrate that after player one leaves the first dollar, two believes that one is rational and two believes that one believes that two is rational. Otherwise (assuming this proof is common knowledge) this would imply that player two believes that rationality is common knowledge. But this is impossible if players' beliefs are restricted to being consistent with the physical description of the game and the current position in it. So, although a rational explanation is available, in a formal sense it cannot be the only available explanation. And it

349

is precisely these (necessarily) available alternatives which make a rational explanation possible at all.

As in TOL(n), one can show that by cooperating in a finitely repeated prisoner's dilemma, the players can create an environment in which rationality can not be common knowledge. As Kreps et al. (1982) have shown, once this is the case both players may be better off. Hence one can also explain, now more fully perhaps, rational cooperation in the finitely repeated prisoner's dilemma.

3. Is TOL(n) an isolated example?

We shall now fulfill a promise made in the introduction and illustrate that there is a large class of games within which the problem of common knowledge of rationality arises. To do so, we shall first restrict our attention to two-person extensive-form games with perfect information, where no player is indifferent between any two endpoints.[10]

Instead of asking which games within this class contain information sets at which it is possible for rationality to be common knowledge, we ask a slightly different question. That is, which games in this class allow rationality to be common knowledge at every information set *simultaneously*? Some clarification is in order.

It turns out that it is possible that rationality is common knowledge at a particular information set so long as at that information set it is also common knowledge that rationality is *not* common knowledge at some other information set. Hence in such cases it may be that taken one at a time, it is possible that rationality is common knowledge at every information set, but taken together, common knowledge of rationality at one information set precludes it at another. In the latter case we say that it is not possible for rationality to be common knowledge at every information set simultaneously. Whenever it is not possible for rationality to be common knowledge at every information set simultaneously, issues such as those described for TOL(n) arise and again, the arguments supporting the traditional equilibrium concepts no longer apply (Reny 1988 contains more details). It can be shown that:[11]

> Any two-person finite extensive-form game with no player indifferent between any two endpoints that allows rationality common knowledge at every information set simultaneously, must be of the form depicted in Figure 3, where the arrows indicate the unique subgame-perfect equilibrium (i.e. every decision node is reached in equilibrium).

In our view, this indicates that the problem of common knowledge of rationality in extensive-form games occurs rather frequently.

Figure 3

4. Conclusion

What are we to make of all of this? We've shown that rationality cannot simply be assumed common knowledge. The physical description of the game and the current position in it may preclude this. But once rationality can no longer be common knowledge, it is not at all clear what the theory should tell the players to believe about one another. One thing, however, is clear. If we insist that the theory itself is common knowledge among the players (as has implicitly been the traditional approach, and even explicitly in Bernheim 1984 and Pearce 1984), then the theory cannot indicate that the players are rational, since this would automatically render rationality common knowledge and this is not always possible. But if the theory cannot indicate that players are rational, then the players may believe just about anything about the behavior of their opponents, since there is no obvious substitute for rational behavior. Clearly, this will lead to a plethora of possible outcomes since rational players who need not believe their opponents are rational may have many strategies which are a best response to *something* their opponents might play. Hence, we can expect a very weak (in terms of predictions) theory to result if we insist that the theory itself is common knowledge among the players at every information set. (For more on this see Reny 1988.)

The most promising avenue to pursue, then, is one which explicitly allows the theory not to be common knowledge at every information set. Indeed one might reach this conclusion simply on the grounds that once a player has deviated it is impossible that the rationality of the players and the equilibrium strategies remain common knowledge. One way to consistently explain the deviation and still hold fast to the rationality of all players (including the deviator) is to postulate that the equilibrium strategies (i.e., the theory) are no longer common knowledge. We close by mentioning that this approach has been recently undertaken (Reny 1987) in a manner that yields relatively strong predictions, indicating that some of the issues raised here can be usefully taken into account.

351

1. In a two-player normal-form game, a pair of strategies (one for each player) is a Nash equilibrium if each player's choice maximizes his expected utility given the choice of the other player.
2. This follows Bernheim (1984) and Pearce (1984).
3. That is, not only must it be that (i) every player has decided what to do in every possible eventuality so that from his perspective at the beginning of the game his choices are best given what the others have planned to do, it must also be the case that (ii) whenever any possible eventuality becomes a reality, no player will wish to change his previously decided upon choice.
4. A physical description of chess, for instance, includes the initial position, the set of all possible first moves and subsequent positions etc. . . . . A subgame is simply the description of what the current position is and which positions are possible during subsequent play. A proper subgame must begin at a point in the game where every player has full knowledge of the past choices made by others. Hence, every subgame in chess is proper. In other games (like poker) one must take a turn in ignorance of what others have done previously. Those aspects known to a player when it is his turn to move are embodied in his "information set." Corresponding to each turn then is an information set, and so "reaching" a particular information set of player one say, indicates that it is a particular turn of player one.
5. One might argue that since Selten's (1975) theory allows players to make "mistakes," one can always preserve the common knowledge of rationality by explaining arrival at any particular information set through a sequence of independent "errors." Our response to this is that players who make "mistakes," independent or not, arbitrarily small or not, are by definition not expected utility maximizers (i.e., not rational). Expected utility maximizers always and everywhere make decisions that maximize their expected utility. Hence any explanation based on mistakes or trembles is one that embraces the lack of common knowledge of rationality.
6. In fact, taking the first dollar is the unique Nash equilibrium outcome.
7. A formal version is contained in Reny (1988). The formal definition given there uses infinite hierarchies of beliefs as in Mertens and Zamir (1985) and Tan and Werlang (1985). It should be pointed out that what here we've called common knowledge should more appropriately be called common belief since we never require any player to actually be rational, only that players believe that one another are rational etc. . . . . Since an event which is common knowledge (in Aumann's 1976 sense) must also be common belief (see Tan and Werlang 1985) we have actually obtained the stronger result that at player two's information set rationality cannot be common belief.
8. The particular kind of irrationality imposed upon the player whose behavior is not completely known is important. Kreps et al. assume in particular that there is a positive probability that this player is one who uses the TIT-FOR-TAT strategy (i.e., cooperate in the first period and in every subsequent period copy the previous period choice of your opponent). TIT-FOR-TAT is not rational since it sometimes dictates cooperation in the last period.
9. Again, the particular kind of irrationality involved is important.
10. Recall that a game is one of perfect information if at every stage of the game both players know all past choices made by their opponent. A player is not

indifferent between any two endpoints if whenever asked to choose which of two ways he would like the game to end, he always strictly prefers one over the other. (Chess does not satisfy the latter condition since, for instance, there are many ways that the game can end with a win for white, and white is indifferent between all of these.) Note that the perfect information restriction rules out the finitely-repeated prisoner's dilemma. The no indifference between endpoints formally rules out TOL(n), but replacing the payoffs of zero at stage k by $1/k$ leaves all relevant features of TOL(n) intact and this new version is a member of the class of games described.

11. For a proof, see Reny (1988, pp. 72–9).

REFERENCES

Aumann, R. (1976), "Agreeing to Disagree." *Annals of Statistics* 4: 1236–9.
Basu, K. (1985), "Strategic Irrationality in Extensive Games." Mimeo, Institute for Advanced Studies, Princeton.
Bernheim, D. (1984), "Rationalizable Strategic Behavior." *Econometrica* 52: 1007–28.
Binmore, K. G. (1985), "Modelling Rational Players." Mimeo, London School of Economics and University of Pennsylvania.
Kreps, D., Milgrom, P., Roberts, J., and Wilson, R. (1982), "Rational Cooperation in the Finitely Repeated Prisoner's Dilemma." *Journal of Economic Theory* 27: 245–52.
Kreps, D., and Wilson, R. (1982), "Sequential Equilibria." *Econometrica* 50: 863–94.
Lewis, D. (1969), *Convention: A Philosophical Study.* Cambridge, MA: Harvard University Press.
Mertens, J. F., and Zamir, S. (1985), "Formulation of Bayesian Analysis for Games with Incomplete Information." *International Journal of Game Theory* 14: 1–29.
Nash, J. (1951), "Noncooperative Games." *Annals of Mathematics* 54: 286–95.
Pearce, D. (1984), "Rationalizable Strategic Behaviour and The Problem of Perfection." *Econometrica* 52: 1008–50.
Reny, P. J. (1987), "Explicable Equilibria." Mimeo, Princeton University and the University of Western Ontario.
Reny, P. J. (1988), "Rationality, Common Knowledge and the Theory of Games." Ph.D. Dissertation, chap. 1, Princeton University.
Rosenthal, R. W. (1981), "Games of Perfect Information, Predatory Pricing and the Chain-Store Paradox." *Journal of Economic Theory* 25: 92–100.
Selten, R. (1965), "Spieltheoretische Behandlung eines Oligopolmodells mit Nachfrageträgheit." *Zeitschrift für die Gesamte Staatswissenschaft* 121: 301–24.
Selten, R. (1975), "Reexamination of the Perfectness Concept for Equilibrium Points in Extensive Games." *International Journal of Game Theory* 4: 25–55.
Tan, T., and Werlang, S. (1985), "On Aumann's Notion of Common Knowledge – An Alternative Approach." Mimeo, University of Chicago Business School and Princeton University.

21

Game solutions and the normal form

John C. Harsanyi

I. Nash equilibria and their refinements

1. Notations

In general, I will denote a strategy (whether pure or mixed) of player i ($i = 1, ..., n$) as s_i. (But for convenience, in discussing specific game examples, I will often use other notations.) The set of all pure and mixed strategies of player i I will denote as S_i. A strategy combination of all n players will be denoted as $s = (s_1, ..., s_n)$. I will write $s_{-i} = (s_1, ..., s_{i-1}, s_{i+1}, ..., s_n)$ to denote the incomplete strategy combination we obtain if we omit player i's strategy s_i from s. I will write $s = (s_i, s_{-i})$. Player i's payoff function will be denoted as U_i.

2. Best replies and Nash equilibria

Suppose the $(n-1)$ players other than player i use the strategy combination s_{-i} whereas player i himself uses the strategy s_i^*. We say that s_i^* is a *best reply* to s_i if s_i^* maximizes player i's payoff when the other players use the strategy combination s_{-i}, that is, if

$$(1) \qquad U_i(s_i^*, s_{-i}) = \max_{s_i \in S_i} U_i(s_i, s_{-i}).$$

We say that a strategy combination $s = (s_1, ..., s_n)$ is an *equilibrium* (or a Nash equilibrium) if the strategy s_i of every player i in s is a best reply to the strategy combination s_{-i} used by the other players.

I wish to thank the National Science Foundation for supporting this research through grant SES-8700454 to the Center for Research in Management, University of California, Berkeley. I am thankful also to MIT Press for permission to use some material from our new book, Harsanyi and Selten (1988). Reprinted from John C. Harsanyi, "Game Solution and Normal Form," copyright © Kluwer Academic Publishers, with permission of the copyright holder.

Nash (1951) was the first to define equilibria. In the same paper, he introduced the distinction between cooperative and noncooperative games, defining the former as games in which the players can make binding agreements, and defining the latter as games in which agreements have no binding force so that each player can break any agreement if he wants to.

Nash also pointed out that the solution of a noncooperative game can be only a strategy combination $s = (s_1, \ldots, s_n)$ that is an equilibrium, in the sense that any theory recommending a nonequilibrium strategy combination as the outcome or solution of a game played by rational players would be *self-defeating*. For suppose such a nonequilibrium strategy combination s would be recommended as the outcome of a given game G, and that each player i in fact expected the other players to use the strategy combination s_{-i} prescribed by s. Since, by assumption, s is not an equilibrium, there would be at least one player i for whom the strategy s_i that s assigned to him would not be a best reply to the strategy combination s_{-i} he would expect the other players to use. As a result, this player i would not use this strategy s_i but would rather use another strategy s_i' that is a best reply to s_{-i}. Thus, any theory recommending use of s as the outcome of the game would be self-defeating in the sense that the very expectation that the other players would act in accordance with s would make some player(s) deviate from the strategy prescribed by s.

3. Perfect and imperfect equilibria

After Nash, it was commonly assumed that, once a strategy combination is a Nash equilibrium, it can be the solution for a noncooperative game. But then Selten (1965, 1975) pointed out that some Nash equilibria do involve *irrational* moves (i.e., moves not maximizing the relevant player's payoff), and that such an equilibrium cannot serve as solution for a game played by rational players. Any such equilibrium Selten called an *imperfect* equilibrium, in contrast to equilibria not involving such irrational moves, which he called *perfect* equilibria.

For example, consider the game whose extensive form is shown by Figure 1, and whose normal form is shown by Figure 2. (In Figure 2, in each cell of the payoff table, the upper-left number is player 1's payoff whereas the lower-left number is player 2's payoff.)

In the normal form, player 1 has two pure strategies, namely A and B. A means "Make move a" whereas B means "Make move b." Player 2 likewise has two pure strategies, namely C and D. C means "If player 1 makes move a then make move c. Otherwise do nothing." D means "If player 1 makes move a then make move d. Otherwise do nothing."

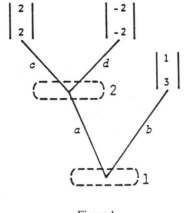

Figure 1

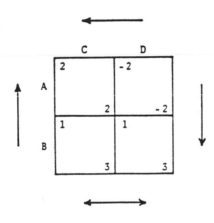

Figure 2

The game has two equilibria in pure strategies, namely $E_1 = (A, C)$ and $E_2 = (B, D)$. Of these, E_1 is a perfect equilibrium whereas E_2 is an imperfect one. It is imperfect because it makes player 1 choose move b on the expectation that if he himself chose move a then player 2 would retaliate by making move d, which would yield both of them the payoff $u_1 = u_2 = -2$. Yet, this is an irrational expectation. For, by definition, game-theoretic analysis is based on the assumption that all players will act rationally, that is, will act in such a way as to maximize their payoffs. Yet, given this assumption, player 1 should know that player 2 will definitely

357

not make move d even if he (player 1) makes move a, because if he did then he would not only reduce player 1's payoff to -2 but would reduce his own payoff to -2 as well. In other words, player 1 cannot rationally expect player 2 to make move d because d would not maximize player 2's own payoff. But, then, player 1 himself cannot rationally make move b because this would make sense only if he did expect player 2 to make move d in case he chose move a.

As I have indicated, a Nash equilibrium is defined as a strategy combination with the property that each player's strategy is a best reply (i.e., a payoff-maximizing strategy) against the strategies used by the other players. How is it then possible that a strategy combination may be an equilibrium fully satisfying this requirement, yet may nevertheless assign to some of the players irrational moves *not* maximizing their payoffs? The answer is that an equilibrium may contain such irrational moves only at information sets that will not be reached (that will be reached with zero probability) if the players follow their equilibrium strategies, so that any move that a player makes at such an information set will not affect his expected payoff.

For instance, in our example, if both players act in accordance with equilibrium E_2 then player 2's information set will not be reached. (It would be reached only if player 1 made move a. But it will not be reached because E_2 requires him to make move b instead.) Consequently, player 2 will never have to make the irrational move d at all. If player 2 did really have to make this move then he would lose 4 units of utility (because he would reduce his payoff from 2 to -2). But as player 1 will actually make move b (if he follows equilibrium E_1) the probability that player 2 will in fact have to make this payoff-losing move is zero.

Accordingly, Selten has suggested that we can recognize imperfect equilibria, and can actually eliminate them from the game, by replacing the original game G by a *perturbed* game G^*. G^* differs from G in the fact that, whenever any player i wants to make some move m, then he will have a very small but positive probability ϵ of making a different and unintended move $m' \neq m$ instead by "mistake." In this perturbed game G^*, owing to these postulated mistakes made with very small yet positive probabilities, all information sets will always be reached with some positive probability. Consequently, the imperfect equilibria of the original game G will no longer be equilibria in the perturbed game G^*. This model involving these mistakes occurring with small probabilities is sometimes called the *trembling-hand* model.

For instance, in our example, it would be enough to assume that player 1 has a "trembling hand." As a result, we would have to replace his pure strategies A and B with the perturbed strategies $A^* = (1-\epsilon)A + \epsilon B$ and

358

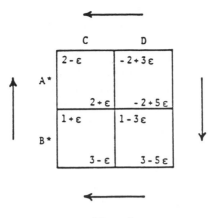

Figure 3

$B^* = \epsilon A + (1 - \epsilon)B$. As a result we would obtain the perturbed payoff table shown in Figure 3.

As this perturbed payoff table shows, the strategy pair (A^*, C), corresponding to the perfect equilibrium $E_1 = (A, C)$, is still an equilibrium in the new perturbed game, but the strategy pair (B^*, D), corresponding to the imperfect equilibrium $E_2 = (B, D)$, is no longer an equilibrium. This is so because even though D was a best reply to the unperturbed strategy B, it is not a best reply to the perturbed strategy B^* since (as can be seen in Figure 3) if player 1 uses B^* then strategy D will yield only $(3 - 5\epsilon)$ to player 2 while strategy C would yield the higher payoff $(3 - \epsilon)$ to him. The difference, 4ϵ, now obtains because in the disturbed game, player 2 would in fact have to implement the payoff-losing move d with probability ϵ if he used strategy D since player 1 would make move a with a positive probability ϵ by "mistake," even if his intended move were b rather than a.

The simplest way of constructing a perturbed game G^* is to assume that, for all players and at all of their information sets, the probability of making a mistaken move is the same very small probability ϵ. Such a perturbed game is called a *uniformly perturbed* game. Any equilibrium that remains an equilibrium even in the uniformly perturbed game is called a *uniformly perfect* equilibrium. The set of all uniformly perfect equilibria is in general a proper subset of the set of all perfect equilibria.

4. Other refinements proposed for Nash equilibria and the question of normal-form dependence

After Selten had proposed the concept of *perfect* equilibria, other game theorists have suggested other refinements to Nash's equilibrium concept.

These include Myerson's (1978) proper equilibria, Kalai and Samet's (1984) persistent equilibria, Kohlberg and Mertens's (1986) stable equilibria, as well as Kreps and Wilson's (1982) sequential equilibria (and some others).

Perfect and sequential equilibria in general cannot be recognized as such in the normal form of the game, and one may have to look at the extensive form (or at the agent normal form, which is in a sense halfway between the extensive and the normal forms) to identify them. In contrast, proper, persistent, and stable equilibria can all be recognized already in the normal form. This property of depending only on the normal form I will call *normal-form dependence*.

Kohlberg and Mertens have in fact argued that normal-form dependence should be regarded as an essential requirement for any acceptable solution concept. In this chapter I will try to show that actually normal-form dependence is not a reasonable requirement and that it often conflicts with the requirement of *sequential rationality* (also called *backward-induction rationality*), which Kohlberg and Mertens themselves recognize as a very important aspect of game-theoretic rationality.

By sequential rationality or backward-induction rationality we mean the following principle. Let G be a sequential game, that is, a game played in two or more stages. Let me call these stages $G_1, ..., G_K$. Then analysis of game G must always start with analysis of stage K, then must proceed to analysis of stage $(K-1)$, then to analysis of stage $(K-2)$, and so on. The reason is this. Even though the rules under which stage G_K will be played may very well depend on what happened at earlier stages, yet *given* these rules, the players' actual behavior in G_K will not depend on what happened at earlier stages. (Nor will it depend on what the players expect to happen at later stages because G_K is, by assumption, the last stage of the game.) More generally, at any stage G_k ($k = 1, ..., K-1$), the players' behavior will not depend on what happened prior to stage G_k. But if the players act rationally then their behavior at stage G_k *will* depend on their expectations on how their behavior at stage G_k will affect their strategic positions at later stages. Therefore, we cannot analyze stage G_k before we have completed our analysis of all later stages.

II. The problem of equilibrium selection

5. Payoff dominance and risk dominance in 2 × 2 games

In the last ten or twenty years, many economists have found that modeling economic problems as noncooperative games has important advantages over modeling them as cooperative games (as was fashionable in the 1950s and 1960s). One reason is that cooperative solution concepts are usually defined only for games in normal form or in characteristic-function form, and cannot be used for analyzing games in extensive form. They cannot be used for analyzing games with incomplete information (Harsanyi 1967/68) either. Moreover, noncooperative models permit more detailed analysis of the strategy problems facing the players, both in bargaining situations and in other game situations, than cooperative models do.

Yet, noncooperative-game models also pose a major problem: Almost all games arising in economic applications of game theory have a very large set, often an infinite set, of very different equilibria. This remains true even if we restrict ourselves to perfect equilibria, or to other classes of equilibria with special properties. Therefore, if all we can tell is that the outcome of the game will be an equilibrium then we are making a rather uninformative statement. In many cases the set of equilibria is so large that by saying that the outcome will be a point in this set, we are saying little more than almost anything can happen in the game.

To overcome this problem, Reinhard Selten and I have developed a *general theory of equilibrium selection* (Harsanyi and Selten 1988), providing rational criteria for selecting one equilibrium – more specifically, one *uniformly perfect* equilibrium – as the solution of any noncooperative game. Our theory can be extended also to cooperative games by remodeling the latter as noncooperative bargaining games. Of course, in this paper I have no space to describe our theory in detail. I can discuss only how it works in the case of a small class of rather simple two-person games.

Consider a two-person game in which each player has only two pure strategies. Such games are called *2×2 games*. Player 1's pure strategies will be called A_1 and A_2 while those of player 2 will be called B_1 and B_2. Moreover, I will assume that the game has two equilibria in pure strategies, namely $E_1 = (A_1, B_1)$ and $E_2 = (A_2, B_2)$. See Figure 4.

To ensure that E_1 and E_2 are in fact equilibria, we have to assume that

$$(2) \qquad a \geq e, \quad b \geq d, \quad g \geq c, \quad \text{and} \quad h \geq f.$$

For the purpose of choosing between the two equilibria E_1 and E_2, we will distinguish two cases.

Case 1: Both players prefer the same equilibrium. For instance, both of them may prefer E_1 to E_2. This will be the case if

$$(3) \qquad a > g \quad \text{and} \quad b > h.$$

In this case, we say that E_1 *payoff-dominates* E_2. More generally, in an n-person game we say that one equilibrium s^* payoff-dominates another equilibrium s if s^* gives every player a strictly higher payoff than s does, that is, if

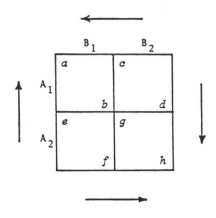

Figure 4

(4) $$U_i(s^*) > U_i(s) \quad \text{for } i = 1, \ldots, n.$$

In a 2×2 game, if one pure strategy equilibrium payoff-dominates the other then our theory selects this payoff-dominant equilibrium as the solution of the game.

Case 2: One player would prefer one pure strategy equilibrium whereas the other player would prefer the other. For instance, suppose that player 1 would prefer E_1, while player 2 would prefer E_2 as outcome because

(5) $$a > g \quad \text{but} \quad h > b.$$

In this case, our theory will select the solution of the game by using the concept of *risk dominance*. Intuitively speaking, we choose as solution the equilibrium associated with strategies less risky for the two players. More formally, in a 2×2 game, risk dominance can be defined as follows. As both $E_1 = (A_1, B_1)$ and $E_2 = (A_2, B_2)$ are equilibria, if player 1 were sure that player 2 would use strategy B_i $(i = 1, 2)$, then he would know that, in order to maximize his payoff, he would have to use strategy A_i. Likewise, if player 2 were sure that player 1 would use strategy A_i, then he would know that, in order to maximize his payoff, he would have to use strategy B_i. But what happens if each player is somewhat uncertain about the other player's strategy?

In particular, we know that player 1 would prefer $E_1 = (A_1, B_1)$ as outcome. Suppose he thinks that player 2 will use strategy B_1 with probability $(1 - p)$ and will use strategy B_2 with probability p. Then, p will be a measure of player 1's uncertainty as to whether player 2 will use his strategy B_1 corresponding to E_1. How large can p be without making player 1 himself switch from strategy A_1 to strategy A_2?

The answer is of course that player 1 will stick to strategy A_1 as long as

(6) $$(1-p)a + pc \geq (1-p)e + pg,$$

that is, as long as

(7) $$p \leq \frac{a-e}{(a-e)+(g-c)}.$$

The highest p value satisfying (7) is

(8) $$p^* = \frac{a-e}{(a-e)+(g-c)}.$$

We can regard p^* as a measure for player 1's insistence on making his favorite equilibrium E_1 the outcome of the game, becaue p^* is the highest uncertainty that player 1 would accept without switching from strategy A_1, prescribed by E_1, to strategy A_2, prescribed by E_2.

By the same token, player 2 would prefer E_2 as outcome. Suppose that he thinks that player 1 will use strategy A_2 with probability $(1-q)$ and will use strategy A_1 with probability q. Now, q will be a measure of player 2's uncertainty as to whether player 1 will use his strategy A_2 corresponding to E_2. How large can q be without making player 2 himself switch from strategy B_2 to strategy B_1?

The answer is of course that player 2 will stick to strategy B_2 as long as

(9) $$qd + (1-q)h \geq qb + (1-q)f,$$

that is, as long as

(10) $$q \leq \frac{h-f}{(h-f)+(b-d)}.$$

The highest q value satisfying (10) is

(11) $$q^* = \frac{h-f}{(h-f)+(b-d)}.$$

We can regard q^* as a measure for player 2's insistence on making his favorite equilibrium E_2 the outcome of the game, because q^* is the highest uncertainty that player 2 would accept without switching from strategy B_2, prescribed by E_2, to strategy B_1, prescribed by E_1.

It is natural to argue that if

(12) $$p^* > q^*,$$

then player 1's pressure to make E_1 the outcome of the game will be stronger than player 2's pressure to make E_2 the outcome and, as a result, in the end both players will accept E_1 as the outcome.

We can argue also as follows. Even before both players accept E_1 as the actual outcome, both of them must realize that E_1 is a more likely outcome than E_2 is (precisely because the pressure to make E_1 the outcome will be stronger than the pressure to make E_2 the outcome). Yet, if this is the case then it will be less risky for each player to use strategy A_1 or B_1 prescribed by E_1 on the assumption that E_1 will be the actual outcome, than it would be to use the strategy A_2 or B_2 prescribed by E_2 on the assumption that E_2 would be the actual outcome.

Accordingly, we say that if inequality (12) holds then E_1 *risk-dominates* E_2 whereas if inequality (12) is reversed then E_2 risk-dominates E_1. Moreover, if one player prefers one equilibrium while the other player prefers the other, then our theory selects the risk-dominant equilibrium as the solution of the game.

Yet, to complete my discussion of risk dominance in 2×2 games, I want to point out that inequality (12) can be written in a more convenient form. First of all, by (8) and (11), (12) is equivalent to

(13) $$\frac{a-e}{(a-e)+(g-c)} > \frac{h-f}{(h-f)+(b-d)}.$$

Yet, the latter can be written as

(14) $$(a-e)(b-d) > (g-c)(h-f).$$

The product on the left-hand side is called the *Nash product* for equilibrium E_1 whereas the product on the right-hand side is called the Nash product for equilibrium E_2. I will denote these two products by $\pi(E_1)$ and $\pi(E_2)$, respectively. (The term "Nash product" is used to indicate that our concept of risk dominance is a generalization of Nash's (1950) two-person bargaining solution, whose definition is based on maximizing a product of two payoff differences, which is often referred to as the Nash product.)

Note that the four payoff differences in (14) have the following intuitive interpretation. $(a-e)$ and $(b-d)$ are the losses that player 1 and player 2 would suffer if they deviated from their strategies A_1 and B_1 prescribed by E_1 to their alternative strategies A_2 and B_2, respectively, assuming that the other player would stick to A_1 or to B_1. Likewise, $(g-c)$ and $(h-f)$ are the losses player 1 and player 2 would suffer if they deviated from their strategies A_2 and B_2 prescribed by E_2 to their alternative strategies A_1 and B_1, respectively, assuming that the other player would stick to A_2 or to B_2. In other words, all four payoff differences occurring in (14) represent the losses the two players would suffer by unilaterally deviating from one of the two equilibria. Thus we can state:

In a 2×2 game, E_1 risk-dominates E_2 if

(15) $$\pi(E_1) = (a-e)(b-d) > \pi(E_2) = (g-c)(h-f),$$

and E_2 risk-dominates E_1 if this inequality is reversed.

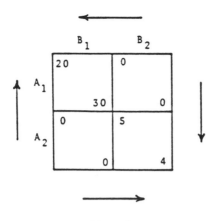

Figure 5

6. The special case of zero-based 2×2 games

A particularly simple class of 2×2 games are those in which the two non-equilibrium strategy pairs (A_2, B_1) and (A_1, B_2) yield zero payoffs to both players so that

(16)
$$c = d = e = f = 0.$$

Such 2×2 games are called *zero-based*. In such games, the two Nash products take the simpler form

(17)
$$\pi(E_1) = ab \quad \text{and} \quad \pi(E_2) = gh.$$

It is easy to see that in such games if either equilibrium E_i payoff-dominates the other equilibrium E_j ($i, j = 1, 2$ and $i \neq j$) then E_i also risk-dominates E_j. For instance, in the case where $i = 1$ and $j = 2$, payoff dominance means that

(18)
$$a > g \quad \text{and} \quad b > h.$$

This immediately implies that

(19)
$$\pi(E_1) = ab > \pi(E_2) = gh.$$

Yet, as we will see presently, this simple relationship in general fails to hold in non–zero-based 2×2 games.

7. The case of general 2×2 games

Consider the zero-based 2×2 game defined by Figure 5. Obviously, in this game, equilibrium $E_1 = (A_1, B_1)$ payoff-dominates equilibrium $E_2 =$

365

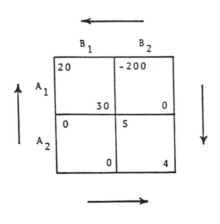

Figure 6

(A_2, B_2) because $20 > 5$ and $30 > 4$. Moreover, risk dominance goes in the same direction because $\pi(E_1) = 20 \times 30 = 600 > \pi(E_2) = 5 \times 4 = 20$.

Yet, suppose we change the payoff $U_1(A_1, B_2)$ from 0 to -200, as shown by Figure 6. E_1 still payoff-dominates E_2. But now risk dominance goes in the opposite direction because now $\pi(E_1) = 20 \times 30 = 600 < \pi(E_2) = [5 - (-200)]4 = 205 \times 4 = 820$. In fact, in this game, the two players face a genuine dilemma. On the one hand, it is still true that both players will be better off by making E_1 the outcome of the game. But, to do so, player 1 must use strategy A_1 and player 2 must use strategy B_1. Yet, for both players it now becomes rather risky to use these strategies. This is clear in the case of player 1: If he uses strategy A_1 then he now risks receiving the very unfavorable payoff, $u_1 = -200$. Yet, this makes it somewhat risky also for player 2 to use strategy B_1. For he can do so only if he can confidently expect player 1 to use strategy A_1; yet, the risk now associated with strategy A_1 from player 1's point of view must raise some doubt in player 2's mind whether player 1 will actually use this strategy.

Nevertheless, in cases like this, our theory always gives precedence to payoff dominance over risk dominance, and selects the payoff-dominant equilibrium E_1 as the solution of the game. The reason is that it is in both players' interest to make E_1 the outcome of the game and, therefore, they will do so if both of them are truly rational and fully trust each other's rationality. Yet, by recognizing that by choosing E_1 over E_2 the two players are making a risky choice, our theory does call attention to the real dilemma facing the two players in such cases.

The dilemma facing the two players becomes even sharper in the game defined by Figure 7. E_1 still payoff-dominates E_2, in the same way as it

366

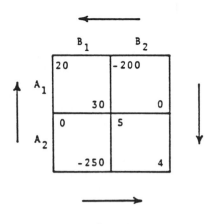

Figure 7

did in the two previous games. But now E_2 has even stronger risk dominance over E_1 because now not only does player 1 risk receiving the very unfavorable payoff $u_1 = -200$ if he uses strategy A_1 but player 2, also, risks receiving the very unfavorable payoff $u_2 = -250$ if he uses strategy B_1. More specifically, now

$$\pi(E_1) = 20 \times 30 = 600 < \pi(E_2) = [5 - (-200)][4 - (-250)]$$
$$= 205 \times 254 = 52{,}070.$$

Yet, even in this case, our theory selects the payoff-dominant equilibrium E_1 as the solution.

8. The intransitivity of risk dominance

Let me mention another problem, arising only in games larger than 2×2 games. This is the fact that, in general, risk dominance is not a transitive relation. Let me write $E_1 > E_2$ to indicate that E_1 risk-dominates E_2. Suppose that E_1, E_2, and E_3 are equilibria of a given game G. Then, it is perfectly possible that $E_1 > E_2$ and $E_2 > E_3$, yet that it is *not* the case that $E_1 > E_3$. In fact, risk dominance can follow a circular path, in the sense that $E_1 > E_2$ and $E_3 > E_3$, yet $E_3 > E_1$. An example for this situation will be discussed in Section 11.

In contrast, payoff dominance is always a transitive relation. Let me write $E_1 \gg E_2$ to indicate that E_1 payoff-dominates E_2. Whenever $E_1 \gg E_2$ and $E_2 \gg E_3$ then we can be sure that also $E_1 \gg E_3$. This is so because if, for all players i, $U_i(E_1) > U_i(E_2)$ and $U_i(E_2) > U_i(E_3)$ then it obviously follows that $U_i(E_1) > U_i(E_3)$.

367

9. The requirements of subgame consistency
and of truncation consistency

As is well known, in the extensive-form, a game is represented by a *game tree*. A *subgame G** is that part of a given game *G* that is represented by a *subtree* of this game tree, with the property that no information set of this subtree extends to nodes not belonging to this subtree. Intuitively, a subgame *G** is a self-contained part of game *G*. (The requirement that no information set of *G** should include nodes not belonging to *G** is meant to ensure that, when any player makes a move, he will always know whether his move will belong to *G** or not.)

Any (proper) subgame *G** is always preceded by information sets not belonging to *G**, and it will depend on the players' moves (and possibly on chance moves) whether any given subgame *G** will be reached or not. If *G** is reached at all then game *G* can be regarded as a two-stage game, with state 1 corresponding to those moves that had been made before *G** was reached, and stage 2 corresponding to the subgame *G** itself.

By the principle of sequential rationality (see Section 4), rational players will act in a subgame *G** in the same way as they would act if *G** were an independent game. In other words, what had happened in game *G* before *G** was reached will have no influence on the players' behavior in *G** itself, once the latter has been reached (though the players' moves and the chance moves occurring before *G** was reached will decide whether *G** will be reached at all). This principle that in any subgame *G** rational players will act in the same way as they would act if *G** were not a subgame but rather were an independent game is called *subgame consistency*. As a corollary of this principle, the payoff vector $u^* = (u_1^*, \ldots, u_n^*)$ the players will obtain in a subgame *G** will be the same payoff vector that the solution of *G** would assign to them if *G** were an independent game.

Let me now define the *truncation* game G^0 corresponding to a subgame *G** of game *G*. G^0 is defined as the game whose game tree can be obtained from the game tree of game *G* by cutting off the subgame *G** and then replacing it with the payoff vector u^* defined by the solution of *G**. Now, it is natural to argue that rational players will act, at all information sets belonging to game *G* but not belonging to subgame *G**, in the same way as they would act in game G^0 if the latter were an independent game. This principle is called *truncation consistency*.

This principle is based on the following considerations. For rational players, games G^0 and G will be equivalent because any given sequence of moves in G^0 will generate the same payoffs for them as the same sequence

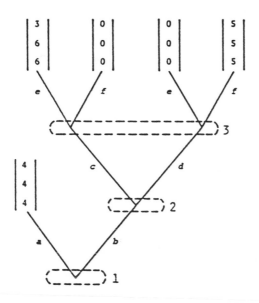

Figure 8

of moves in G would. The only difference is that the sequence of moves in G that would steer them into subgame G^* and would yield the payoff vector u^* to them in G^* will in game G^0 yield this payoff vector u^* directly, without a prior participation in subgame G^*. But since rational players will be interested only in the payoffs they will receive,[1] they will not care whether they will obtain this payoff vector u^* by participating in subgame G^* or without participating in G^*.

Both subgame consistency and truncation consistency are essential components of sequential rationality.

10. An example involving payoff dominance

Consider the following three-person game G in extensive form (see Figure 8). In this game, the information sets of players 2 and 3 form a subgame, to be called G^*, in which only players 2 and 3 are active players whereas player 1 is a passive player who does obtain a payoff but has no move in the subgame.

If player 1 makes move a then the game will end at once and the three players will obtain the payoff vector $(4, 4, 4)$. In contrast, if he makes move b then players 2 and 3 will have to play the subgame G^*. The latter has two equilibria in pure strategies, namely $E_1 = (c, e)$ and $E_2 = (d, f)$.

369

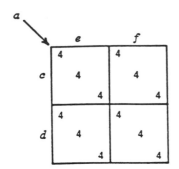
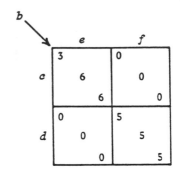

Figure 9

From the point of view of players 2 and 3, E_1 payoff-dominates E_2 because E_1 yields the payoffs $u_2 = u_3 = 6$ to them whereas E_2 would yield only the payoffs $u_2 = u_3 = 5$. Consequently, E_1 is the solution of the subgame. Therefore, player 1 will know that if he steers the game *into* this subgame, then the other two players will choose E_1 and he himself will obtain only the payoff $u_1 = U_1(E_1) = 3$. Knowing this, he will steer the game away from the subgame by making move a, so as to obtain the payoff $u_1 = 4 > 3$.

Yet, this fact could not be discovered by looking merely at the normal form of the game, shown by Figure 9. In this normal form,[2] player 1 chooses the payoff matrix, whereas player 2 chooses the row, and player 3 chooses the column. As this normal form shows, the game as a whole has only two really distinct equilibria, namely $\mathcal{E}_1 = (a, \cdot, \cdot)$ and $\mathcal{E}_2 = (b, d, f)$. The two dots in the definition of $\mathcal{E}_1$ indicate the fact that if player 1 chooses strategy a then it does not matter what strategies the other two players choose. Now $\mathcal{E}_1$ yields the payoff vector $(4, 4, 4)$ whereas $\mathcal{E}_2$ yields the payoff vector $(5, 5, 5)$. Thus, $\mathcal{E}_2$ clearly payoff-dominates $\mathcal{E}_1$. Hence, as far as one can tell from the normal form of the game, $\mathcal{E}_2$ should be the solution.

Yet, this is clearly an absurd conclusion. For the equilibrium $\mathcal{E}_2 = (b, d, f)$ cannot be reached at all, because if player 1 did choose strategy b then players 2 and 3 would not choose strategies d and f but, as we have seen and as the extensive form of the game clearly shows, they would rather choose strategies c and e instead. (To be sure, (b, c, e) is not an equilibrium of game G as a whole, but (c, e) is an equilibrium and, indeed, is the payoff-dominant equilibrium of subgame G^*.[3])

370

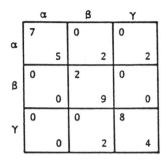

Figure 10. Game Γ.

11. An example involving risk dominance

Consider the normal-form two-person game Γ defined by Figure 10. Γ has three equilibria in pure strategies, namely (α, α), (β, β), and (γ, γ), which for convenience I will call simply α, β, and γ, respectively. None of these payoff-dominates any of the others. On the other hand, risk dominance is circular in that $\alpha > \beta > \gamma > \alpha$. To verify this, consider the three 2×2 games $\Gamma_{\alpha\beta}$, $\Gamma_{\beta\gamma}$, and $\Gamma_{\alpha\gamma}$ obtained from game Γ, shown by Figures 11, 12, and 13, respectively. Game $\Gamma_{\alpha\beta}$ is obtained from game Γ by omitting both row γ and column γ. Game $\Gamma_{\beta\gamma}$ is obtained from game Γ by omitting both row α and column α. Finally, game $\Gamma_{\alpha\gamma}$ is obtained from game Γ by omitting both row β and column β.

To verify that $\alpha > \beta$, we can apply definition (14) to game $\Gamma_{\alpha\beta}$. This gives $\pi(\alpha) = (7 - 0)(5 - 2) = 21 > \pi(\beta) = (2 - 0)(9 - 2) = 14$, indicating that indeed $\alpha > \beta$. To verify that $\beta > \gamma$, we can apply (14) to game $\Gamma_{\beta\gamma}$, which gives $\pi(\beta) = (2 - 0)(9 - 0) = 18 > \pi(\gamma) = (8 - 0)(4 - 2) = 16$, indicating that indeed $\beta > \gamma$. Finally, to verify that $\gamma > \alpha$, we apply (14) to game $\Gamma_{\alpha\gamma}$, which gives $\pi(\alpha) = (7 - 0)(5 - 3) = 14 < \pi(\gamma) = (8 - 0)(4 - 0) = 32$, verifying that $\gamma > \alpha$.[4]

Next consider the two two-person games G_1 and G_2 in extensive form, defined by Figures 14 and 15. As we will see, both games G_1 and G_2 have as normal form game Γ shown by Figure 10. In both games, the two players have to choose among the three outcomes α, β, and γ. The two games, however, differ in the order in which they are asked to choose among these three possible outcomes.

More particularly, in game G_1, at stage 1 of the game they have to decide whether they want to accept outcome β or to reject it. Acceptance is indicated by choosing move β whereas rejection is indicated by choosing

371

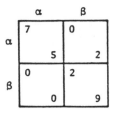

Figure 11. Game $\Gamma_{\alpha\beta}$.

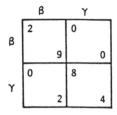

Figure 12. Game $\Gamma_{\beta\gamma}$.

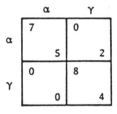

Figure 13. Game $\Gamma_{\alpha\gamma}$.

move $-\beta$. If one or both players vote for acceptance of β then the game ends already at stage 1. (If both vote for acceptance then the outcome will be β, yielding the payoffs $u_1 = 2$ and $u_2 = 9$ to players 1 and 2. If only one of them votes for acceptance then they obtain the lower payoffs $(0, 2)$ and $(0, 0)$, as indicated in Figure 14.) On the other hand, if both players vote for rejection of β then the game proceeds to stage 2, where they can choose between outcomes α and γ. In Figure 14, stage 1 of G_1 is represented by the lower information sets of both players whereas stage 2 of G_1 is represented by their upper information sets. As is easy to see, stage 2 of G_1 has the nature of a subgame, which I will call G_1^*.

372

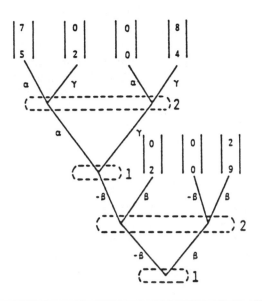

Figure 14. Game G_1.

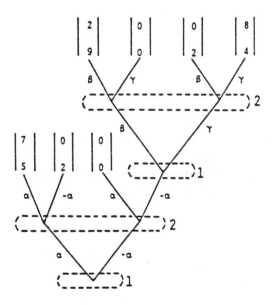

Figure 15. Game G_2.

In this game G_1, both players have the four strategies $\beta\alpha$, $\beta\gamma$, $-\beta\alpha$, and $-\beta\gamma$. (Here the first letter indicates the relevant player's move at stage 1 while the second letter indicates his move at stage 2.) Yet, strategies $\beta\alpha$ and $\beta\gamma$ are equivalent because if either player votes for β at stage 1 then the game will not proceed to stage 2 at all, so that it does not matter whether he would make move α or γ at stage 2 if there were a stage 2. One can easily verify that game G_1 does in fact have the normal form Γ shown by Figure 10 if one identifies each player's strategies $\beta\alpha$, $\beta\gamma$, $-\beta\alpha$, and $-\beta\gamma$ in game G_1 with his strategies α, β, and γ in game Γ as follows: $\beta\alpha = \beta\gamma = \beta$, $-\beta\alpha = \alpha$, and $-\beta\gamma = \gamma$.

In contrast, in game G_2, at stage 1 of the game the two players have to decide whether they want to accept or to reject outcome α by choosing between moves α and $-\alpha$. Again, if one or both players vote for acceptance then the game ends already at stage 1, with the payoffs shown by Figure 15. On the other hand, if both players vote for rejection of α then the game will proceed to stage 2, where they can choose between outcomes β and γ. Once more, in Figure 15, stage 1 of game G_2 corresponds to the lower information sets of players 1 and 2 whereas stage 2 of the game corresponds to their upper information sets. Again, stage 2 of G_2 has the nature of a subgame, to be called G_2^*.

In game G_2, each player has the four strategies $\alpha\beta$, $\alpha\gamma$, $-\alpha\beta$, and $-\alpha\gamma$, but of these $\alpha\beta$ and $\alpha\gamma$ are equivalent. One can easily verify that G_2 has the normal form Γ shown by Figure 10 if one identifies the strategies $\alpha\beta$, $\alpha\gamma$, $-\alpha\beta$, and $-\alpha\gamma$ of game G_2 with the strategies α, β, and γ of game Γ as follows: $\alpha\beta = \alpha\gamma = \alpha$, $-\alpha\beta = \beta$, and $-\alpha\gamma = \gamma$.

Yet, even though games G_1 and G_2 have the same normal form Γ, they have different solutions. This is so because of the circular risk dominance relation $\alpha > \beta > \gamma > \alpha$, which makes the outcome dependent on the order in which the players have to choose among the three possible outcomes. To find the solution of game G_1, we must first find the solution of subgame G_1^*. This subgame has the normal form $\Gamma_{\alpha\gamma}$ as shown by Figure 13. Because $\gamma > \alpha$, the solution of $\Gamma_{\alpha\gamma}$ is γ. Therefore, at stage 1 of game G_1, the two players will know that a vote $-\beta$ for rejecting β really amounts to a vote for γ. Consequently, as $\beta > \gamma$, at stage 1 they will both vote for accepting β. Therefore, the solution of the entire game G_1 is β.

In contrast, the solution of game G_2 is α. To verify this, we must first find the solution of the subgame G_2^*. This subgame has the normal form $\Gamma_{\beta\gamma}$ as shown by Figure 12. Because $\beta > \gamma$, the solution of $\Gamma_{\beta\gamma}$ is β. Therefore, at stage 1 of game G_2, the two players will know that a vote $-\alpha$ for rejecting α is really a vote for β. Consequently, as $\alpha > \beta$, at stage 1 they will both vote for accepting α. Therefore, the solution of the entire game G_2 is in fact α.

This example shows once more that, contrary to Kohlberg and Mertens's suggestion, we cannot expect the solution of a game to satisfy the requirement of normal-form dependence, that is, to be identifiable by looking merely at the normal form of the game. For games G_1 and G_2 have the same normal form, yet have different solutions. If we assigned the solution α of game G_2 to game G_1, or if we assigned the solution β of game G_1 to game G_2, we would violate the principles of sequential rationality.

12. Conclusion

In modeling economic or other real-life situations by noncooperative-game models, deciding how to select one particular Nash equilibrium as the solution from the often very large set of equilibria in the game is an important problem. Our theory uses the concepts of *payoff dominance* and of *risk dominance* as our main criteria for choosing a solution for the game. (For a more detailed discussion of our theory, see Harsanyi and Selten 1988.)

Kohlberg and Mertens (1986) recently argued that any acceptable solution concept should depend only on the normal form of the game. We have tried to show by means of two numerical examples that this requirement of normal-form dependence is not a reasonable requirement, and that it cannot be satisfied without violating the basic principles of sequential rationality (backward-induction rationality), which Kohlberg and Mertens themselves recognize as an important aspect of game-theoretic rationality.

NOTES

1. This is true by definition. For anything of any interest to a given player is modeled as something increasing or decreasing his payoff.
2. The normal form specified by Figure 9 is not the conventional normal form of the game. In the latter, player 2 would have four strategies, namely cc, cd, dc, and dd, rather than the two strategies c and d. (In the two-letter strategies, the first letter refers to the strategy that player 2 would use in the first matrix whereas the second letter refers to the strategy he would use in the second matrix.) Likewise, player 3 would have the four strategies ee, ef, fe, and ff, rather than the two strategies e and f. Yet, for our purposes, this conventional normal form would be needlessly complicated.
3. The strategy combination (b, c, e) is not an equilibrium in the entire game G because strategy b of player 1 is not a best reply to the strategies c and e of players 2 and 3. In other words, if player 1 expects the other two players to use strategies c and e then he will not use strategy b.
4. It is now easy to see why, in general, risk dominance is an intransitive relation. The reason is that the Nash product $\pi(E)$ associated with a given equilibrium E will in general depend on what other equilibrium E' is compared with E. Thus,

375

in our example, when α is compared to β then $\pi(\alpha) = 21$; whereas when α is compared with γ then $\pi(\alpha) = 14$.

REFERENCES

Harsanyi, J. C. (1967/68), "Games with Incomplete Information Played by Bayesian Players," Parts I, II, III. *Management Science* 14: 159–82, 320–34, 486–502.
Harsanyi, J. C., and Selten, R. (1988), *A General Theory of Equilibrium Selection in Games*. Cambridge, MA: MIT Press.
Kalai, E., and Samet, D. (1984), "Persistent Equilibria in Strategic Games." *International Journal of Game Theory* 13: 129–44.
Kohlberg, E., and Mertens, J. F. (1986), "On the Strategic Stability of Equilibria." *Econometrica* 54: 1003–37.
Kreps, D., and Wilson, R. (1982), "Sequential Equilibria." *Econometrica* 50: 863–94.
Myerson, R. B. (1978), "Refinements of the Nash Equilibrium Concept." *International Journal of Game Theory* 7: 73–80.
Nash, J. F. (1950), "The Bargaining Problem." *Econometrica* 18: 155–62.
Nash, J. F. (1951), "Noncooperative Games." *Annals of Mathematics* 54: 286–95.
Selten, R. (1965), "Spieltheoretische Behandlung eines Oligopolmodells mit Nachfrageträgheit," Parts I-II. *Zeitschrift für die Gesamte Staatswissenschaft* 121: 301–24, 667–89.
Selten, R. (1975), "Reexamination of the Perfectness Concept for Equilibrium Points in Extensive Games." *International Journal of Game Theory* 4: 25–55.

22

The dynamics of belief systems: Foundations versus coherence theories

PETER GÄRDENFORS

1. THE PROBLEM OF BELIEF REVISION

In this chapter I want to discuss some philosophical problems one encounters when trying to model the dynamics of epistemic states. Apart from being of interest in themselves, I believe that solutions to these problems will be crucial for any attempt to use computers to handle *changes* of knowledge systems. Problems concerning knowledge representation and the updating of such representations have become the focus of much recent research in artificial intelligence (AI).

Human beings perpetually change their states of knowledge and belief in response to various cognitive demands. There are several different kinds of belief changes. The most common type occurs when we learn something new, either by perception or by accepting the information provided by other people. This kind of change will be called an *expansion* of a belief state. Sometimes we also have to revise our beliefs in the light of evidence that contradicts what we had earlier (mistakenly) accepted, a process that will here be called a *revision* of a belief state. And sometimes – for example, when a measuring instrument is malfunctioning – we discover that the reasons for some of our beliefs are invalid, and so we must give up those beliefs. This kind of change will be called a *contraction* of an epistemic state. Note that when a state of belief is revised, it is also necessary to give up some of the old beliefs in order to maintain consistency. Some of the changes of states of belief are made in a rational way, others not. The main problem to be treated here is how to characterize rational changes of belief.

But before we can attack the problems related to the dynamics of epistemic states, we must know something about their statics: the properties

I want to thank Paul Gochet, Sven Ove Hansson, and David Makinson for several helpful comments. Research for this article has been supported by the Swedish Council for Research in the Humanities and Social Sciences.

of single states of belief. To this purpose, one can formulate criteria for what should count as a *rational belief state*. There are two main approaches to modeling epistemic states. One is the "foundations" theory, which holds that one needs to keep track of the justifications for one's beliefs: Propositions that have no justification should not be accepted as beliefs. The other is the "coherence" theory, which holds that one need not consider the pedigree of one's beliefs; the focus is instead on the logical structure of the beliefs – what matters is how a belief coheres with the other beliefs that are accepted in the present state. Each of these theories will be presented in greater detail in the following two sections. After that, I shall also introduce some empirical evidence relating to how people in fact do update their epistemic states, and discuss its relevance for the two theories.

It should be obvious that the foundations and the coherence theories have very different implications for what should count as rational changes of belief systems. According to the foundations theory, belief revision should consist, first, in giving up all beliefs that no longer have a satisfactory justification and, second, in adding new beliefs that have become justified. On the other hand, according to the coherence theory the objectives are, first, to maintain consistency in the revised epistemic state and, second, to make minimal changes of the old state that guarantee sufficient overall coherence. Thus, the two theories of belief revision are based on conflicting ideas of what constitutes rational changes of belief. The choice of underlying theory is, of course, also crucial for how an AI researcher will attack the problem of implementing a belief revision system on a computer.

In order to illustrate the more abstract aspects of these theories, I shall present two modelings of belief revisions. The first is Doyle's (1979) Truth Maintenance System (TMS), which is a system for keeping track of justifications in belief revisions; thus it directly follows the foundations theory.[1] The second is the theory of belief revision that has been developed by myself in collaboration with Carlos Alchourrón and David Makinson.[2] The theory is often referred to as the "AGM model" of belief revision. This modeling operates essentially with minimal changes of epistemic states and is thus, in its simplest form, in accordance with the coherence theory. The advantages and drawbacks of each modeling will be discussed.

One of the main criticisms directed against coherence models of belief states is that they cannot be used to express that some beliefs are justifications or reasons for other beliefs. This is true, of course, for single states of belief that only contain information about which beliefs are accepted and which are not. However, in Section 8 I shall argue that if one has further information about how such a state will potentially be revised

378

under various forms of input, then it is possible to formulate criteria for considering one belief to be a reason for another. In particular, the notion of the "epistemic entrenchment" of beliefs, which plays a central role in a development of the AGM model, is useful here. Adding this kind of information about the beliefs in an epistemic state will thus produce a model of belief revision that satisfies most of the requirements of the foundations as well as the coherence theory.

2. The foundations theory of belief revision

The basic principle of the foundations theory is that one must keep track of the reasons for the beliefs that are accepted in an epistemic state. This means that an epistemic state has a justificational structure, so that some beliefs serve as justifications for others.

Justifications come in two kinds. The standard kind of justification is that a belief is justified by one or several other beliefs, but not justified by itself. However, since all beliefs should be justified and since infinite regresses are disallowed, some beliefs must be justified by themselves. Harman (1986, p. 31) calls such beliefs *foundational*. One finds ideas like this, for example, in the epistemology of the positivists: Observational statements need no further justification – they are self-evident. Another requirement on the set of justifications is that it be noncircular, so that we do not find a situation where a belief in A justifies B, a belief in B justifies C, while C in turn justifies A. Sosa (1980) describes the justificational structure of an epistemic state in the following way: "For the foundationalist, every piece of knowledge stands at the apex of a pyramid that rests on stable and secure foundations whose identity and security does not derive from the upper stories or sections."

Another feature of the justification relation is that a belief A may be justified by several independent beliefs, so that even if some of the justifications for A are removed, the belief may be retained because it is supported by other beliefs.

Probably the most common models of epistemic states used in cognitive science are called *semantic networks*. A semantic network typically consists of a set of *nodes* representing some objects of belief and, connecting the nodes, a set of *links* representing relations between the nodes. The networks are then complemented by some implicit or explicit interpretation rules that make it possible to extract beliefs and epistemic attitudes. Changing a semantic network consists in adding or deleting nodes or links. If nodes represent beliefs and links represent justifications, semantic networks seem to be ideal tools for representing epistemic states according to the foundational theory.

379

However, not all nodes in semantic networks represent beliefs, and not all links represent justifications. Different networks have different types of objects as nodes and different kinds of relations as links. In fact, the diversity is so large that it is difficult to see what the various networks have in common. It seems that any kind of object can serve as a node in the networks, and that any type of relation or connection between nodes can be used as a link between nodes. This diversity seems to undermine the claims that semantic networks represent epistemic states. In his excellent methodological article, Woods (1975, p. 36) admits that "we must begin with the realization that there is currently no 'theory' of semantic networks." As a preliminary to such a theory, Woods formulates requirements for an adequate notation of semantic networks and explicit interpretation rules for such a notation. My aim here is not a presentation of his ideas, but only to give a brief outline of semantic networks as models of epistemic states suitable for the foundational theory.

If we now turn to the implications of the foundational theory for how belief revisions should be performed, the general principle formulated by Harman (1986, p. 39) is of interest:

Principle of Negative Undermining: One should stop believing P whenever one does not associate one's belief in P with an adequate justification (either intrinsic or extrinsic).

This means that if a state of belief containing A is revised so that the negation of A becomes accepted, not only should A be given up in the revised state, but also all beliefs that depend on A for their justification. Consequently, if one believes that B and all the justifications for believing B are given up, continued belief in B is no longer justified, so it should be rejected. A drawback of this principle, from an implementational point of view, is that it may lead to chain reactions and thus to severe bookkeeping problems. A specific example of this type of process (namely, TMS) will be presented in Section 5.

3. THE COHERENCE THEORY OF BELIEF REVISION

According to the coherence theory, beliefs do not usually require any justification; the beliefs are justified just as they are. A basic criterion for a coherent epistemic state is that it should be *logically consistent*. Other coherent criteria for the statics of belief systems are less precise. Harman (1986, pp. 32–3) says that coherence

includes not only consistency but also a network of relations among one's beliefs, especially relations of implication and explanation. . . . According to the coherence theory, the assessment of a challenged belief is always holistic. Whether such a belief is justified depends on how well it fits together with everything else one

380

believes. If one's beliefs are coherent, they are mutually supporting. All one's beliefs are, in a sense, equally fundamental.

Sosa (1980) compares the coherence theory with a raft: "For the coherentist, a body of knowledge is a free-floating raft every plank of which helps directly or indirectly to keep all the others in place, and no plank of which would retain its status with no help from the others."

A more far-reaching criterion on an epistemic state is that it be closed under logical consequences; that is, if A is believed in an epistemic state **K**, and B follows logically from A, then B is believed in **K** too. This criterion is presumed in the AGM model of belief revision that will be presented in Section 6. An immediate objection to this criterion is that human beings can handle only a finite number of beliefs and there are infinitely many logical consequences of any belief, so a human epistemic state cannot be closed under logical consequences. This objection will be discussed in Sections 7 and 8.

Turning now to the problem of belief revision, the central coherentist criterion is that changes of belief systems should be minimal. This criterion provides the motivation for many of the postulates formulated for expansions, contractions, and revisions of belief systems in the AGM model. The postulates capture some general properties of minimal changes of belief. However, from an implementational point of view, they leave the main problem unsolved: What is a reasonable metric for comparing different epistemic states that can be used in a computer program to update such states?

According to the coherence theory, belief revision is a "conservative" process, in the following sense.[3]

Principle of conservation. When changing beliefs in response to new evidence, you should continue to believe as many of the old beliefs as possible.

It should be noted that, in general, it is not meaningful to actually count the number of beliefs changed, but other comparisons of minimality must be applied.[4]

There is an economic side to rationality that is the main motivation for the principle of conservation. When we change our beliefs, we want to retain as much as possible of our old beliefs; information is in general not gratuitous, and unnecessary losses of information are therefore to be avoided. We thus have a criterion of *informational economy* motivating the coherentist approach.

The main drawback of coherence models of belief is that they cannot be used to directly express that some beliefs may be justifications for other beliefs. And, intuitively, when we judge the similarity of different epistemic states, we want the structure of justifications to count as well. In such a case we may end up contradicting the principle of conservation. The following example, adopted from Tichy (1976), illustrates this point:[5] Suppose that, in my present state **K** of belief, I accept as known that Oscar always wears his hat when it rains, but when it does not rain, he wears his hat completely at random (about 50% of the days). I also believe that it rained today and that he wore his hat. Let A be the proposition "It rains today" and B "Oscar wears his hat." Thus both A and B are accepted in **K**. What can we say about the beliefs in the state where **K** has been revised to include $\neg A$? Let us denote this state by $\mathbf{K}^*_{\neg A}$. According to the principle of conservation, B should still be accepted in $\mathbf{K}^*_{\neg A}$ because the addition of $\neg A$ does not conflict with B. However, I no longer have any *justification* for believing B, so intuitively neither B nor $\neg B$ should be accepted in $\mathbf{K}^*_{\neg A}$. This and similar examples will be discussed in Section 8, where a way out of the problem (remaining within the coherentist framework) will also be presented.

4. SOME EMPIRICAL EVIDENCE

Levi (1980, p. 1) criticizes the foundations theory of knowledge as follows:

Knowledge is widely taken to be a matter of pedigree. To qualify as knowledge, beliefs must be both true and justified. Sometimes justification is alleged to require tracing of the biological, psychological, or social causes of belief to legitimating sources. Another view denies that causal antecedents are crucial. Beliefs become knowledge only if they can be derived from impeccable first principles. But whether pedigree is traced to origins or fundamental reasons, centuries of criticism suggest that our beliefs are born on the wrong side of the blanket. There are no immaculate preconceptions.

If we consider how people in fact change their beliefs, Levi seems to be right. Although the foundations theory gives intuitively satisfying recommendations about what one *ought* to do when revising one's beliefs, the coherence theory is more in accord with what people *actually* do.

In a survey article, Ross and Anderson (1982) present some relevant empirical evidence.[6] Some experiments have been designed to explore "the phenomenon of belief perseverance in the face of evidential discrediting." For example, in one experiment,

[s]ubjects first received continuous false feedback as they performed a novel discrimination task (i.e., distinguishing authentic suicide notes from fictitious ones) . . . [Each subject then] received a standard debriefing session in which he learned that his putative outcome had been predetermined and that his feedback had been totally unrelated to actual performance. Before dependent variable measures were introduced, in fact, every subject was led to explicitly acknowledge his understanding of the nature and purpose of the experimental deception. (pp. 147–9)

Following this total discrediting of the original information, the subjects completed a dependent variable questionnaire dealing with [their] performance and abilities. The evidence for postdebriefing impression perseverance was unmistakable . . . . On virtually every measure . . . the totally discredited initial outcome manipulation produced significant "residual" effects . . . .

A recent series of experiments . . . first manipulated and then attempted to undermine subjects' theories about the functional relationship between two measured variables: the adequacy of firefighters' professional performances and their prior scores on a paper and pencil test of risk preference. . . . [S]uch theories survived the revelations that the cases in question had been totally fictitious and the different subjects had, in fact, received opposite pairings of riskiness scores and job outcomes.

Ross and Anderson conclude from these and other experiments that

it is clear that beliefs can survive potent logical or empirical challenges. They can survive and even be bolstered by evidence that most uncommitted observers would agree logically demands some weakening of such beliefs. They can even survive the total destruction of their original evidential basis. (p. 149)

Harman (1986, p. 38) has the following comments on the experiments:

In fact, what the debriefing studies show is that people simply do not keep track of the justification relations among their beliefs. They continue to believe things after the evidence for them has been discredited because they do not realize what they are doing. They do not understand that the discredited evidence was the *sole* reason why they believe as they do. They do not see they would not have been justified in forming those beliefs in the absence of the now discredited evidence. They do not realize these beliefs have been undermined. It is this, rather than the difficulty of giving up bad habits, that is responsible for belief perseverence.

So these empirical findings concerning belief perseverance mean that people are irrational? Shouldn't they try to keep track of the justifications of their beliefs? I think not. The main reason is that it is intellectually extremely costly to keep track of the sources of beliefs, and the benefits are, by far, outweighed by the costs.[7] A principle of intellectual economy would entail that it *is* rational to neglect the pedigree of one's beliefs. To be sure, we will sometimes hold on to unjustified beliefs, but the erroneous decisions caused by this negligence must be weighed against the cost of remembering all reasons for one's beliefs. The balance will certainly be in favor of forgetting reasons. After all, it is not very often that a justification for a belief is actually withdrawn and, as long as we do not introduce new beliefs without justification, the vast majority of our beliefs will hence remain justified.[8]

5. DOYLE'S TRUTH MAINTENANCE SYSTEM

Doyle's (1979) Truth Maintenance System (TMS) is an attempt to model changes of belief within the setting of the foundational theory. As Doyle remarks (p. 232), the name "truth maintenance system" not only sounds like Orwellian Newspeak but is also a misnomer, because what is maintained is the consistency of beliefs and reasons for belief. Doyle (1983) later changed the name to "reason maintenance system." In a broad sense TMS can be said to be a semantic network model, but its belief structure and its technique for handling changes of belief are more sophisticated than in other semantic network models.

There are two basic types of entities in TMS: *nodes* representing propositional beliefs and *justifications* representing reasons for beliefs. These justifications may be other beliefs from which the current belief is derived. A node may be *in* or *out,* which corresponds to the epistemic attitudes of accepting and not accepting the belief represented by the node. As should be expected, if a certain belief is *out* in the system, this does not entail that its negation is *in.* On the other hand, as a rationality requirement, if both a belief and its negation are *in,* then the system will start a revision of the sets of nodes and their justifications in order to reestablish consistency.

A justification consists of a pair of lists: an "inlist" and an "outlist." A node is *in* if and only if it has some justification (there may be several for the same node), the inlist of which contains only nodes that are *in* and the outlist of which contains only nodes that are *out.* A particular type of justification, called "nonmonotonic justification," is used to make tentative guesses within the system. For example, a belief in A can be justified simply by the fact that the belief in $\neg A$ is *out.* Beliefs that are justified in this way are called *assumptions.* This technique gives us a way of representing commonsense "default" expectations. It also leads to nonmonotonic reasoning in the following sense: If belief in A is justified only by the absence of any justification for $\neg A$, then a later addition of a justification for $\neg A$ will lead to a retraction of the belief in A.

The basic concepts of TMS are best illustrated by an example, as follows.

| | Justification | | |
Node	Inlist	Outlist	Status
(N1) Oscar is not guilty of defamation	(N2)	(N3)	*in*
(N2) The accused should have the benefit of the doubt	–	–	*in*
(N3) Oscar called the queen a harlot	(N4), (N5)	–	*out*
(N4) It may be assumed that the witness's report is correct	–	–	*in*
(N5) The witness says he heard Oscar call the queen a harlot	–	–	*out*

In this situation (N1) is *in* because (N2) is *in* and (N3) is *out*. Node (N3) is *out* because not both of (N4) and (N5) are *in*. If (N5) changes status to *in* (this may be assumed to be beyond the control of system), then (N3) will become *in* and consequently assumption (N1) is *out*.

Apart from the representation of nodes and justifications as presented here, TMS contains techniques for handling various problems that arise when the system of beliefs is adjusted to accommodate the addition of a new node or justification. In particular, when a contradiction is found, the system uses a form of backtracking to find the fundamental assumptions that directly or indirectly give support to the contradiction. One of these assumptions is chosen as the culprit and is given the status *out*. This process sometimes needs to be iterated, but it is beyond the scope of this chapter to give a full description of the mechanics of TMS.

However, the TMS representation of beliefs is not without epistemological problems. The example can be used to illustrate some of the drawbacks of TMS. In the handling of beliefs and justifications, TMS takes no notice of what the nodes happen to stand for. The sentences that I have added to the node names are not interpreted in any way by the system. This means that TMS completely lacks a semantic theory. As a consequence, much of the logic of propositions is lost in the TMS representation of beliefs. All forms of logical inferences that are to be used by the system must be reintroduced as special systems of justifications. Doyle discusses conditional proofs, but the process for handling such inferences seems extremely complex.

Furthermore, TMS leaves much of the work to the programmer. The programmer produces the nodes and their justifications; she must organize the information in levels, and she must also decide on how contradictions are to be engineered.[9]

6. The AGM model

An example of a model of belief revision that agrees with the coherence theory is the AGM model outlined in this section.[10]

Epistemic states are modeled by *belief sets,* which are sets of sentences from a given language. Belief sets are assumed to be closed under logical consequences (classical logic is generally presumed), which means that if **K** is a belief set and **K** logically entails B, then B is an element in **K**. A belief set can be seen as a partial description of the world, partial because in general there are sentences A such that neither A nor $\neg A$ are in **K**.

Belief sets model the statics of epistemic states; I now turn to their dynamics. What we need are methods for updating belief sets. Three kinds of updates will be discussed here.

(i) Expansion: A new sentence together with its logical consequences is added to a belief set **K**. The belief set that results from expanding **K** by a sentence A will be denoted $\mathbf{K}_A^+$.

(ii) Revision: A new sentence that is inconsistent with a belief set **K** is added, but in order that the resulting belief set be consistent, some of the old sentences of **K** are deleted. The result of revising **K** by a sentence A will be denoted $\mathbf{K}_A^*$.

(iii) Contraction: Some sentence in **K** is retracted without adding any new beliefs. In order that the resulting belief set be closed under logical consequences, some other sentences from **K** must be given up. The result of contracting **K** with respect to A will be denoted $\mathbf{K}_A^-$.

Expansions of belief sets can be handled with comparative ease. $\mathbf{K}_A^+$ can simply be defined as the logical consequences of **K** together with A:

(Def +) $$\mathbf{K}_A^+ = \{B : \mathbf{K} \cup \{A\} \vdash B\}.$$

Here $\vdash$ is the relation of logical entailment. As is easily shown, $\mathbf{K}_A^+$ defined in this way is closed under logical consequences, and will be consistent when A is consistent with **K**.

It is not possible to give a similar explicit definition of revisions and contractions in logical and set-theoretical notions only. To see the problem for revisions, consider a belief set **K** that contains the sentences A, B, $A \& B \rightarrow C$, and their logical consequences (among which is C). Suppose that we want to revise **K** by adding $\neg C$. Of course, C must be deleted from **K** when forming $\mathbf{K}_{\neg C}^*$, but at least one of the sentences A, B, or $A \& B \rightarrow C$ must also be given up in order to maintain consistency. There is no purely logical reason for making one choice rather than the other, so we must rely on additional information about these sentences. Thus, from a logical point of view, there are several ways of specifying the revision of a belief set. What is needed here is a (computationally well-defined) method of determining the revision.

As should be easily seen, the contraction process faces parallel problems. In fact, the problems of revision and contraction are closely related, being two sides of the same coin. To establish this more explicitly, we note that a revision can be seen as a composition of a contraction and an expansion. Formally, in order to construct the revision $\mathbf{K}_A^*$, one first contracts **K** with respect to $\neg A$ and then expands $\mathbf{K}_{\neg A}^-$ by A, which amounts to the following definition:

(Def *) $$\mathbf{K}_A^* = (\mathbf{K}_{\neg A}^-)_A^+.$$

Conversely, contractions can be defined in terms of revisions. The idea is that a sentence B is accepted in the contraction $\mathbf{K}_A^-$ if and only if B is accepted in both **K** and $\mathbf{K}_A^*$. Formally:

(Def −) $K_A^- = K \cap K_{\neg A}^*.$

These definitions indicate that revisions and contractions are interchangeable, and a method for explicitly constructing one of the processes would automatically yield a construction of the other.

There are two methods of attacking the problem of specifying revision and contraction operations. One is to present "rationality" postulates for the processes. Such postulates are introduced in Gärdenfors (1984) and Alchourrón et al. (1985), and are discussed extensively in Gärdenfors (1988); they will not be repeated here. In these works the connections between various postulates are also investigated.

The second method of solving the problems of revision and contraction is to adopt a more constructive approach. A central idea here is that the sentences that are accepted in a given belief set **K** have different degrees of *epistemic entrenchment;* not all sentences that are believed to be true are of equal value for planning or problem-solving purposes, but certain pieces of our knowledge and beliefs about the world are more important than others when planning future actions, conducting scientific investigations, or reasoning in general.[11] It should be noted that the ordering of epistemic entrenchment is not prima facie motivated by justificational considerations. However, the connection between entrenchment and justification will be discussed in Section 8.

The degrees of epistemic entrenchment of the sentences in a belief set will have a bearing on what is abandoned and what is retained when a contraction or a revision is carried out. The guiding idea for the construction is that when a belief set **K** is revised or contracted, the sentences in **K** that are given up are those having the lowest degree of epistemic entrenchment. Fagin, Ullman, and Vardi (1983, pp. 358ff.) introduce the notion of "database priorities," which is closely related to the concept of epistemic entrenchment and is used in a similar way to update belief sets.

I do not not assume that degrees of epistemic entrenchment can be quantitatively measured, but work only with qualitative properties of this notion. One reason for this is that I want to emphasize that the problem of uniquely specifying a revision method (or a contraction method) can be solved assuming very little structure on the belief sets apart from their logical properties.

If A and B are sentences, the notation $A \leq B$ will be used as shorthand for "B is at least as epistemically entrenched as A." The strict relation $A < B$ is defined in the usual way. Note that the relation $\leq$ is defined only in relation to a given **K**; different belief sets may be associated with different orderings of epistemic entrenchment.

The following postulates for epistemic entrenchment will be assumed:[12]

(EE1) Transitivity: If $A \leq B$ and $B \leq C$, then $A \leq C$.

(EE2) Dominance: If $A \vdash B$, then $A \leq B$.

(EE3) Conjunctiveness: For any A and B, $A \leq A \& B$ or $B \leq A \& B$.

(EE4) Minimality: If **K** is consistent, then $A \notin$ **K** if and only if $A \leq B$ for all B.

(EE5) Maximality: $B \leq A$ for all B, only if A is logically valid.

The motivation for (EE2) is that if A logically entails B, and either A or B must be retracted from **K**, then it will be a smaller change to give up A and retain B rather than to give up B, which would require that A be retracted too if we want the revised knowledge set to be closed under logical consequences. The rationale for (EE3) is as follows: If one wants to retract $A \& B$ from **K**, this can be achieved only by giving up either A or B, and consequently the informational loss incurred by giving up $A \& B$ will be the same as the loss incurred by giving up A or that incurred by giving up B. Note that it follows already from (EE2) that $A \& B \leq A$ and $A \& B \leq B$. The postulates (EE4) and (EE5) only take care of limiting cases: (EE4) requires that sentences already not in **K** have minimal epistemic entrenchment in relation to **K**; (EE5) states that only logically valid sentences can be maximal in $\leq$.

After these technicalities, we can now return to the main problem of constructing a contraction (or revision) method. The central idea is that if we want to contract **K** with respect to A, then the sentences that should be retained in $\mathbf{K}_A^-$ are those which have a higher degree of epistemic entrenchment than A. For technical reasons (see Gärdenfors and Makinson 1988, pp. 89–90), the comparisons $A < B$ do not always give the desired result, but in general one must work with comparisons of the form $A < A \vee B$. The appropriate definition is the following:

(C−) $B \in \mathbf{K}_A^-$ if and only if $B \in$ **K** and
   (i) A is logically valid (in which case $\mathbf{K}_A^- = \mathbf{K}$) or
   (ii) $A < A \vee B$.

It can be shown (Gärdenfors and Makinson 1988, Thm. 4) that if the ordering $\leq$ satisfies (EE1)–(EE5) then the contraction method just defined satisfies all the appropriate postulates. By way of (Def *), the definition (C−) can also be used to construct a revision method with the desired properties.

From an epistemological point of view, this result suggests that the problem of constructing appropriate contraction and revision methods can be reduced to the problem of providing an appropriate ordering of epistemic entrenchment. Furthermore, condition (C−) gives an explicit answer to which sentences are included in the contracted belief set, given the initial belief set and an ordering of epistemic entrenchment. Computationally, applying (C−) is trivial, once the ordering ≤ of the elements of **K** is given.

7. FINITE BASES FOR INFINITE BELIEF SETS

I now turn to some of the problems of the AGM model. Although the model and the results presented in the previous section provide some insights into the problem of revising epistemic states, it seems doubtful whether the approach could really be used in a computational context (e.g., in updating databases). First of all, belief sets cannot be handled directly in a database because they contain an infinite number of logical consequences; what is needed is a finite representation of belief sets. Such a representation will be called a *finite base* for a belief set. Second, when we contract or revise a belief set it can be argued that what we operate on is not the belief set *in toto* but rather a finite base for the belief set. Nebel (1988, p. 165) writes that

[p]ropositions in finite bases usually represent something like facts, observations, rules, laws, etc., and when we are forced to change the theory we would like to stay as close as possible to the original formulation of the finite base.

And Makinson (1985, p. 357) observes that

. . . in real life when we perform a contraction or a derogation, we *never do it to the theory itself* (in the sense of a set of propositions closed under consequence) but rather on some finite or recursive or at least recursively enumerable *base* for the theory. . . . In other words, contrary to casual impressions, the *intuitive* processes of contraction and revision are always applied to more or less clearly identified finite or otherwise manageable *bases* for theories, which will in general be either irredundant or reasonably close to irredundant.

If **B** is a finite base, let Cn(**B**) be the set of all logical consequences of **B**. A consequence of using finite bases instead of belief sets to represent epistemic states is that there may be two different bases **B** and **B**' such that Cn(**B**) = Cn(**B**') but where revisions or contractions performed on these bases may lead to different new states. Hansson (1989) gives the following illustrative example:

. . . suppose that on a public holiday you are standing in the street of a town that has two hamburger restaurants. Let us consider the subset of your belief set that represents your beliefs about whether or not each of these two restaurants is open.

When you meet me, eating a hamburger, you draw the conclusion that at least one of the restaurants is open ($A \vee B$). Further seeing from a distance that one of the two restaurants has its lights on, you believe that this particular restaurant is open (A). This situation can be represented by the set of beliefs $\{A, A \vee B\}$.

When you have reached the restaurant, however, you find a sign saying that it is closed all day. The lights are only turned on for the purpose of cleaning. You now have to include the negation of A, i.e. $\neg A$ into your belief set. The revision of $\{A, A \vee B\}$ to include $\neg A$ should still contain $A \vee B$, since you still have reasons to believe that one of the two restaurants is open.

In contrast, suppose you had not met me or anyone else eating a hamburger. Then your only clue would have been the lights from the restaurant. The original belief system in this case can be represented by the set $\{A\}$. After finding out that this restaurant was closed, the resulting set should not contain $A \vee B$, since in this case you have no reason to believe that one of the restaurants is open.

This example illustrates the need to differentiate in some epistemic contexts between the set $\{A\}$ and the set $\{A, A \vee B\}$. The closure of $\{A\}$ is identical to the closure of $\{A, A \vee B\}$. Therefore, if all sets are assumed to be closed under consequence, the distinction between $\{A\}$ and $\{A, A \vee B\}$ cannot be made.

This example will be analyzed in Section 8.

8. REPRESENTING JUSTIFICATIONS WITH THE AID OF EPISTEMIC ENTRENCHMENT

It is now, finally, time to return to the challenge from the foundations theory that models of epistemic states based on the coherence theory cannot express that some beliefs are justifications or reasons for other beliefs. I want to show that if we take into account the information provided by the ordering of epistemic entrenchment, the belief sets of the AGM model can, at least to some extent, handle this problem.

As an illustration of the problems, let us return to the example in Section 3 where the belief set **K** was generated from the finite base consisting of the beliefs A "It rains today" and $A \rightarrow B$ "If it rains, Oscar wears his hat." A logical consequence of these beliefs is B "Oscar wears his hat." Now if it turns out that the belief in A was in fact false, so that it does not rain after all, then together with A we would like to get rid of B since A presumably is the only reason to believe in B. However, it seems that if we follow the principle of conservation, B should be retained in the revised state since that principle requires that we keep as many of the old beliefs as possible.

My reply is that this is far from obvious; in fact, *the result of the contraction depends on the underlying ordering of epistemic entrenchment.* The belief set **K** is the set of all logical consequences of the beliefs A and $A \rightarrow B$ – that is, $\mathrm{Cn}(\{A, A \rightarrow B\})$, which is identical to $\mathrm{Cn}(\{A \& B\})$. The logical relations between the elements of this set can be divided into equivalence classes that can be described as an 8-element Boolean algebra. This algebra can be depicted by the Hesse diagram shown in Figure 1.

Here, upward lines indicate logical implication and $\top$ is the logical tautology. If we want to contract **K** by A (in order to be able to later add $\neg A$) then we must give up the elements A and $A \& B$, as well as some of

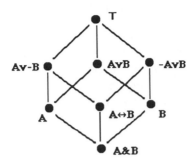

Figure 1

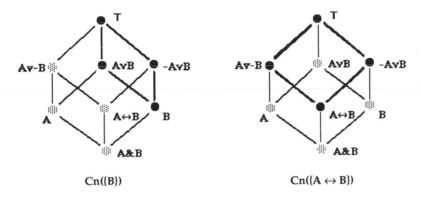

Cn({B})   Cn({A ↔ B})

Figure 2

the elements the conjunction of which entail A (e.g., we cannot have both $A \vee \neg B$ and $A \vee B$ in $\mathbf{K}_A^-$ because A would then also be in $\mathbf{K}_A^-$). If we require the principle of recovery, that $(\mathbf{K}_A^-)_A^+ = \mathbf{K}$,[13] then we are left with three alternatives for $\mathbf{K}_A^-$, namely the belief sets $\mathrm{Cn}(\{B\})$, $\mathrm{Cn}(\{A \leftrightarrow B\})$, and $\mathrm{Cn}(\{\neg A \vee B\})$. The first two of these are depicted (the black dots and heavy lines) as Hesse diagrams in Figure 2.

Which of these possibilities for $\mathbf{K}_A^-$ is the correct one is determined by the underlying ordering of epistemic entrenchment. This amounts to a choice between whether B or $A \leftrightarrow B$ should be included in $\mathbf{K}_A^-$ (we cannot have both). According to the criterion (C−), we should include in $\mathbf{K}_A^-$ exactly those elements C for which $A < A \vee C$. The proposition $A \vee (A \leftrightarrow B)$ is equivalent to $A \vee \neg B$, so ultimately we must decide which of $A \vee \neg B$ and $A \vee B$ has the highest degree of epistemic entrenchment.

To do this we should return to the example and look at the meaning of the formulas. $A \vee \neg B$ is equivalent to $\neg A \rightarrow \neg B$, which amounts roughly to "If it does not rain then Oscar does not wear his hat." On the other hand, $A \vee B$ is equivalent to $\neg A \rightarrow B$; in words, "If it does not rain then Oscar wears his hat." (Both formulas are material conditionals, and are included in $\mathbf{K}$ since they are logical consequences of A.) If rain is the *only* reason for Oscar to wear his hat, then $A \vee \neg B$ is epistemically more entrenched than $A \vee B$, and $\mathbf{K}_A^-$ will then be identical to $Cn(\{A \leftrightarrow B\})$. On the other hand, if Oscar wears his hat even if it does not rain, then $A \vee B$ is the more entrenched proposition and in this case we have $\mathbf{K}_A^- = Cn(\{B\})$. Finally, if they are regarded epistemically as being equally entrenched then we have the third case where neither of the two disjunctions $A \vee \neg B$ and $A \vee B$ will be included, so $\mathbf{K}_A^- = Cn(\{\neg A \vee B\})$.

The point of this fairly technical exercise is to show that if we consider the extra information provided by the ordering of epistemic entrenchment, then this information suffices to account for much of what is required by a foundations theory of belief revision. A belief set in itself does not contain any justifications of the beliefs included. However, if an ordering of epistemic entrenchment is added to such a belief set, then at least some justifications for the beliefs can be reconstructed from this ordering.

A similar analysis can be applied to the hamburger example of the previous section. Let A be the proposition that the restaurant with the lights on is open, and let B be the proposition that the other restaurant is open. Hansson claims that if all sets are assumed to be closed under logical consequence, the distinction between $\{A\}$ and $\{A, A \vee B\}$ cannot be made. This is true enough if we consider belief sets without the ordering of epistemic entrenchment. However, the relevant distinction can be made if such an ordering is introduced. If $\mathbf{K} = Cn(\{A\}) = Cn(\{A, A \vee B\})$ and if we want to form the contraction $\mathbf{K}_A^-$, we must decide which of the disjunctions $A \vee B$ and $A \vee \neg B$ is epistemically more entrenched. If you have met a man eating hamburgers so that you believe that at least one restaurant is open, this information is captured by the fact that $A \vee B$ is more entrenched than $A \vee \neg B$. Consequently, $\mathbf{K}_A^-$ will in this case be identical to $Cn(\{A \vee B\})$ as desired. Without this information, however, you have no reason why $A \vee B$ should be more entrenched than $A \vee \neg B$. If they are equally entrenched then $\mathbf{K}_A^-$ will be $Cn(\emptyset)$ – that is, will contain only logically valid formulas. (The third case, where $A \vee \neg B$ is more entrenched than $A \vee B$, does not occur in Hansson's example. In this case $\mathbf{K}_A^-$ would have been $Cn(\{A \vee \neg B\})$.) In conclusion, the distinction between $\{A\}$ and $\{A, A \vee B\}$ *can* be made without resorting to finite bases, if one considers the epistemic entrenchment of the beliefs in a belief set.

The upshot of the analyses of the two examples presented here is that the ordering of epistemic entrenchment of propositions, if added to a belief set, contains information about which beliefs are *reasons* for other beliefs. From a computational point of view, keeping track of the entrenchment of propositions in a belief set is much simpler than the techniques used in justification-based systems like TMS.[14]

However, I have not provided a definition, in terms of epistemic entrenchment, of what constitutes a reason. The only analysis (that I know of) of reasons and causes in terms of belief revisions is that of Spohn (1983). I shall now show how Spohn's analysis relates to the one presented here, and also indicate some problems connected with his proposal.

The guiding idea for Spohn is that "my belief in a reason strengthens my belief in what it is a reason for" (p. 372). More technically, he formulates this as follows: A is a reason for B for the person X (being in an epistemic state $\mathbf{K}$) if and only if X's believing A would raise the epistemic rank of B.[15] In the terminology of this chapter, the notion of "raising the epistemic rank" can be defined as follows:

(R) A is a reason for B in $\mathbf{K}$ if and only if
    (a) $B \in \mathbf{K}_A^*$ but $B \notin \mathbf{K}_{\neg A}^*$, or
    (b) $\neg B \in \mathbf{K}_{\neg A}^*$ but $\neg B \notin \mathbf{K}_A^*$.

If we now apply (Def *) and (Def +), (R) can be rewritten as

(R′) A is a reason for B in $\mathbf{K}$ if and only if
    (a) $A \to B \in \mathbf{K}_{\neg A}^-$ but $\neg A \to B \notin \mathbf{K}_A^-$, or
    (b) $\neg A \to \neg B \in \mathbf{K}_A^-$ but $A \to \neg B \notin \mathbf{K}_{\neg A}^-$.

And if (C−) is applied in the case when both A and B are in $\mathbf{K}$, case (a) can in turn be formulated as

(R″) A is a reason for B in $\mathbf{K}$ if
    $\neg A < \neg A \vee B$ and $A \vee B \leq A$.

The clause that $\neg A < \neg A \vee B$ is automatically fulfilled according to the postulate (EE4), since in the assumed case $\neg A \notin \mathbf{K}$ but $\neg A \vee B \in \mathbf{K}$. So what remains is the following simple test:

(R‴) If A and B are in $\mathbf{K}$ and $A \vee B \leq A$, then A is a reason for B in $\mathbf{K}$.

(A similar analysis can be applied to case (b) of (R′).) Note that in the case assumed in (R‴) we also have $A \to B \in \mathbf{K}$.

(R‴) means that if we want to determine whether A is a reason for B, this can be done by comparing the epistemic entrenchment of $A \vee B$ with that of A. For example, if A is "It rains today" and B is "Oscar wears his hat" then A is a reason for B if $A \vee B \leq A$, that is, if it is at least as easy to

393

give up a belief in "It rains today or Oscar wears his hat" as it is to give up a belief in "It rains today." Note that to give up $A \vee B$ we must give up both A and B. Thus $A \vee B \leq A$ obtains only if B is retracted as soon as A is. However, this is just another way of saying that A is the only justification for B!

What this series of transformations of Spohn's condition (R) shows is that his analysis of "being a reason for" can be translated into an analysis in terms of epistemic entrenchment. However, Spohn's definition suffers from some defects.[16] We would like to require that a relation of "being a reason for" be noncircular, in the sense that we never have that A_1 is a reason for A_2 is a reason for . . . is a reason for A_1, where the A_i's are logically different propositions. But Spohn's definition does not satisfy this requirement.

To see this, let A be a proposition in a belief set $\mathbf{K}$ and let B be a proposition that is not equivalent to A or $\neg A$. According to (EE3) we have either $A \vee B \leq A$ or $A \vee \neg B \leq A$. Assume the first case; the second is similar. Since $A \vee B$ is in $\mathbf{K}$, it follows that $A \vee (A \vee B) \leq A$ and hence from (R''') that A is a reason for $A \vee B$. On the other hand, it follows from (EE2) that $A \vee (A \vee B) \leq A \vee B$, so $A \vee B$ is a reason for A according to (R'''). I have here used (R''') as a basis for my criticism, but an analogous example can be provided showing that Spohn's definition (R) will lead to the same form of circularity. Thus the relation of "being a reason for" is circular, in an undesirable way. This means that Spohn's definition must, at least, be supplemented with other criteria before it can be accepted as an analysis of the justification relation.

9. Conclusion

The foundational and the coherence theories of knowledge representation have traditionally been presented as being in conflict with each other. The foundational theory stresses justifications of beliefs and tones down logical relations between beliefs; the coherence theory emphasizes logical relations, but cannot keep track of justifications. I have presented one paradigm example of each theory, namely TMS and the AGM model. The main drawback of TMS is precisely that it has severe problems handling logical relations between beliefs.

In the AGM model one can distinguish two levels of representing epistemic states. At the first level an epistemic state is represented by a belief set that contains only information about whether the relevant propositions are accepted or not. Belief sets are construed to mirror logical relations between propositions, but there is no way of representing reasons or justifications at this level.

However, at the second level – where a belief set is supplemented with an ordering of epistemic entrenchment of the propositions – the situation is quite different. In this chapter I have aimed at showing that it is possible to express many forms of reasons or justifications on the basis of this additional information about the beliefs in an epistemic state. The upshot is that representing epistemic states as belief sets, together with an ordering of epistemic entrenchment, provides us with a way to handle most of what is desired both by the foundations theory and by the coherence theory. However, as the criticism of Spohn's analysis shows, a conclusive definition of "reason for" within this setting is still lacking.

NOTES

1. For newer developments of the theory see, e.g., Goodwin (1987).
2. See in particular Alchourrón, Gärdenfors, and Makinson (1985), Gärdenfors (1988), and Gärdenfors and Makinson (1988).
3. Gärdenfors (1988, p. 67); cf. also Harman (1986, p. 46).
4. This problem is discussed in Gärdenfors (1988, §3.5).
5. Cf. also Stalnaker (1984, pp. 127-9).
6. Cf. also Harman (1986, pp. 35-7).
7. The same seems to apply to computer programs based on the foundations theory. As will be seen in Section 5, one of the main drawbacks of the TMS system is that it is inefficient.
8. Harman (1986, pp. 41-2) expresses much the same point when he writes about "clutter avoidance."
9. A later development of the TMS system is presented in Goodwin (1987). A system with a slightly different methodology, but still clearly within the foundations paradigm, is de Kleer's (1986) Assumption-based TMS (ATMS). For further comparison between ATMS and TMS, see Rao (1988).
10. The main references are Gärdenfors (1984), Alchourrón et al. (1985), Gärdenfors (1988), and Gärdenfors and Makinson (1988).
11. The epistemological significance of the notion of epistemic entrenchment is further spelled out in Gärdenfors (1988, §§4.6-4.7).
12. Cf. Gärdenfors (1988, §4.6) and Gärdenfors and Makinson (1988).
13. For a presentation and discussion of this postulate, see Gärdenfors (1988, p. 62).
14. Some computational aspects of the entrenchment relation are studied in Gärdenfors and Makinson (1988). Foo and Rao (1988) also show how an ordering of epistemic entrenchment can be used very fruitfully to solve problems in planning programs. Their principle (P<) is clearly related to the analysis presented here.
15. In his paper, Spohn (1983) also discussed probabilistic reasons and causes. For this case, the notion of epistemic rank receives a different analysis.
16. This has been pointed out to me by David Makinson.

REFERENCES

Alchourrón, C. E., Gärdenfors, P., and Makinson, D. (1985), "On the Logic of Theory Change: Partial Meet Functions for Contraction and Revision." *Journal of Symbolic Logic* 50: 510-30.

De Kleer, J. (1986), "An Assumption-based TMS." *Artificial Intelligence* 28: 127–62.

Doyle, J. (1979), "A Truth Maintenance System." *Artificial Intelligence* 12: 231–72.

Doyle, J. (1983), "The Ins and Outs of Reason Maintenance." Report CMU-CS-83-126, Department of Computer Science, Carnegie-Mellon University.

Fagin, R., Ullman, J. D., and Vardi, M. Y. (1983), "On the Semantics of Updates in Databases." In *Proceedings of Second ACM SIGACT-SIGMOD*, Association for Computing Machinery, New York, pp. 352–65.

Foo, N. Y., and Rao, A. S. (1988), "The Gärdenfors Postulates and Nonlinear Planning." Mimeo, Basser Department of Computer Science, University of Sydney, Australia.

Gärdenfors, P. (1984), "Epistemic Importance and Minimal Changes of Belief." *Australasian Journal of Philosophy* 62: 136–57.

Gärdenfors, P. (1988), *Knowledge in Flux: Modeling the Dynamics of Epistemic States.* Cambridge, MA: MIT Press.

Gärdenfors, P., and Makinson, D. (1988), "Revisions of Knowledge Systems Using Epistemic Entrenchment." In M. Y. Vardi (ed.), *Proceedings of the Second Conference on Theoretical Aspects of Reasoning about Knowledge.* Los Altos, CA: Morgan Kaufmann, pp. 83–95.

Goodwin, J. W. (1987), "A Theory and System for Non-monotonic Reasoning." Dissertation No. 165, Linköping Studies in Science and Technology, Linköping University, Sweden.

Hansson, S. O. (1989), "New Operators for Theory Change." *Theoria* 55: 114–39.

Harman, G. (1986), *Change in View.* Cambridge, MA: MIT Press.

Levi, I. (1980), *The Enterprise of Knowledge.* Cambridge, MA: MIT Press.

Makinson, D. (1985), "How to Give It Up: A Survey of Some Formal Aspects of the Logic of Theory Change." *Synthese* 62: 347–63.

Nebel, B. (1988), "Reasoning and Revision in Hybrid Representation Systems." Lecture Notes in Computer Science, 422. Berlin: Springer.

Rao, A. S. (1988), "Dynamics of Belief Systems: A Philosophical, Logical and AI Perspective." Ph.D. dissertation, Basser Department of Computer Science, University of Sydney, Australia.

Ross, L., and Anderson, C. A. (1982), "Shortcomings in the Attribution Process: On the Origins and Maintenance of Erroneous Social Assessments." In D. Kahneman, P. Slovic, and A. Tversky (eds.), *Judgement under Uncertainty: Heuristics and Biases.* Cambridge: Cambridge University Press, pp. 129–52.

Sosa, E. (1980), "The Raft and the Pyramid: Coherence versus Foundations in the Theory of Knowledge." *Midwest Studies in Philosophy* 5: 3–25.

Spohn, W. (1983), "Deterministic and Probabilistic Reasons and Causes." *Erkenntnis* 19: 371–96.

Stalnaker, R. (1984), *Inquiry.* Cambridge, MA: MIT Press.

Tichy, P. (1976), "A Counterexample to the Stalnaker–Lewis Analysis of Counterfactuals." *Philosophical Studies* 29: 271–3.

Woods, W. A. (1975), "What's in a Link." In D. G. Bobrow and A. Collins (eds.), *Representation and Understanding.* New York: Academic Press, pp. 35–81.

23

Counterfactuals and a theory of equilibrium in games

HYUN SONG SHIN

1. INTRODUCTION

In this chapter, I shall develop a theory of equilibrium in normal-form games based on a formalization of counterfactuals. This theory has its starting point in the decision-theoretic framework of Jeffrey (1965), but develops this framework in ways that diverge from the spirit of Jeffrey's own theory. In order to motivate the theory, and to explain why it diverges from Jeffrey's, let us start by describing the main features of Jeffrey's framework for decisions.

In abolishing Savage's (1954) distinctions between "acts," "states of the world," and "consequences," Jeffrey (1965) was able to construct a unified framework for decisions in which the consequences of an individual's action and the action itself is as much a part of the description of the world as any other feature of the world. To choose an act in Jeffrey's framework is to make a certain proposition true. Thus, when Ω is the *state space* consisting of all states of the world ω, an *act* can be seen as a subset of Ω. To choose an act a_k is to ensure that the true state of the world is an element of a_k.

Let $\{a_1, a_2, \ldots, a_m\}$ be the set of acts for the decision maker. This set partitions Ω, reflecting the condition that one (and only one) act is chosen. It is assumed that the decision maker has a probability distribution p over Ω and that $u(\omega)$, the desirability of the state ω, is thus the value of a random variable u at ω. Denote by $E(u \mid a_k)$ the conditional expectation of u on a_k. Then, it might seem that the following is a natural criterion of rationality:

I am grateful to the participants at the workshop in Castiglioncello for their comments, and in particular to Michael Bacharach, Ken Binmore, Cristina Bicchieri, Bill Harper, Peter Gärdenfors, and Brian Skyrms. I also thank participants in seminars at Oxford and Essex for their comments.

(1.1) a_k is rational whenever $E(u\,|\,a_k) \geq E(u\,|\,a_l)$ for all actions a_l
for which the payoff $E(u\,|\,a_l)$ is defined.

We have supposed that the decision maker has a probability distribution p over Ω. An act is a subset of Ω, and so the decision maker attaches a probability to each of his own acts. Moreover, let us suppose that the decision maker is aware that he holds the beliefs implied by p. In particular, suppose p is the probability distribution obtained after full deliberation by the decision maker about his choice problem. In this case, the probability attached to an act is the probability attached by the decision maker to his choosing that act. Among other things, this would mean that when the decision maker decides *against* an act, zero probability is attached to choosing this act. That is:

(1.2) If a_k is not chosen then $p(a_k) = 0$.

This would seem to be uncontroversial in the extreme. It nevertheless generates a problem for our initial criterion for rationality given by (1.1). When a_k is chosen, $p(a_l) = 0$ for all $l \neq k$, so that $E(u\,|\,a_l)$ is undefined for all $l \neq k$. Because $E(u\,|\,a_k)$ is equal to itself, (1.1) is satisfied trivially, and *any* act that is chosen is rational. At the root of this difficulty is the fact that, when we abolish Savage's distinction between states of the world and consequences, the distribution p over Ω cannot be specified independently of the act chosen.

The problem is that, in order to assess the desirability of an act, the conditional expectation of the random variable u must be defined, and for this, the particular act must be given positive probability. This, in turn, implies that the act is chosen by the decision maker with nonzero probability. Clearly, what is required is a way of assessing the desirability of those acts that are *not* chosen. What is required is a device enabling the decision maker to use the sort of reasoning contained in the statement:

(1.3) "I choose x, since if I were to choose y, the outcome would
be z."

Such statements are counterfactuals, and we are led to an investigation of how such statements may be formalized in the context of Jeffrey's framework for decisions.

I shall employ the theory of counterfactuals developed by Stalnaker (1968) and Lewis (1973), based on the notion of a set of *possible worlds*. We shall start by constructing a possible-worlds space for a given normal-form game, and formalize the notion of "similarity" of possible worlds by means of a *metric* on this space. Thus, two possible worlds are *similar* if the distance between them is small. This approach to the analysis of

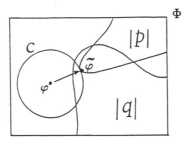

Figure 1

counterfactuals seems particularly suited to game theory, since the subject matter of game theory (strategies, payoffs, and probabilities) supplies some very natural metrics on the space of natural worlds.

The flavor of our approach can be conveyed by Figure 1, where Φ is the set of possible worlds, $|p|$ is the subset of Φ at which the proposition p is true, and $|q|$ is the subset of Φ at which the proposition q is true. Suppose φ is the true world, and we are concerned with the truth or falsity of the counterfactual "if p were the case then q would be the case." We denote this counterfactual by $p \,\square\!\!\rightarrow q$.

We implement the Stalnaker–Lewis criterion by introducing a metric m on Φ. We pick out the closest possible world to φ in this metric in which p is true, and see whether q is also true there. If so, then $p \,\square\!\!\rightarrow q$ is true at φ. Otherwise, $p \,\square\!\!\rightarrow q$ is false at φ. In Figure 1, the closest possible world to φ in which p is true is $\tilde{\varphi}$. But q is also true at $\tilde{\varphi}$. Thus, we conclude that $p \,\square\!\!\rightarrow q$ is true at φ.

The counterfactuals of particular interest to use are those of the form "if player i were to play strategy x, his payoff would be higher." When the truth values of these counterfactuals are known, a natural rationality criterion suggests itself. Namely, a player should never find himself at a possible world at which, according to his metric m, his payoff would be higher if he were to deviate. This is the principle that motivates the rationality criterion in this chapter.

For the particular metric adopted here, the rationality criterion just sketched turns out to be identical to that proposed by Aumann (1987). As Aumann starts from quite different premises, the discussion in this chapter should have some value in casting Aumann's framework in a different light. Before presenting the general theory, we start with an example to illustrate the central ideas.

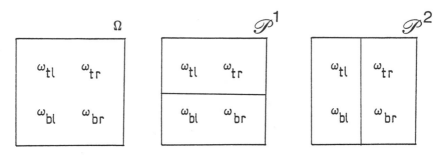

Figure 2

2. An example

Consider the following 2×2 game, the familiar "chicken" game:

	L	R
T	6, 6	2, 7
B	7, 2	0, 0

We define the state space Ω for this game to be the set $\{\omega_{tl}, \omega_{tr}, \omega_{bl}, \omega_{br}\}$, where ω_{tl} is the state in which 1 plays T and 2 plays L, ω_{tr} is the state in which 1 plays T and 2 plays R, and so on. Notice that each state specifies the actions of both players. Each player has a partition of Ω, so that to choose an act corresponds to choosing an element of this partition. Player 1's partition is denoted by $\mathcal{P}^1$ and player 2's partition by $\mathcal{P}^2$; $\mathcal{P}^1$ and $\mathcal{P}^2$ are depicted in Figure 2.

We suppose that each player has a prior probability distribution p over Ω, and forms beliefs by conditioning on the element of the partition that contains the true state. In particular, a player attaches probabilities to propositions of the form "player i plays strategy s." For example, if player 1 has the prior distribution p given by

$$p(\omega_{tl}) = p(\omega_{tr}) = p(\omega_{bl}) = \frac{1}{3},$$

then at ω_{tl} he holds the following beliefs:

(i) T is played with probability 1;
(ii) L is played with probability $\frac{1}{2}$;
(iii) R is played with probability $\frac{1}{2}$.

More succinctly, we can combine (ii) and (iii) to give the proposition

(ii)' Player 2 randomizes $(\frac{1}{2}, \frac{1}{2})$ over her strategy set.

400

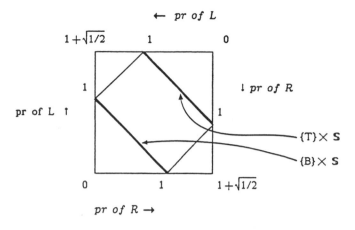

Figure 3

In general, at any state $\omega \in \Omega$, each player attaches probability 1 to one of his own actions, and believes that his opponent randomizes with some pair of probabilities. For player i, we represent these beliefs by the pair $\langle x, y \rangle$, where x is a pure strategy of i and y is a probability distribution over the strategies of i's opponent. The interpretation of the pair $\langle x, y \rangle$ is that player i attaches probability 1 to his own strategy x and believes that his opponent randomizes with distribution y. For example, player 1's beliefs at the state ω_{ll} are represented by the pair $\langle T, x \rangle$, where x is the randomization $\langle \frac{1}{2}, \frac{1}{2} \rangle$. The set of all such pairs for player 1 is given by the product $\{T, B\} \times S$, where S is the one-dimensional unit simplex representing the set of all probability distributions over $\{L, R\}$.

We shall define $\{T, B\} \times S$ to be player 1's *possible-worlds space*, denoted Φ^1. We denote by φ a typical element of this set, calling it a *possible world* for player 1. Geometrically, we can represent Φ^1 as in Figure 3; Φ^1 is represented by the two parallel bold lines. The upper line is the set $\{T\} \times S$ and the lower line is $\{B\} \times S$. As we move toward the top left-hand corner, the probability of L increases; as we move toward the bottom right-hand corner, the probability of R increases. The reason for this particular representation will become clear when we introduce a particular metric on this space.

Given a prior probability distribution p for player 1, we can define a function $\beta^1: \Omega \to \{T, B\} \times S$ such that $\beta^1(\omega)$ represents the beliefs held by player 1 at ω. When $\beta^1(\omega) = \varphi$, we shall use the metaphor "at ω, player 1 believes he is at the possible world φ." Figure 4 depicts the image of the function β^1.

401

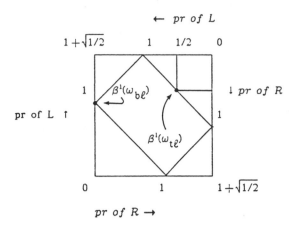

Figure 4

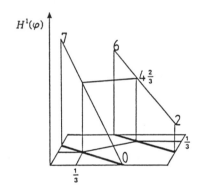

Figure 5

Each possible world φ in Φ^1 determines a unique action for player 1 and a unique probability distribution over 2's actions. Hence, each φ determines an expected payoff for player 1. This is shown in Figure 5, where $H^1(\varphi)$ denotes the expected payoff of player 1 at φ.

Having thus defined the possible-worlds space for player 1, we introduce a metric on this space that will serve as player 1's theory of the similarity between possible worlds. Each point in Φ^1 is a pair $\langle x, y \rangle$, where x is either T or B and $y = \langle y_1, y_2 \rangle \in \mathbf{R}^2$, where $y_1 + y_2 = 1$. For any two points $\langle x, y \rangle$ and $\langle \bar{x}, \bar{y} \rangle$ in Φ^1, define the distance between them as

402

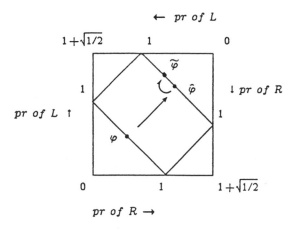

Figure 6

$$(2.1) \quad \begin{cases} \left[\sum_{i=1}^{2} |y_i - \tilde{y}_i|^2 \right]^{1/2} & \text{if } x = \tilde{x}, \\[2mm] \left[\sum_{i=1}^{2} |y_i - \tilde{y}_i|^2 \right]^{1/2} + 1 & \text{if } x \neq \tilde{x}. \end{cases}$$

Geometrically, this metric could be dubbed the "library-stack metric." Refer to Figure 6. Imagine the two bold lines representing Φ^1 to be two parallel corridors in a library. There are stacks of shelves at right angles to the corridors. In order to get from one point in the library to another, one must follow the corridors and the spaces between the stacks. Thus, for example, the distance between φ and $\tilde{\varphi}$ in Figure 6 is the sum of two distances – between φ and $\hat{\varphi}$ (which is 1) and between $\hat{\varphi}$ and $\tilde{\varphi}$ (which is the Euclidean distance between $\hat{\varphi}$ and $\tilde{\varphi}$). We denote this metric by λ.

Given a player's possible-worlds space and his metric, we have the apparatus to analyze the counterfactuals entertained by that player. Refer to Figure 7. Suppose player 1 believes that he is at the possible world φ. At φ, 1 plays T, and his payoff is 4. We are interested in the following counterfactual:

(2.2) "If player 1 were to play B, his payoff would be higher."

To evaluate this counterfactual at φ, we find the closest possible world to φ at which player 1 plays B, and see whether 1's payoff is higher here than at φ. As we see in Figure 7, there is a unique closest possible world in which 1 plays B – namely, $\tilde{\varphi}$. But at $\tilde{\varphi}$, 1's payoff is 3.5. Thus, according to our criterion for analyzing counterfactuals, (2.2) is false at φ.

403

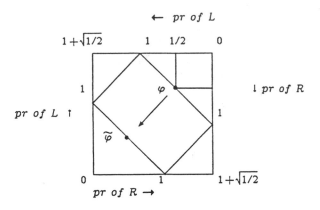

Figure 7

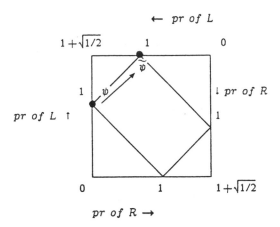

Figure 8

Next, refer to Figure 8. Suppose player 1 believes he is at the possible world ψ. His action at ψ is B, and his payoff at ψ is 7. The closest possible world to ψ in which 1 plays T is $\tilde{\psi}$, and his payoff there is 6. Thus, the following counterfactual is false at ψ:

(2.3)   "If player 1 were to play T, his payoff would be higher."

This suggests a very natural rationality criterion for player 1. Namely, that he should never find himself at a possible world at which, according to his metric λ, he would be strictly better off if he were to deviate from

404

his current action. We leave the formal definition of this rationality to the next section. For now, notice that this criterion is satisfied by player 1 in our example, since

$$\beta^1(\omega_{tl}) = \beta^1(\omega_{tr}) = \varphi \quad \text{and} \quad \beta^1(\omega_{bl}) = \beta^1(\omega_{br}) = \psi,$$

and we showed that both φ and ψ satisfy the rationality requirement sketched previously.

It should be emphasized that the counterfactuals (2.2) and (2.3) are evaluated with reference to one particular metric λ. If player 1 held a different metric over his space of possible worlds, our conclusions could be very different. Nevertheless, the metric λ plays a prominent role in our chapter. This is because a player who acts rationally according to λ satisfies Aumann's (1987) criterion of Bayes-rationality. The converse is also true. Moreover, since Aumann's notion of Bayes-rationality leads directly to the standard solution concepts of Nash and correlated equilibria, this result provides a point of contact between our approach and other (more familiar) treatments of equilibrium in normal-form games.

3. The general case

In this section, we make precise the concepts introduced previously. The order of the discussion follows that of Section 2, except that we consider the general n-person case.

3.1. The game G

Let G be a game in normal form between n players, $G = \langle S, h \rangle$, where $S = \times_{i=1}^{n} S^i$ and $h = \langle h^1, \ldots, h^n \rangle$. S^i is player i's strategy set, and h^i is his payoff function $h^i : S \to \mathbf{R}$. We assume that each S^i is finite and has K^i elements. Denote by s_j^i the jth strategy of player i. Let $K := \prod_{i=1}^{n} K^i$, so that S has K elements. Denote by S^{-i} the product $S^1 \times S^2 \times \cdots \times S^{i-1} \times S^{i+1} \times \cdots \times S^n$. S^{-i} has K/K^i elements. We order this set in some well-defined manner by the index set $\{1, 2, \ldots, K^{-i}\}$, where $K^{-i} = K/K^i$. Thus $S^{-i} = \{s_1^{-i}, s_2^{-i}, \ldots, s_{K^{-i}}^{-i}\}$.

3.2. The state space Ω

With each strategy n-tuple $s \in S$, we associate a unique state ω. The set of all such states is the state space Ω. Then we can define a function $\underline{s} : \Omega \to S$ such that $\underline{s}(\omega)$ is the strategy n-tuple associated with ω. The function $\underline{s}$ is therefore the n-tuple $\langle \underline{s}^1, \underline{s}^2, \ldots, \underline{s}^n \rangle$, where $\underline{s}^i$ is the function $\underline{s}^i : \Omega \to S^i$ such that $\underline{s}^i(\omega)$ is the strategy of player i associated with the state ω. We

405

denote by $\underline{s}^{-i}$ the $(n-1)$-tuple $\langle \underline{s}^1, \ldots, \underline{s}^{i-1}, \underline{s}^{i+1}, \ldots, \underline{s}^n \rangle$. Let $\mathcal{P}^i$ be the partition of Ω generated by the equivalence relation $\equiv^i$, defined as

$$\omega \equiv^i \omega' \Leftrightarrow \underline{s}^i(\omega) = \underline{s}^i(\omega').$$

We denote by $P^i(\omega)$ the element of $\mathcal{P}^i$ containing the state ω. Finally, each player i has a probability distribution p^i over Ω.

3.3. The possible-worlds space Φ^i

For each player i, we define the possible-worlds space Φ^i as the product

(3.1) $$\Phi^i := S^i \times \Delta(S^{-i}),$$

where $\Delta(S^{-i})$ is the unit simplex of dimension $K^{-i} - 1$, representing the set of all probability distributions over the set S^{-i}. An element of Φ^i will be called a possible world and be denoted by φ. In turn, $\varphi = \langle \varphi^i, \varphi^{-i} \rangle$, where φ^i is the projection of φ into S^i (so that $\varphi^i \in S^i$) and φ^{-i} is the projection of φ into $\Delta(S^{-i})$ (so that $\varphi^{-i} \in \mathbf{R}^{K^{-i}}$). In particular, φ^{-i} will be denoted by the K^{-i}-dimensional vector

(3.2) $$\langle \varphi^{-i}[s_1^{-i}], \varphi^{-i}[s_2^{-i}], \ldots, \varphi^{-i}[s_{K^{-i}}^{-i}] \rangle,$$

where $\varphi^{-i}[s_k^{-i}]$ has the interpretation of the probability weight given to s_k^{-i} in φ^{-i}. When no confusion is likely, we shall abbreviate (3.2) as

(3.3) $$\langle \varphi_1^{-i}, \varphi_2^{-i}, \ldots, \varphi_{K^{-i}}^{-i} \rangle.$$

Finally, with each possible worlds space Φ^i we associate a metric m on Φ^i. This metric has the interpretation of the theory with which player i forms counterfactual beliefs. Call the pair $\langle \Phi^i, m \rangle$ player i's *theory space*.

3.4. The belief function β^i

Consider the posterior probability attached to s_k^{-i} by player i at ω, obtained by conditioning on his partition $\mathcal{P}^i$. We denote this probability by $\varphi_k^{-i}(i, \omega)$. More precisely,

(3.4) $$\varphi_k^{-i}(i, \omega) := p^i(\{\omega \mid \underline{s}^{-i}(\omega) = s_k^{-i}\} \mid P^i(\omega)).$$

Denote by Ω_+^i the set $\{\omega \mid p^i(\omega) > 0\}$, and define the function $\underline{t}^i : \Omega_+^i \to \Delta(S^{-i})$ as follows:

(3.5) $$\underline{t}^i(\omega) := \langle \varphi_1^{-i}(i, \omega), \varphi_2^{-i}(i, \omega), \ldots, \varphi_{K^{-i}}^{-i}(i, \omega) \rangle.$$

From this, we define player i's *belief function* $\beta^i : \Omega_+^i \to \Phi^i$ as

(3.6) $$\beta^i := \langle \underline{s}^i, \underline{t}^i \rangle.$$

We can give the following interpretation to this function. Player i forms beliefs by conditioning on the partition $\mathcal{P}^i$. Thus, at state ω, i computes posterior probabilities by conditioning on $P^i(\omega)$. Thus, at state ω, i attaches probability 1 to one of his own actions and attaches probabilities to his opponents' strategy combinations s^{-i}. Each point in the possible-worlds space Φ^i constitutes a possible set of beliefs for player i obtained in this manner. The belief function β^i is constructed so that, at ω, player i has the set of probability beliefs given by the possible world $\beta^i(\omega)$. More figuratively we say that, at state ω, player i believes that he is "at" the possible world $\beta^i(\omega)$.

3.5. The payoff function H^i

Each possible world $\varphi \in \Phi^i$ determines a probability distribution over S. By taking the weighted sum of $h^i(s)$ over $s \in S$ with the weights given by φ, we arrive at the expected payoff of player i at the possible world φ. We shall denote player i's payoff at the possible world φ as $H^i(\varphi)$. To define this formally, let I_j^i be the indicator function $I_j^i : S^i \to \{0, 1\}$ such that

$$(3.7) \qquad I_j^i(s^i) = \begin{cases} 1 & \text{if } s^i = s_j^i, \\ 0 & \text{otherwise.} \end{cases}$$

The function $H^i : \Phi^i \to \mathbf{R}$ is defined as follows:

$$(3.8) \qquad H^i(\varphi) = \sum_{j=1}^{K^i} \sum_{k=1}^{K^{-i}} I_j^i(\varphi^i) \varphi_k^{-i} h^i(s_j^i, s_k^{-i}).$$

To verify that H^i is in accordance with the intuition outlined above, note that from (3.4) and the definition of β^i, when $\beta^i(\omega) = \varphi$, $I_j^i(\varphi^i)\varphi_k^{-i}$ is the probability of the event $\{\omega \mid \underline{s}(\omega) = \langle s_j^i, s_k^{-i}\rangle\}$ according to p^i, conditional on $P^i(\omega)$.

3.6. The proposition set Ψ^i

We shall associate with each player i a set of *propositions* Ψ^i. This set is defined by the following three clauses.

(i) The following are elements of Ψ^i:
   $\sigma_j^i :=$ "player i plays s_j^i" for all $j \in \{1, \ldots, K^i\}$;
   $\sigma_\mu^{-i} :=$ "the set of players except i play the (possibly correlated) strategy μ" for all $\mu \in \Delta(S^{-i})$; and
   $\pi_r^i :=$ "player i's payoff does not exceed r" for all $r \in \mathbf{R}$.
(ii) Suppose $\psi, \chi \in \Psi^i$. Then $\psi \,\square\!\!\rightarrow \chi$ is an element of Ψ^i.
(iii) Ψ^i is the smallest set satisfying (i) and (ii).

3.7. Truth conditions and events

Given player i's theory space $\langle \Phi^i, m \rangle$, consider a function $e: \Phi^i \times \Psi^i \to \{0, 1\}$. For any such function, denote by $|\psi|_e$ the set $\{\varphi \in \Phi^i \mid e(\varphi, \psi) = 1\}$. We say that the function e is the *truth function* relative to the metric m if, for all $\varphi \in \Phi^i$ and $\psi \in \Psi^i$,

(i) $e(\varphi, \sigma^i_j) = 1 \Leftrightarrow \varphi^i = s^i_j \; \forall j \in \{1, \ldots, K^i\}$;

(ii) $e(\varphi, \sigma^{-i}_\mu) = 1 \Leftrightarrow \varphi^{-i} = \mu \; \forall \mu \in \Delta(S^{-i})$;

(iii) $e(\varphi, \pi^i_r) = 1 \Leftrightarrow H^i(\varphi) \leq r \; \forall r \in \mathbf{R}$;

(iv) $e(\varphi, \psi \mathbin{\Box\!\!\to} \chi) = 1 \Leftrightarrow$ there is a closed sphere C around φ in the metric m such that $C \cap |\psi|_e$ is nonempty and $C \cap |\psi|_e \subseteq |\chi|_e$.

When e is a truth function, we say that ψ is *true at* φ if $e(\varphi, \psi) = 1$; ψ is *false at* φ otherwise. Worthy of note is clause (iv), formalizing the truth condition for counterfactual propositions. It states that $\psi \mathbin{\Box\!\!\to} \chi$ is true at the possible world φ if and only if, in the closest possible world(s) to φ in which ψ is true, χ is also true.

When e is a truth function, we shall drop the subscript e from $|\psi|_e$. In this case, we call $|\psi|$ the *event* corresponding to ψ.

3.8. Rationality

Suppose $\beta^i(\omega) = \varphi$, and that player i holds the metric m. We say that i is *m-rational at* ω if

(3.9) for all $j \in \{1, \ldots, K^i\}$, there is $r \leq H^i(\varphi)$ such that $\sigma^i_j \mathbin{\Box\!\!\to} \pi^i_r$ is true at φ.

In other words, i is m-rational at ω if i believes that he is at a possible world at which, according to his metric m, he would not gain if he were to deviate. In general, we say that i is m-rational if i is m-rational at all states on which β^i is defined. Notice that the notion of m-rationality is a stipulation on a player's probability distribution p. Namely, a player's probability distribution should be such that, given his metric m, he will not find himself in a situation in which he believes that he would do better if he were to deviate. Thus, m-rationality describes a type of consistency, the consistency of a player's probability distribution with his metric.

3.9. The metric λ

Thus far, we have left open the exact form of a player's metric over his space of possible worlds. We shall explore the equilibrium structure arising from the metric λ. This is the so-called library-stack metric for the n-player case. This metric will be the sum of two metrics on the two component sets of Φ^i. Denote by λ^i the discrete metric on S^i and by λ^{-i} the Euclidean norm on $\Delta(S^{-i})$. That is,

(3.10)
$$\lambda^i(s_j^i, s_q^i) = \begin{cases} 0 & \text{if } j = q, \\ 1 & \text{otherwise}; \end{cases}$$

(3.11)
$$\lambda^{-i}(\varphi^{-i}, \hat{\varphi}^{-i}) = \|\varphi^{-i} - \hat{\varphi}^{-i}\|$$
$$= \left(\sum_{k=1}^{K^{-i}} |\varphi_k^{-i} - \hat{\varphi}_k^{-i}|^2 \right)^{1/2}.$$

Thus, suppose $\varphi = \langle \varphi^i, \varphi^{-i} \rangle$ and $\hat{\varphi} = \langle \hat{\varphi}^i, \hat{\varphi}^{-i} \rangle$. We define λ as the sum of λ^i and λ^{-i}:

(3.12)
$$\lambda(\varphi, \hat{\varphi}) := \lambda^i(\varphi^i, \hat{\varphi}^i) + \lambda^{-i}(\varphi^{-i}, \hat{\varphi}^{-i}).$$

3.10. Rationality

We now come to our central result. We shall demonstrate that a player who is rational with respect to the metric λ will act in accordance with the criterion of rationality as set out by Aumann (1987), and conversely. We use the term "Aumann-rationality" to refer to Aumann's criterion of Bayes-rationality, in order to avoid confusion with other formulations of Bayes-rationality used in different contexts.

Denote by c_q^i the constant function $c_q^i : \Omega \to S^i$ such that $c_q^i(\omega) = s_q^i$ for all ω. We say that player i is *Aumann-rational* at the state ω if

(3.13) $\quad E(h^i(\underline{s}) | P^i(\omega)) \geq E(h^i(c_q^i, \underline{s}^{-i}) | P^i(\omega)) \quad$ for all $q \in \{1, ..., K^i\}$,

where E denotes the expectation with respect to the distribution p^i.

3.11. The theorem

Theorem. *Player i is λ-rational at ω if and only if i is Aumann-rational at ω.*

This theorem is an immediate consequence of the following pair of lemmas. Denote by $\varphi \backslash s_q^i$ the vector $\langle s_q^i, \varphi^{-i} \rangle$. In other words, $\varphi \backslash s_q^i$ is the possible world obtained from φ by replacing φ^i with s_q^i. We also use the following abbreviations:

$$[s_j^i] = \{\omega \mid \underline{s}^i(\omega) = s_j^i\};$$
$$[s_k^{-i}] = \{\omega \mid \underline{s}^{-i}(\omega) = s_k^{-i}\};$$
$$[s] = \{\omega \mid \underline{s}(\omega) = s\}.$$

A final piece of notation: We define $S_{il} = \{s \in S \mid s^i = s_l^i\}$. That is, S_{il} is the set of all strategy combinations in which player i plays s_l^i.

Lemma 1. *Suppose* $\beta^i(\omega) = \varphi$. *Then:*
(i) $H^i(\varphi) = E(h^i(\underline{s}) \mid P^i(\omega))$;
(ii) $H^i(\varphi \setminus s_q^i) = E(h^i(c_q^i, \underline{s}^{-i}) \mid P^i(\omega))$.

Proof. (i) $\underline{s}^i(\omega) = s_l^i$ for some $l \in \{1, \dots, K^i\}$. Then

$$H^i(\varphi) = \sum_{j=1}^{K^i} \sum_{k=1}^{K^{-i}} I_j^i(\varphi^i) \varphi_k^{-i} h^i(s_j^i, s_k^{-i})$$

$$= \sum_k \varphi_k^{-i} h^i(s_l^i, s_k^{-i}) \quad [\text{since } \varphi^i = s_l^i]$$

$$= \sum_k p^i([s_k^{-i}] \mid P^i(\omega)) h^i(s_l^i, s_k^{-i}) \quad [\text{from (3.4)}]$$

$$= \sum_k p^i([s_k^{-i}] \cap [s_l^i] \mid P^i(\omega)) h^i(s_l^i, s_k^{-i}) \quad [\text{since } P^i(\omega) = [s_l^i]]$$

$$= \sum_k p^i(\{\omega \mid \underline{s}(\omega) = \langle s_l^i, s_k^{-i} \rangle\} \mid P^i(\omega)) h^i(s_l^i, s_k^{-i})$$

$$= \sum_{s \in S_{il}} p^i([s] \mid P^i(\omega)) h^i(s)$$

$$= \sum_{s \in S} p^i([s] \mid P^i(\omega)) h^i(s) \quad [\text{since } [s] \cap P^i(\omega) \text{ is null for all } s \notin S_{il}]$$

$$= E(h^i(\underline{s}) \mid P^i(\omega)).$$

(ii) We have

$$H^i(\varphi \setminus s_q^i) = \sum_{j=1}^{K^i} \sum_{k=1}^{K^{-i}} I_j^i(s_q^i) \varphi_k^{-i} h^i(s_j^i, s_k^{-i})$$

$$= \sum_k \varphi_k^{-i} h^i(s_q^i, s_k^{-i})$$

$$= \sum_k p^i([s_k^{-i}] \mid P^i(\omega)) h^i(s_q^i, s_k^{-i}) \quad [\text{from (3.4)}]$$

$$= \sum_{s^{-i} \in S^{-i}} p^i([s^{-i}] \mid P^i(\omega)) h^i(s_q^i, s^{-i})$$

$$= E(h^i(c_q^i, \underline{s}^{-i}) \mid P^i(\omega)). \qquad \square$$

Lemma 2. *Suppose* $\beta^i(\omega) = \varphi$. *Then* i *is* λ-*rational at* ω *if and only if*

(3.14) $$H^i(\varphi) \geq H^i(\varphi \setminus s_q^i) \quad \forall q.$$

Proof. (Sufficiency) Suppose $H^i(\varphi) \geq H^i(\varphi \setminus s_q^i)$. Define C to be the closed sphere around φ in the metric λ with radius $\lambda(\varphi, \varphi \setminus s_q^i)$. We claim that $C \cap |\sigma_q^i|$ is the singleton $\{\varphi \setminus s_q^i\}$. Suppose not; then there is $\hat{\varphi} \in C \cap |\sigma_q^i|$ such that $\hat{\varphi} \neq \varphi \setminus s_q^i$. However, then

410

$$\lambda(\varphi, \hat{\varphi}) = \lambda^i(\varphi^i, \hat{\varphi}^i) + \lambda^{-i}(\varphi^{-i}, \hat{\varphi}^{-i})$$
$$= \lambda^i(\varphi^i, s_q^i) + \lambda^{-i}(\varphi^{-i}, \hat{\varphi}^{-i}) \quad [\text{since } \hat{\varphi} \in |\sigma_q^i|]$$
$$> \lambda^i(\varphi^i, s_q^i) \quad [\text{since } \hat{\varphi}^{-i} \neq \varphi^{-i}].$$

But this contradicts the supposition that $\hat{\varphi} \in C \cap |\sigma_q^i|$, since the radius of C is given by $\lambda(\varphi, \varphi \backslash s_q^i) = \lambda^i(\varphi^i, s_q^i)$, thereby establishing the claim.

Denote by r the number $H^i(\varphi \backslash s_q^i)$. Then $|\sigma_q^i| \cap C \subseteq |\pi_r^i|$, so that $\sigma_q^i \square \rightarrow \pi_r^i$ is true at φ. However, $r \leq H^i(\varphi)$ by supposition. In other words, for all q there is $r \leq H^i(\varphi)$ such that $\sigma_q^i \square \rightarrow \pi_r^i$ is true at φ. This is the definition of λ-rationality at ω.

(Necessity) By λ-rationality at ω, for all q there is $r \leq H^i(\varphi)$ such that $\sigma_q^i \square \rightarrow \pi_r^i$ is true at φ. By the truth condition for $\square \rightarrow$, there is a closed sphere C around φ such that $C \cap |\sigma_q^i|$ is nonempty and $C \cap |\sigma_q^i| \subseteq |\pi_r^i|$. Let $\bar{\varphi} \in C \cap |\sigma_q^i|$. Then

$$\lambda(\varphi, \bar{\varphi}) = \lambda^i(\varphi^i, \bar{\varphi}^i) + \lambda^{-i}(\varphi^{-i}, \bar{\varphi}^{-i})$$
$$\geq \lambda^i(\varphi^i, \bar{\varphi}^i)$$
$$= \lambda^i(\varphi^i, s_q^i);$$
$$\lambda(\varphi, \varphi \backslash s_q^i) = \lambda^i(\varphi^i, s_q^i) + \lambda^{-i}(\varphi^{-i}, \varphi^{-i})$$
$$= \lambda^i(\varphi^i, s_q^i).$$

Because the radius of C is no less than $\lambda(\varphi, \bar{\varphi})$ and $\lambda(\varphi, \bar{\varphi}) \geq \lambda(\varphi, \varphi \backslash s_q^i)$, we have $\varphi \backslash s_q^i \in C$; also, $\varphi \backslash s_q^i \in |\sigma_q^i|$. By supposition, $C \cap |\sigma_q^i| \subseteq |\pi_r^i|$, so that $\varphi \backslash s_q^i \in |\pi_r^i|$. Hence $H^i(\varphi \backslash s_q^i) \leq r$. But we know that $r \leq H^i(\varphi)$; thus, $H^i(\varphi) \geq H^i(\varphi \backslash s_q^i)$. Moreover, this is the case for all q. $\square$

Proof of Theorem. Player i is λ-rational at $\omega \leftrightarrow H^i(\varphi) \geq H^i(\varphi \backslash s_q^i) \, \forall q$ (by Lemma 2) $\leftrightarrow i$ is Aumann-rational at ω (by Lemma 1). $\square$

3.12. Corollaries

We can reap the following corollaries by virtue of Aumann's (1987) main theorem and its converse.

Corollary 3.1. *Suppose $p^i = p \, \forall i$. Then all players are λ-rational if and only if p is a correlated equilibrium distribution.*

Corollary 3.2. *Suppose $p^i = p \, \forall i$ and that p is independent. Then all players are λ-rational if and only if the mixed strategies given by p constitute a Nash equilibrium.*

411

Thus, taken together with existing results in the literature, our main theorem constitutes a point of contact between our approach (based on the notion of counterfactuals) and the standard solution concepts in the literature (which make no explicit reference to counterfactuals). Moreover, since the theorem describes the consequences arising from one particular metric λ, we may legitimately claim that our framework is a generalization of these conventional concepts.

4. CONCLUSIONS AND RELATED WORK

Counterfactuals and game theory are inseparable. If counterfactuals are not explicitly invoked, it is because the assumptions are buried implicitly in the discussion. The main objective of this chapter has been to demonstrate that the standard solution concepts of Nash and correlated equilibria can be reconstructed within the decision-theoretic framework of Jeffrey, by invoking counterfactual beliefs generated by the metric λ.

The analysis in this chapter is closely related to two strands in the literature. The first is the analysis of extensive-form games by modeling the counterfactual beliefs of players along the off-equilibrium paths of the game. Bicchieri (1988) has provided the most explicit development of this line of inquiry, building on the theories of belief change of Levi (1977) and Gärdenfors (1984).

In many ways, the role of counterfactuals is more conspicuous in the analysis of extensive-form games, as any theory of equilibrium in extensive-form games must specify off-equilibrium beliefs. The main difference between Bicchieri's work and this chapter (apart from the difference in subject matter) is in the approach to counterfactuals. The theories of belief revision (due to Levi and Gärdenfors) that Bicchieri uses are couched in terms of sets of sentences, and this has the advantage of making belief revisions more explicit. This is essential in analyzing extensive-form games. For normal-form games, however, the belief revisions are of a much simpler kind, and the possible-worlds approach provides a geometric intuition that is lacking in the Levi–Gärdenfors approach.

The second strand in the literature to which this chapter is related is that of "ratifiability." Ratifiability is the notion developed by Jeffrey in the second edition of his monograph (1983), and developed in subsequent papers such as Harper (1986), which is based on the idea that a decision maker may not be able to perform an action that was chosen. In Jeffrey's own words:

The notion of ratifiability is applicable only where, during deliberation, the agent finds it conceivable that he will not manage to perform the act he finally decides to peform, but will find himself performing one of the other available acts instead. (1983, p. 18)

One perspective on this problem (which may be controversial) is to regard the framework surrounding ratifiability as a way of generating an implicit set of counterfactual beliefs. By allowing for the possibility that a decision maker may not be able to carry out a decision, the framework is capable of accommodating the sort of reasoning contained in statements such as

(4.1) "I play x, but if I were to tremble and play y instead, the outcome would be z."

Elsewhere (Shin 1989) I show that, for a particular formalization of ratifiability, the restrictions on decisions obtained are identical to those obtained in this chapter. This seems to bear out the intuition that ratifiability is a way of introducing an implicit set of counterfactual beliefs. However, this interpretation may prove to be a controversial one, since counterfactuals do not sit well within the overall Bayesian view of decision theory espoused by Jeffrey. It remains to be seen to what extent the two approaches can be reconciled.

REFERENCES

Aumann, R. J. (1987), "Correlated Equilibrium as an Expression of Bayesian Rationality." *Econometrica* 55: 1–19.
Bicchieri, C. (1988), "Strategic Behavior and Counterfactuals." *Synthese* 76: 135–69.
Harper, W. (1986), "Mixed Strategies and Ratifiability in Causal Decision Theory." *Erkenntnis* 24: 25–36.
Gärdenfors, P. (1984), "Epistemic Importance and Minimal Changes of Belief." *Australasian Journal of Philosophy* 62: 136–57.
Jeffrey, R. C. (1965), *The Logic of Decision.* New York: McGraw-Hill.
Jeffrey, R. C. (1983), *The Logic of Decision,* 2nd ed. Chicago: Chicago University Press.
Lewis, D. K. (1973), *Counterfactuals.* Oxford: Basil Blackwell.
Levi, I. (1977), "Subjunctives, Dispositions and Chances." *Synthese* 34: 423–55.
Savage, L. J. (1954), *The Foundations of Statistics.* New York: Wiley.
Shin, H. S. (1989). "Two Notions of Ratifiability and Equilibrium in Games." in M. Bacharach and S. Hurley (eds.), *Foundations of Decision Theory.* Oxford: Basil Blackwell.
Stalnaker, R. (1968), "A Theory of Conditionals." In N. Rescher (ed.), *Studies in Logical Theory.* Oxford: Basil Blackwell.